Climate Change in Human History

Climate Change in Human History

Prehistory to the Present

Second Edition

BENJAMIN LIEBERMAN
AND ELIZABETH GORDON

BLOOMSBURY ACADEMIC
LONDON • NEW YORK • OXFORD • NEW DELHI • SYDNEY

BLOOMSBURY ACADEMIC
Bloomsbury Publishing Plc
50 Bedford Square, London, WC1B 3DP, UK
1385 Broadway, New York, NY 10018, USA
29 Earlsfort Terrace, Dublin 2, Ireland

BLOOMSBURY, BLOOMSBURY ACADEMIC and the Diana logo are trademarks of Bloomsbury Publishing Plc

First published in Great Britain 2018

This edition published 2022

Cover design: Anna Berzovan
Cover image © Bushfires below Stacks Bluff, Tasmania, Australia. Photo by Matt Palmer on Unsplash

A catalogue record for this book is available from the British Library.

Library of Congress Cataloging-in-Publication Data
Names: Lieberman, Benjamin David, 1962- author. | Gordon, Elizabeth (Professor), author.
Title: Climate change in human history: prehistory to the present / Benjamin Lieberman and Elizabeth Gordon.
Description: Second edition. | New York: Bloomsbury Academic, 2022. | Includes bibliographical references and index.
Identifiers: LCCN 2021028667 (print) | LCCN 2021028668 (ebook) | ISBN 9781350170339 (paperback) | ISBN 9781350170346 (hardback) | ISBN 9781350170360 (pdf) | ISBN 9781350170353 (epub)
Subjects: LCSH: Climatic changes–Social aspects–History. | Human beings–Effect of climate on–History. | Climate and civilization. | Prehistoric peoples–Climatic factors. | Human ecology–History.
Classification: LCC QC903 .L52 2021 (print) | LCC QC903 (ebook) | DDC 304.2/5–dc23
LC record available at https://lccn.loc.gov/2021028667
LC ebook record available at https://lccn.loc.gov/2021028668

ISBN: HB: 978-1-3501-7034-6
PB: 978-1-3501-7033-9
ePDF: 978-1-3501-7036-0
eBook: 978-1-3501-7035-3

Typeset by Deanta Global Publishing Services, Chennai, India

To find out more about our authors and books visit www.bloomsbury.com and sign up for our newsletters.

CONTENTS

FIGURES

ACKNOWLEDGMENTS

The authors wish to thank our students in our classes on climate change in human history. This book owes a great deal to their enthusiasm and to their questions.

We would like to thank Academic Affairs at Fitchburg State University for helping us advance this project through sabbaticals.

We would also like to thank the staff of the Amelia V. Gallucci-Cirio Library at Fitchburg State University for their professional support in obtaining materials.

We would also like to thank Daniel Lieberman for comments and advice.

Thanks, as well, to Isabel Lieberman for the photos. We received very useful comments from external review that helped us to expand the scope of this book.

Finally, we thank Abigail Lane, Maddie Holder, Emma Goode and the editorial and production staff at Bloomsbury for their help at many stages.

ACRONYMS

ACR Antarctic Cold Reversal
AMH anatomically modern humans
BCE Before Common Era
Be-10 Beryllium-10
C Centigrade or Celsius
C-14 Carbon-14
CE Common Era
CFCs chlorofluorocarbons
CH_4 Methane
CO_2 Carbon Dioxide
D-O Dansgaard-Oeschger
ENSO El Niño-Southern Oscillation
EOR enhanced oil recovery
EU European Union
F Fahrenheit
FEMA Federal Emergency Management Agency
G-7 Group of 7
GCMs Global Climate Models
GDP Gross Domestic Product
GHGs Greenhouse Gasses
HCFCs Hydrochlorofluorocarbons
IPCC Intergovernmental Panel on Climate Change
ITCZ intertropical convergence zone
k thousand as in thousand years
LALIA Late Antique Little Ice Age
LGM Last Glacial Maximum
LIA Little Ice Age
MCA Medieval Climate Anomaly
Ma Mega-annum (million years)
mya million years ago
NAO North Atlantic Oscillation
NASA National Aeronautics and Space Administration
NOAA National Oceanic and Atmospheric Administration
O_3 Ozone
PDO Pacific Decadal Oscillation

ppm parts per million
RCPs representative concentration pathways
Tg Terragrams
TW Terrawatt
XR Extinction Rebellion
y years
ybp years before present

Introduction

Beginning 10,000 years ago, in a warming climate with ample water, farmers domesticated grains and expanded their fields. The growing surplus sustained a rise in population. Societies grew more complex: new towns and cities boosted demand for a wide range of artisans and specialists. The thriving economy enabled political and religious leaders to build elaborate palaces and monuments: pyramids, ziggurats, and even a sphinx. Complex societies and civilizations in other regions of the world followed much the same pattern as they emerged during a long warm phase after the last glacial maximum, popularly termed the "ice age."

Consider a different climate trend: a once-fertile region dried out as prevailing winds shifted, taking away once reliable rains. Salinity in soil rose. The supply of food shrank. The surplus that once supported large cities vanished. The population moved away and, in doing so, abandoned urban complexes and the culture and society associated with their lost cities. Prolonged, severe droughts have already produced such outcomes, leading either to the collapse or to the retreat of civilizations during the historic past, and megadroughts associated with climate change threaten to confront human societies of the present and future with severe challenges.

Yet another complex society faced a shift in climate. Heavier storms with more intense rains caused more frequent flooding and exacerbated the risk of famine. Colder winters lowered crop yields at higher elevations. Those living on the extreme margins in some cases abandoned their villages, but others turned to more efficient heating sources, and political authorities improved their ability to provide relief from dearth. Societies confronting past regional climate change along these lines have suffered losses, but in some cases have also adapted to shifts in climate.

Each of these scenarios displays the interaction between climate change and human history. In the first, a climate with favorable conditions for farming contributed to the rise of a flourishing civilization dependent on extracting a large surplus from agriculture. In the second scenario, a sharp shift in precipitation was so severe as to persuade people to migrate away from their homes. The third scenario exemplifies both the challenges that climate shifts can pose and resiliency and the ability to adapt to changes in climate. From before the start of human civilization to the present, climate has influenced human history in many ways. This book introduces and outlines the vital, complex, and often changing interaction between climate and human societies.

Scientific and historical methods

Historians have traditionally focused on a wide range of important historical phenomena to a much greater extent than climate. A conventional top-down approach to history describes the achievements and failures of leaders and elites; of prophets, emperors, kings, military commanders, presidents; or of towering figures who marshaled forces of protest. Modifying this focus on leaders, historians have adopted multiple approaches to recast history from other perspectives. Thus, entire fields focus on social history, economic history, and the history of gender. Some historians have reversed the top-down approach entirely to look at history from the grassroots or from the perspective of oppressed or marginalized groups. With a very few notable exceptions, however, historians until recently seldom emphasized the role of climate, instead implicitly assuming that the climate provided a general frame or base for other historical events and trends. With the surge in interest in climate research, historians over the last few decades have increasingly included climate as a significant influence on history.

With very few exceptions no historical event can be attributed to a single cause. To take a few major historical events, neither the French Revolution, nor the rise of Hitler, nor the collapse of the Soviet Union stemmed from any single factor. Short of the most extreme natural disaster or catastrophe of warfare, almost any major event or trend arises from multiple causes. As this book shows, climate change has influenced human history in many fundamental ways, but it is also important to keep in mind that climate has interacted with other factors in affecting history. Thus we should focus closely on the effects of climate, but we do not presume that any shift in climate has determined a particular historical outcome without interacting with other causes.

Both history and climate science are evolving fields. This may seem more obvious in the case of climate science, but written sources remain scant or nonexistent for many human societies. Archaeology provides additional information, but in many cases the study of climate and human history has yielded more than one possible scenario. One well-known example is that of the demise of Viking settlements on the vast North Atlantic island of Greenland. Rather than present only one interpretation, this book seeks to make clear when research into climate and human history has yielded more than one interpretation or scenario. In many such cases, including the case of Greenland, ongoing research may strengthen or weaken scenarios or may create new models of the interaction between climate and human history.

Both climate science and historical research can build consensus while leaving or creating new areas for research and debate. Questions about the interaction between multiple powerful causes allow for the possibility of more than one plausible answer. However, both history and climate science have established many firm conclusions. On the side of history, we possess increasingly firm information on the timing of human migration. We know when major civilizations and human societies emerged, and in many cases we know when their power eroded. We can in many cases make good estimates of population, and we have a good grasp on the use of different fuel sources and on the emergence of new technologies. We also possess detailed political chronologies for many civilizations.

The historical record for investigating the effects of climate change is paradoxically both dense but also scant, depending on the time and place. Thus, we have far more direct evidence about certain societies, usually those that left writing and abundant physical sites, than about human societies that lacked writing or complex governments.

From the climate science perspective, we have integrated knowledge from several fields, such as astronomy, geology, and climatology, to understand the growth and retreat of ice sheets throughout the past 3 million years, and have established methods that allow us to measure air samples that date about half as far back. We have routine proxies from various geological records that lock in climate conditions at various times in Earth's past, and emerging new trends in the field that will help address unanswered questions. We still see the Holocene, the last 11,700 years of Earth's history, as a generally stable period with respect to climate, but have also begun to recognize short-term instabilities in the climate system that can influence human civilization. There is also an overwhelming consensus in the scientific community about the impact that human activities are having on our current climate—a warming trend that is unprecedented since our ancestors first roamed the planet.

Scientists employ a number of tools to examine past climate. Our first direct measurements of carbon dioxide began in 1958, at the Mauna Loa Observatory in Hawaii. One way to track the changes of carbon dioxide in our atmosphere before direct measurements is by drilling into glacial ice and extracting cores. Small bubbles of air are trapped at the time the ice is forming, providing a time capsule of the atmosphere from the past. Scientists can determine the age of the ice through counting the annual layers—summer and winter ice have different appearances, so bands that include summer and winter ice represent a single year—or from other distinct markers, and age models help recreate a timeline in deep layers that may not have these easily identifiable annual bands. The ancient air bubbles can then be analyzed for gas concentration, and the chemical signature of the ice can be used to infer air temperature. Other indicators, such as dust and volcanic deposits, can be used to reconstruct other events.

Reliable continuous ice core records date back around 1 million years. To study climate further back in time relies on geological samples such as rocks and sediment. Fossils of ancient plankton, microscopic organisms that are contained in the sediment, have been used extensively to track changes in the amount of ice that exists globally, as well as atmospheric CO_2 and ocean temperature. Ancient pollen grains and other vegetation markers that are buried in ocean or lake sediment can similarly detect climatic changes on land. Lake levels help constrain precipitation changes over time. High-resolution records of climate can also be found in tree rings, corals, and cave formations, or speleothems. As with ice cores, these climate proxies, or indirect measures of climate, contain annual bands that provide a timeline for change. Analyzing chemical signatures of tree rings and speleothems helps scientists reconstruct temperature and precipitation on land, while the calcium carbonate skeleton of coral samples provides information about climate conditions in the ocean.

Historical records are also used to track past changes in climate. Documentary evidence such as farmers' journals can provide a written record of frost dates and harvest records.

Overview of Earth's climate system

The greenhouse effect, while sometimes painted solely in an unfavorable light, is necessary for life on Earth. Greenhouse gases (GHGs) make up less than 1 percent of our atmosphere, yet their heat-absorbing capabilities maintain Earth at a habitable temperature for plants and animals, including humans. Energy moves between the sun and the planets in the form of waves—electromagnetic radiation—and the wavelength of that energy dictates how it behaves when it encounters Earth's atmosphere. This is because our atmosphere is composed of various gases, some of which absorb radiation. For example, ozone (O_3) is a gas that resides in relatively large abundance about 20–25 kilometers above Earth's surface—making up our ozone layer—and absorbs ultraviolet radiation efficiently. Energy with shorter waves, such as gamma rays, X-rays, and ultraviolet radiation, can be harmful to life on Earth, but most of the radiation, including radio waves, microwaves, and infrared energy, is not. Although the sun emits energy that spans the electromagnetic spectrum, most of it falls within a small range that includes ultraviolet light, visible light, and infrared energy, or heat. Earth's upper atmosphere absorbs much of the shorter-wave radiation, including ultraviolet light, thanks to the ozone layer; in all, about 20 percent of the energy coming from the sun is absorbed by Earth's atmosphere en route to the surface. The visible light—the wavelengths of light that can be seen by human eyes—passes through our atmosphere to reach Earth's surface, and about 30 percent of that is reflected back to space. The 50 percent of incoming solar radiation, or insolation, that reaches Earth's surface warms the Earth, which in turn emits radiation. Because the Earth is not nearly as warm as the sun, the energy emitted by Earth's surface is not visible to our eyes—it's emitted as heat.

The greenhouse gases in our atmosphere, such as carbon dioxide (CO_2), methane (CH_4), and water vapor, among others, absorb much of that infrared radiation, warming the atmosphere as a whole. The warm atmosphere gives off its own heat, emitting radiation in all directions, mostly toward the ground but some upward back to space. This process is our greenhouse effect, a natural process that keeps the Earth at an average temperature of 15°C (59°F), rather than the −18°C (0°F) temperature it would be without it: a planet bound in ice. It therefore supports most forms of life on Earth, but it is also subject to alteration with changes in greenhouse gas concentration. As the abundance of greenhouse gases increases, so does Earth's temperature, and this warming sets off chain reactions within the Earth system that amplify the warming further. Such internal processes that either magnify or counteract an initial change in climate are referred to as climate feedbacks.

Because these two pathways of energy reaching the planet—coming from the sun or from our own atmosphere—are responsible for heating the planet, temperatures on Earth change globally either when the amount of insolation received by Earth's system changes, or when the energy coming from our atmosphere changes. We therefore look to changes in insolation and changes in greenhouse gas concentration in the study of what causes climate to change, or climate forcings.

Timescales of climate change

We consider climate and human interactions by focusing on various timescales, from the long-term changes in global climate that influenced human evolution and early innovation, to short-term oscillations that have more regional footprints and consequences. Both climate forcings and feedbacks interact to cause our climate to change. Other processes redistribute energy around the globe but do not have a long-lasting impact on global temperature—these represent climate variability rather than climate change. Both climate variability and overall change have influenced human history.

Over most of the 4.5 billion years of Earth's history, we can see how the climate system functions to maintain temperatures within a habitable range. When Earth first formed, the sun was only about 70 percent as bright as it is today, and therefore less solar energy was received by our planet. At that level of insolation, any water on Earth would have been frozen even with the greenhouse effect—our preindustrial atmosphere would have warmed Earth above the freezing point only in the latter half of Earth's existence after about 2 billion years ago. Evidence, mostly in the form of sedimentary rocks dating back 3.8 billion years, lets scientists know that the Earth was not frozen at this time—sedimentary rocks only form where liquid water exists. This conundrum, known as the Faint Young Sun Paradox, has been the subject of research for decades. How could liquid water have existed with a weaker sun? Although still a matter of scientific debate, the likeliest explanation is that our greenhouse effect was stronger on ancient Earth, with a greater concentration of GHGs, especially carbon dioxide and methane. As the sun became brighter over time, the GHGs were removed from the atmosphere through geological and biological processes, concurrently weakening the greenhouse effect. This illustrates the importance of feedback mechanisms in the climate system, and how they have operated throughout Earth's history to create our goldilocks planet.

There are several processes that cause climate to change on geological timescales, mostly connected to the shifting of Earth's tectonics plates. The rate of volcanic activity and its associated release of CO_2 influences the strength of Earth's greenhouse over these long periods of time, as does the creation of mountains that increase chemical weathering of Earth's surface. An increase in weathering associated with the uplift of the Himalayas beginning around 50 million years ago, for example, has been hypothesized as the reason for the overall cooling trend observed since that time. These competing processes—the addition of CO_2 through volcanism and removal of CO_2 through weathering—have kept climate within a habitable range for billions of years. Even within this range, Earth has fluctuated between warmer, largely ice-free periods, known as greenhouse states, and cooler periods that saw the growth of polar ice sheets, known as icehouse states. We entered our current icehouse state around 34 million years ago with the initiation of the Antarctic ice sheet; ice growth came later in the Northern Hemisphere, beginning around 3 million years ago.

The shifting of Earth's tectonic plates affects climate in other ways, including the various changes that occur when continents are located in different positions, forcing a rerouting of ocean currents and shifting wind patterns. These geological processes are a fascinating aspect of Earth's history that receives only brief mention in the book because, although plate motion plays an important role in climate change, it operates slowly and over such long time periods that it factors minimally into human history (Figure 0.1).

On shorter timescales, during the millions of years during which ancestors of humans existed, climate shifts influenced the availability of food and helped drive evolution. Climate change on this timescale, tens to hundreds of thousands of years, is mostly associated with glacial-interglacial cycles driven by changes in Earth's orbit. Milutin Milankovitch, a Serbian astrophysicist, proposed in the 1920s that these climatic changes were linked to changes in Earth's orbital relationship to the sun. While these Milankovitch cycles, as they are known, have primarily been connected to

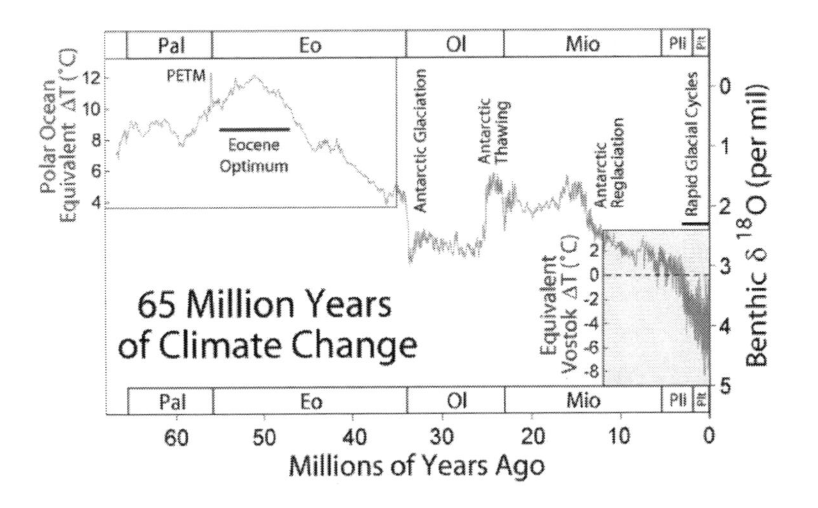

FIGURE 0.1 *Climate change over the last 65 million years. Source: Prepared by Robert A. Rohde for the Global Warming Art project. Accessed via Wikimedia Commons, https://commons.wikimedia.org/wiki/File:65_ Myr_Climate_Change.png.*

the growth and retreat of large ice sheets over the past few million years, they have also influenced human history within the Holocene by affecting the strength of monsoons.

Most Holocene climate forcings can be attributed to volcanic activity, solar variability, and changes in greenhouse gas concentration. Solar variability within the Holocene is primarily related to sunspot activity, which currently follows an eleven-year cycle. The small changes in solar output associated with sunspots can have compounding effects on climate when they trigger internal feedbacks of the climate system. Similarly, short-term cooling initiated by major volcanic eruptions can have longer-lasting effects when amplified by Earth's own processes. Internal variations in the climate system become more prominent on these millennial and shorter timescales. For example, alteration of deep ocean circulation has been linked with rapid climate changes in Earth's past, such as the abrupt cooling of the Younger Dryas approximately 12,000 years ago. Other perturbations to Earth's climate system, such as the El Niño-Southern Oscillation and the North Atlantic Oscillation, affect climate and weather worldwide on various timescales—from seasonal to annual to decadal.

Collapse and resiliency

The interaction between climate and human history can lead in numerous directions. At a base level, suitable climate is indispensable for the existence of humans. A short thought experiment makes this clear: it is hard to imagine thriving human societies during the extreme cold and heat many hundreds of millions of years ago in the distant geologic past. On shorter timescales, during the millions of years when ancestors of humans existed, climate shifts influenced the availability of food and helped drive evolution. On a shorter timescale still, the era since the New Stone Age, or Neolithic, climate at one extreme could contribute to thriving human societies being able to extract plentiful food supplies from their environment, or, at another extreme, could undercut and undermine complex societies. The study of collapse has garnered historians' interest for centuries, in particular for cases such as the end of the Roman Empire. As climate history has emerged, research into collapse has cited climate change as one major contributing factor.[1]

Theories of collapse, in turn, have met with criticism. A frequent counterargument has stressed that what we may see after the fact as a collapse can better be described as a longer, slower, and more complex transition. Roman history again has furnished a major example of this approach. Instead of a sharp collapse produced by dramatic invasions, historians of transition argue that Roman culture persisted in some regions long after the supposed collapse, and that elements of Rome endured through political changes.

Much the same debate that contrasts collapse with transition can be found in climate history. Thus, in place of looking for causes of collapse, a different approach emphasizes the resiliency and capacity to adapt displayed by human societies. It is worth considering that too sharp a focus on collapse can lead to identifying any major trend or event as a cause of collapse, but the study of resilience faces the same risk. Human societies have endured through times of crisis, but they do not possess unlimited resilience. This book therefore considers both human resilience and capacity to adapt and cases of crisis and even collapse.

Organization of the book

Chapter 1 focuses on natural causes of climate variability on timescales tens of thousands of years and longer, along with shorter cycles of change, and describes how climate change influenced the prehistory of *Homo sapiens*. The major climate drivers during this time period were the Milankovitch cycles embedded in the long-term cooling brought on by declining CO_2. The orbital variations caused warming and cooling patterns that resulted in expansion and retreat of continental ice sheets. Climate conditions during the last glacial maximum (LGM) were characterized by large volumes of ocean water trapped in large ice sheets, resulting in lowered sea level and exposed continental lands. Climate change during this time period affected how and where humans lived. Periods of aridity and moist phases influenced the dispersal of human ancestors. Climate change, in particular glacial maxima, also created challenges for human populations. The chapter discusses the emergence and dispersal of human ancestors and of early humans, the fate of closely related humans such as Neanderthals, and the dispersal of our species, *Homo sapiens*.

Chapter 2 describes climate change during deglaciation, the further dispersal of humans, and the emergence of farming. As the Earth began to warm and emerge from maximum glacial conditions, a sudden, but short-lived, return to nearly glacial conditions, referred to as the Younger Dryas, occurred around 12,000 years ago. Climate system feedbacks that are also relevant to our own time amplified the return to cold conditions. By 10,000 years ago, global climate had entered a period of relative stability.

The warming trend after the LGM provided greater opportunities for hunter-gatherers. With the end of the Younger Dryas, the resumption of the warming trend created hospitable conditions for the expansion of farming and the rise of agricultural societies that for the most part provided the basic model for long-term expansion of human population and the emergence of multiple complex societies.

In Chapter 3 we consider climate and human interactions by focusing on various timescales, from the long-term changes in global climate that influenced human evolution and early innovation, to short-term oscillations that have more regional footprints and consequences. Oscillations with a regional, rather than a global, footprint have dominated millennial- and centennial-scale climate change. This chapter discusses drought, focusing on the climate conditions that characterize and lead to extensive aridity. This was a period of generally favorable climate for humans, but the chapter also notes how climate fluctuations pressured and at times appear to have undermined human societies. Greater aridity around 4,000 years ago, for example, contributed to the demise of the Indus River Valley civilization. The chapter discusses the challenges created by climate shifts in the late Bronze Age, and concludes by outlining the interaction between climate and Rome and the Han dynasty in China.

Chapter 4 introduces regional climate shifts from 500 to 1300 CE and the effects of climate fluctuations during this time period. The chapter describes a period of relative warmth from 900 to 1300 CE, referred to as the Medieval Climate Anomaly (MCA), which shows up in various climate and historical records from the North Atlantic region. Many European societies expanded during this period. The chapter also outlines regional climate fluctuations, particularly droughts in Asia and in the Americas, and the interaction between climate shifts and societies in China, Southeast Asia, and North and Central America, including the Maya.

Chapter 5 provides an overview of the climate fluctuation often described as the Little Ice Age. The cause of the Little Ice Age is still a matter of scientific debate. Reduced solar activity, caused by a dearth of sunspots during this time period, may have played a role in the cooling. A series of volcanic eruptions coincident with the cooling onset has also been proposed as a potential cause, as has an alteration in deep ocean circulation. Cooling during the Little Ice Age presented the greatest threats to human societies on the margins of areas of cultivation. In other regions, however, such as the Netherlands, human societies adapted to the cooling.

Chapter 6 outlines key historical changes that made humans the chief agents of climate change. With the Industrial Revolution, Britain in the late eighteenth and early nineteenth centuries created a new path for production that broke all previous constraints on growth. The critical use of fossil fuels provided the unprecedented capacity to exploit resources to generate power, creating a startling pace of change and a shift toward urban living. The further spread of industrialization through several phases of globalization transplanted the industrial model across ever-wider regions of the globe during the nineteenth and twentieth centuries. The spread of industry powered by fossil fuel, in turn, measurably altered the makeup of Earth's atmosphere. This chapter also provides an overview of early scientific research into the greenhouse effect and global warming. As far back as the nineteenth century, scientists such as John Tyndall and Svante Arrhenius described the warming effect of greenhouse gases in the atmosphere.

Chapter 7 takes the discussion of climate change and human societies up to the present. Modern records of climate change, such as global and regional temperature trends, changes in precipitation, sea-level rise, and ice retreat, highlight the extensive climatic changes that have already been observed. The impacts of climate change on human societies include rising sea levels that most immediately threaten coastal areas and changes in precipitation that affect agriculture and water supply. The chapter outlines how climate change is transforming biomes, magnifying threats including wildfires, contributing to sea-level rise, and precipitation extremes and interacting with other factors to increase stress on human societies.

Chapter 8 reviews projections for future climate change and areas of uncertainty. Climate models are introduced, along with descriptions of future scenarios that are used to run the models and various model outputs. A host of human factors, including debate and negotiation at the international level, politics, economic interest, and cognitive biases, affect understanding of climate change and its effects.

The final chapter introduces responses to climate change. The chapter discusses responses to intensifying climate change, including new social and political movements. The chapter also discusses approaches to adapt to climate change, as well as strategies to reduce the human impact on warming. Along with decarbonization, the chapter outlines the controversy over geoengineering.

This book draws on both history and a growing body of climate research to introduce the relationship between climate change and human history in multiple settings over many thousands of years. At the same time, ongoing research continues to add to the knowledge of climate change and history in multiple regions. Thus the history of climate change and human history promises to become ever more global.

CHAPTER ONE

A fragile start

- Global cooling
- Forest habitats
- From woodlands to savannas
- Hunter-gatherers and dispersal
- Climate change and *Homo sapiens*
- Neanderthals and *Homo sapiens*
- The last glacial maximum and *Homo sapiens*
- Summary

Long before it influenced the history of modern humans, climate change over hundreds of thousands and even millions of years helped shape the evolution of our species. Multiple factors influenced human evolution and, indeed, all of human prehistory and history. Just as neither politics, nor economics, nor culture, nor religion by itself determines history, neither does climate change by itself make a particular historical outcome inevitable. However, climate change was a pivotal driving force for human evolution.

Climate change took several forms during the period of human evolution. Human ancestors evolved during a period of general cooling. Within Africa, the opening of the Great Rift Valley led to increasing aridity in East Africa, where most hominin species originated. Cycles of glaciation, often termed ice ages, sharply affected habitats starting 2.58 million years ago during the period called the Quaternary. These cycles of glaciation periodically shifted or moved the habitats in which our human ancestors lived. All of these climate trends affected the evolution of human ancestors, and the cycles of glaciation were key factors that led to the dispersals of human ancestors and eventually to that of humans.

Global cooling

An overall cooling trend prevailed during the millions of years when the ancestors of gorillas, chimpanzees, and humans diverged from their common ancestor and began to evolve on their own. Evidence of a long-term global cooling for at least the past 50 million years comes from remains of microscopic marine organisms (fossilized plankton) buried as sediment on the ocean floor. Fossil vegetation also suggests a warmer climate prior to 50 million years. Our largest ice-covered continent, Antarctica, was ice-free until about 35 million years ago, while the formation of large ice sheets in the Northern Hemisphere came later, around 3 million years ago. Embedded within this long-term cooling trend are shorter-term variations in temperature, but the overall trend has been toward a cooler climate.

Several ideas have been proposed to explain the growth of ice sheets on Antarctica and later those in the Northern Hemisphere. On these long timescales—millions of years—the shifting of continents plays an important role in climate change. The isolation of Antarctica is one example: Australia became separated from Antarctica around 35 million years ago, followed by the opening of the Drake Passage between South America and Antarctica 25 to 20 million years ago. This continental rearrangement allowed for the formation and strengthening of an ocean current that encircles Antarctica and isolates it from the warming effects of tropical currents. Some climate modeling efforts indicate that the opening of the Drake Passage could have led to cooling in the southern high latitudes, initiating ice-sheet growth on the polar continent.[1]

Global cooling associated with a decrease in atmospheric carbon dioxide (CO_2)—the chemical effects produced by the rise of the Himalayan mountains—provides an alternative explanation for Antarctic glaciation.[2] Proxies for atmospheric CO_2 buried in deep-sea sediments, including carbon isotopes of organic molecules[3] and boron isotopes used to infer ocean pH[4], indicate a decline in CO_2. The decrease in atmospheric CO_2 at this time was most likely the result of enhanced chemical weathering associated with the uplift of the Himalayas and the Tibetan plateau, which began around 50 million years ago.[5] The overall drawdown of CO_2 since then helps explain the long-term global cooling trend that continued into the Quaternary, influencing the formation of ice sheets in the Northern Hemisphere around 3 million years ago (mya) and subsequent changes that drove human evolution.

Further shifts in the alignment of continents may have contributed to the Northern Hemisphere glaciation around 3 million years ago with the formation of the Isthmus of Panama and the closure of the Central American seaway.[6] Closure of the seaway enhanced the delivery of warm salty water to the North Atlantic via the Gulf Stream, which in turn strengthened deep-water formation in the North Atlantic. An intensified overturning circulation increased moisture supply to the atmosphere and, along with the cooler temperatures, set the stage for glaciation. Cooler Northern Hemisphere summers brought on by changes in Earth's tilt may have provided the ultimate trigger for ice-sheet formation.[7] As in the case for Antarctic glaciation, climate feedbacks then helped sustain ice-sheet growth.

The expansion of Northern Hemisphere ice sheets coincided with a major transition in Earth's climate, from a relatively warm period characterized by an ice-free Arctic before 3 million years ago, to one characterized by the cyclical expansion and retreat of ice sheets governed by changes in Earth's orbit. Temperatures just before this transition were, on average, 3°C warmer than today. This also represents the most recent time in Earth's history when CO_2 levels matched our current level around 400 ppm. With the cooling that followed, many regions became drier, particularly in Africa, and this drying of the continent led to a shift in vegetation that influenced human evolution.

Within Africa, the shifting of tectonic plates further altered the climate, in particular in East Africa along a series of rifts where the Earth is spreading apart. Although the development of the East African Rift System may have begun as early as 45 million years ago, uplift and notable landscape changes accelerated around 10 million years ago. The formation of the rift valley transformed the region from a relatively flat tropical forest biome to one with varied topography and vegetation. The terrain is now composed of deep rift valleys as well as tall mountains—the most famous are Mount Kilimanjaro and Mount Kenya—and lake basins. The mountains of the rift system intensified the overall drying trend as the high peaks blocked moisture from the Indian Ocean, creating a rain shadow desert in this region. The increasing aridity may have led to the expansion of grasslands and savannas, and may have created an environment increasingly sensitive to hydrological changes. The overall decline in atmospheric CO_2 during this time period may also have contributed to the expansion of savanna vegetation. Recent research indicates that savanna grasses became more abundant beginning 10 million years ago, but before the onset of drier conditions around 7 million years ago. Because the grasses are adapted to low CO_2 conditions, this vegetation shift may have been triggered by the global decrease in CO_2.[8]

Forest habitats

The cooling and drying trend brought significant changes in the habitat in which human ancestors lived—the rain forests of Africa. We are *Homo sapiens*, also described as anatomically modern humans, or AMH for short, and we are the last survivors of a host of related species. *Homo sapiens* are the only living species of hominins. Chimpanzees and anatomically modern humans shared a last common ancestor approximately 6 to 7 million years ago. In comparison to any other species, chimpanzees are the closest living relatives to *Homo sapiens*—we share more than 98 percent of our DNA. Gorillas, the second most closely related species to humans, share some 98 percent of their DNA with us. Gorillas and modern humans shared a common ancestor approximately 9 to 12 million years ago, though these dates are more open to debate.

Although we are more similar to chimpanzees than to any other living species, the differences between humans and chimpanzees stand out today. *Homo sapiens* have colonized most of the globe and have pushed into regions where they were once unknown. In contrast, chimpanzees and gorillas persist in often-shrinking habitats in Africa. Some subspecies such as the mountain gorilla currently survive in small zones in East Africa. Population attests to the current dominance of humans. As of 2020, total human population exceeded 7.7 billion people. In contrast, the chimpanzee population in the wild in Africa has been estimated at somewhere between 150,000 and 250,000, and estimates of the gorilla population in the wild range from 100,000 to 150,000, mostly western lowland gorillas. We also use incomparably more energy than do our closest relatives. The average carbon footprint, the amount of CO_2 emitted per year by the activity of an individual, is about four tons per person per year for humans, with an average three to five times that for residents of comparatively wealthy countries. Chimpanzees and gorillas have essentially no carbon footprint.

How and why did the ancestors of modern humans evolve on a different path from their most closely related species? Humans in no way descended directly from either chimpanzees or gorillas, or indeed, from any other living animal. For that matter, chimpanzees did not descend directly from gorillas. However, we do share a common ancestor. The study of chimpanzees, in particular, provides vital clues for reconstructing where and how our shared common ancestor lived and for how climate change affected the emergence of human ancestors. Chimpanzees live in the greatest density and numbers in the rain forests of tropical Africa. They also survive in woodland forests, and a few populations make excursions into grasslands, but chimpanzees generally stick to areas dominated by trees with fruit. Their preferred habitat is very closely linked to their diet. More than 90 percent of their diet consists of fruit, and they fall back on other plants for the rest. Male chimpanzees eat very small amounts of meat. Even more than chimpanzees, gorillas are restricted to forested areas. They are split up into several subspecies in the rain forests of Africa. By far the largest number of gorillas live in the western lowland areas of Africa in a range that crosses through countries including Cameroon, Gabon, the Central African Republic and the Republic of the Congo, and the Democratic Republic of the Congo. Gorillas obtain their nutrition almost entirely from plants. Like chimpanzees, they prefer fruit.

Observation of our closest relatives among all other species helps shape a picture of our common ancestors. The common ancestors of *Homo sapiens*, chimpanzees, and gorillas lived millions of years ago in the forests of Africa, and in particular in the tropics. These rain forests provided the copious quantities of fruit that sustained them. We should imagine an animal more closely resembling the apes than us living in or near trees and foraging for fruit. If the diet preferred by chimpanzees and gorillas is a guide, they ate massive quantities of wild figs. This kind of habitat, then as now, required a high constant temperature with abundant rainfall.

Given their adaptation to life in the rain forest and to eating a diet heavy in fruit, why did an ancestral species to humans evolve along a separate path that led far beyond the rain forest? Chimpanzees and gorillas, after all, did not, so it was still certainly possible to survive and thrive as a species by maintaining the same basic way of life. What, then, propelled the ancestors of humans to begin the process of leaving the rain forests of Africa and moving to a much wider range of habitats? Tracking changes in climate helps answer that question. A shift in climate that led to the expansion of the rain forest would provide a greater range with abundant food, but a shift that reduced the area of the rainforest would diminish food supply. Climate change over millions of years that broke up the edges of rainforest therefore helped drive the evolution of human forebears as distinct and separate from their closest relatives.

From woodlands to savannas

The next group of hominin species, which emerged about 4 million years ago, have been dubbed the australopiths. The australopiths, if we could meet them, would not closely resemble modern humans, but would instead look more like bipedal chimpanzees. The most famous of all the australopiths, discovered in 1974 and given the name Lucy, lived 3.2 million years ago. She stood about 3 1/2 feet tall and weighed about 65 pounds. The males were larger, standing a little under 5 feet tall on average and weighing about 110 pounds or more. Lucy's species, *Australopithecus afarensis*, existed between around 3 to 4 million years ago. We might not find them well suited to life in our modern society, but they were better adapted to walking than their ancestors who had once dwelled in the rain forest, though with much shorter legs they had a far shorter stride than humans.

A shift in climate was likely a key factor in the emergence of australopiths. In a cooling climate, forests broke up and areas of woodland and savannas grew, reducing the supply of the preferred food: fruit. This shift in habitat created selective pressure to consume fallback foods, including tubers, harder and tougher than fruit. Australopiths were able to walk and to dig in search of additional calories. They developed large teeth and jaws that better enabled them to chew hard foods for long periods of time.

Continuing climate change was most likely a pivotal factor in a second key phase in human evolution long before the emergence of *Homo sapiens*. To obtain food in a cooling Africa, human ancestors continued to diversify their diet by engaging in greater food processing and by eating more meat.

Natural selection in a cooling Africa favored traits apparent in the best-known early species of the human genus: *Homo erectus*. First appearing almost 2 million years ago, *Homo erectus* far more closely resembled us than did any of the australopiths. There was much variation, but many of them were taller and had longer limbs and larger brains than australopiths who continued to exist in Africa until approximately 2 million years ago. With a larger brain, *Homo erectus* required more energy, which further gave incentive to procure energy-rich foods such as meat, and placed even more of a premium on walking and running. The tallest *Homo erectus* eventually stood as much as 6 feet tall and weighed up to 150 pounds. *Homo erectus* was far better at walking and running than australopiths. A foot race between an average healthy individual from these two species would not even be close. If we could meet our *Homo erectus* ancestors the experience would be disconcerting: they would far more closely resemble us than any species alive today, but differences would still be readily apparent.

Homo erectus were hunter-gatherers. Their ability to run and to dissipate heat by sweating made them capable persistence hunters. Hunter-gatherers on the African savannas or grasslands could cover long distances in pursuit of meat. Individual animals could still run at greater speeds than could *Homo erectus*. The same is true of the gap in maximum speed between human hunters and much of their prey today. However, *Homo erectus* could maintain pursuit for far longer, sometimes walking, until some of their prey tired and overheated.

Homo erectus also created and employed more complex tools than earlier species—the oldest stone tools date to 3.3 million years. These included hand axes and other tools for hunting and butchering animals. Several sites from East Africa show the remains of animals consumed by humans, and the distribution of prey suggests that humans were hunting rather than simply scavenging off the oldest and youngest animals. Specialized tools enabled butchering. Excavated sites such as Olduvai Gorge in Tanzania and Olorgesailie in Kenya contain large numbers of sharp, flaked tools next to the bones of large mammals such as elephants and giraffes.

Hunter-gatherers and dispersal

As hunter-gatherers, *Homo erectus* and other related humans in the *Homo* genus faced limits on population density. A given area of savanna could only provide enough food to support a small population. Demographic growth over time therefore led to dispersal as hunter-gatherers expanded their range without increasing their population density. A larger population could support itself only if *Homo erectus* groups lived across a wider area.

Climate change during the Pleistocene, the period between 2.6 million and 11.7 thousand years ago, shaped patterns of human dispersal. Early humans experienced significant shifts in climate during the Pleistocene, namely, a series of cold glacial periods that saw the expansion of ice sheets, interspersed with warmer interglacial periods and ice sheet retreat.

During the Pleistocene, shifts in Earth's orbit shaped the growth and retreat of ice sheets. These orbital variations, known as Milankovitch cycles after the Serbian astrophysicist who outlined their effects, include changes in Earth's eccentricity, obliquity, and precession. The shape of Earth's orbit around the sun, or eccentricity, influences how close Earth approaches the sun during its orbit. A perfect circle would place the Earth at the same distance from the sun year-round, while a more elliptical orbit—greater eccentricity—places Earth farther from the sun during part of the year and closer to the sun about six months later. The slow changes to eccentricity, from nearly circular to more elliptical, occur at 100,000 and 413,000y timescales. Obliquity, the angle of Earth's rotational axis, drives the strength of seasonal change—Earth's tilt is why we have seasons at all. Our current 23.5° angle is about midway in the range of values (~22–24.5°) observed over this 41,000-year cycle. The larger the tilt, the greater the difference between winter and summer seasons. Finally, the "wobble" of Earth's axis like a spinning top, along with the rotational shift of Earth's orbit itself, creates climate variations approximately every 22,000y. This precession of the equinoxes, as it is known, shifts the season that Earth experiences during its closest approach to the sun, its perihelion. The Northern Hemisphere currently experiences winter during perihelion, but 11,000y ago, that season was summer.

Beginning around 2.7 million years ago, these three cycles interacted to drive our glacial-interglacial cycles. Milankovitch proposed that the amount of incoming solar radiation, or insolation, during the summer season at high latitudes of the Northern Hemisphere initiated the growth and decay of large ice sheets. In this scenario, a minimum in summer insolation would allow ice sheets to persist into the winter season, and over time, ice sheets would expand. A configuration that created cooler summers (minimal tilt), with the summer season farthest from the sun, enhanced by a more elliptical orbit (greater eccentricity), would yield ideal conditions for ice-sheet growth. Conversely, a maximum in summer insolation would initiate ice melt and the transition to an interglacial period.

Support for the Milankovitch theory first came in the 1970s from shells of microscopic organisms buried in ocean sediments. Oxygen isotopes of the calcium carbonate that make up these shells provide a record of past ocean temperature and global ice volume. The variations in the isotope

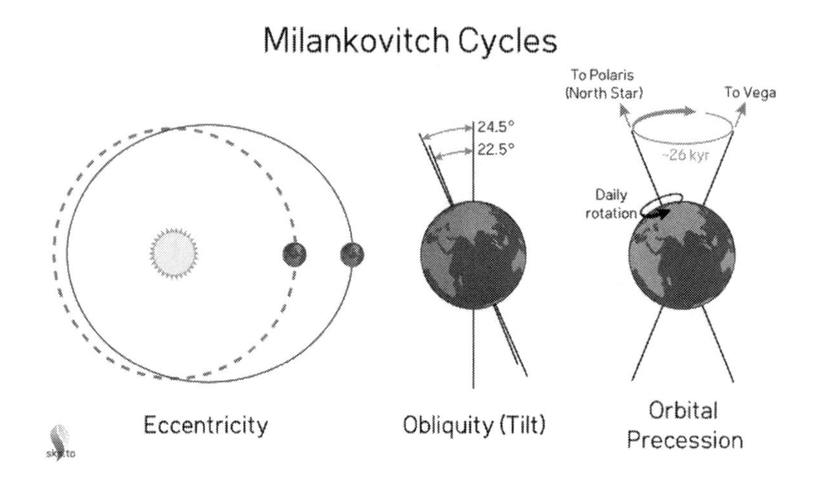

FIGURE 1.1 *Illustration of Milankovitch cycles: Eccentricity, obliquity, and precession. Source: Skeptical Science, https://skepticalscience.com/graphics.php?g=342.*

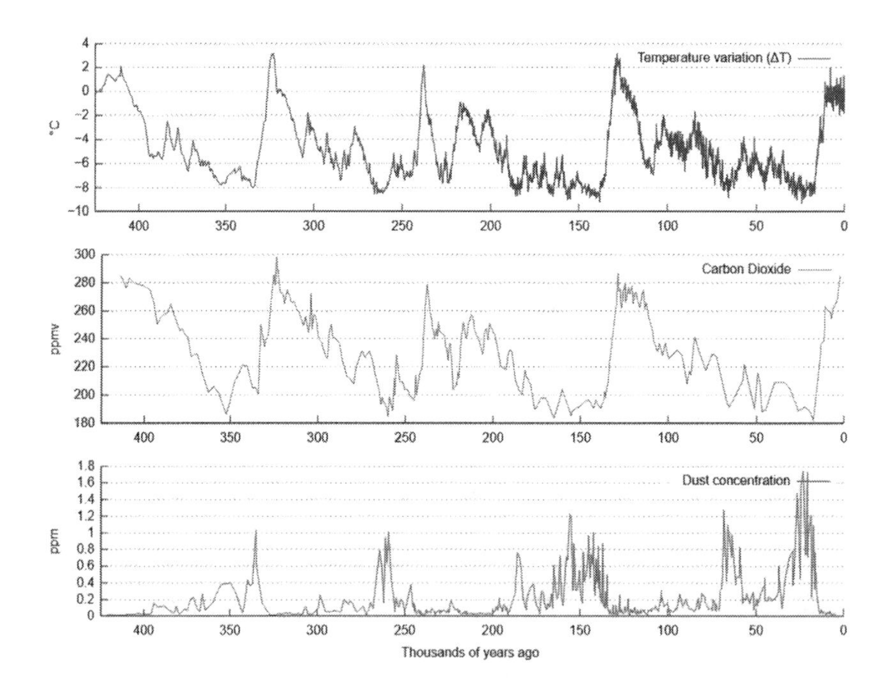

FIGURE 1.2 *Graph of CO_2 (top graph), temperature (middle graph), and dust concentration (bottom graph) measured from the Vostok, Antarctica, ice core as reported by Petit et al., 1999. Higher dust levels are believed to be caused by cold, dry periods. Source: NOAA, https://upload.wikimedia.org/wikipedia/commons/b/b8/Vostok_ Petit_data.svg.*

record were found to be consistent with expected glaciations according to the Milankovitch theory;[9] numerous studies based on a variety of geological records have since corroborated this finding. These records reveal that the 41,000y obliquity cycle dominated climate swings between 2.7 and 0.9Ma. Since that time, glacial-interglacial periods followed a 100,000y cycle, with sharper shifts between warm and cold periods than before 0.9Ma. The glacial maxima, which took up water in the form of ice, lowered sea level.

Sea-level shifts driven by change in global ice volume altered the dispersal routes available to *Homo erectus*. Long before *Homo sapiens* colonized most of the globe, *Homo erectus* dispersed out of Africa to populate sites across much of southern Eurasia. The first discovery of a skeleton that turned out to be from a *Homo erectus* individual was made in 1890, nowhere near Africa but in the Dutch East Indies, in what is now Indonesia. Eugene Dubois, a physician and anatomist inspired by the work of the German scientist Ernst Haeckel, set out to find a fossil that would fill a gap between modern humans and apes. In 1890, he discovered a fragment from a jaw, and digging in 1891 yielded the top of a skull. Because of their place of discovery, the specimens became known as Java Man. As early as 1.6 million years ago *Homo erectus* reached China and Indonesia: the finds from China were dubbed Peking Man. In 1950, Ernst Mayr, a preeminent evolutionary biologist, identified both Java Man and Peking Man as *Homo erectus*. The oldest *Homo erectus* remains were later discovered in East Africa.

Very limited evidence from large parts of the world in the form of fossil remains makes it difficult to reconstruct all the phases of dispersal. *Homo erectus* most likely followed routes along or near the coast to disperse across wide stretches of Eurasia. Periods of abrupt sea-level change could create a broken coastline that impeded or slowed long-distance travel along the coast. However, lower seas revealed or expanded land bridges. Such land bridges would have aided *Homo erectus* in reaching areas of Indonesia. Thus, *Homo erectus* may have traveled along an emergent land, later dubbed Sundaland, to cross into areas of Indonesia such as Java. Today, the Sunda Shelf region of Sundaland is a submerged section of Asian continental shelf lying at a depth of less than 100 meters below the ocean, with many even shallower regions.

Within Eurasia, the expansion of ice sheets reduced the range of the most favorable living areas available. *Homo erectus* survived several glacial maxima, but may have done so in smaller zones or refugia. The constriction of population could have led to the isolation of populations, to the further evolution of humans, but also to extinction.

Several forms of humans evolved from *Homo erectus*. These included our species, *Homo sapiens*, along with *Homo heidelbergensis*, *Homo neanderthalensis* (Neanderthals), a population dubbed the Denisovans, and *Homo floresiensis*, a population that likely evolved from *Homo erectus*.

Homo heidelbergensis dates back some 700,000 years. Neanderthals, found mainly in Europe, date back around 200,000 to 300,000 years. The Denisovans were identified only in 2010 from genetic analyses of bones found in a cave in southern Siberia, and the extent to which populations such as the Denisovans were distinct from *Homo heidelbergensis* remains under discussion. The small skeletons found on the island of Flores in Indonesia provide an example of another human. These fossils are a separate species, *Homo floresiensis*, popularized under the name Hobbit.

Glacial maxima created major challenges for human populations, in particular near the northern range of habitation, but humans also developed greater resiliency to climate shifts during the Pleistocene. Humans started to exploit fire. Evidence for the first use of fire dates back as far as 800,000 or even 1 million years ago, but by 400,000 years ago humans made regular use of fire. Survival in colder areas of Europe suggests that *Homo heidelbergensis* employed some kind of clothing, though no direct evidence survives from this time. They also created sharper points for spears. Greater technological capacity and the ability to manipulate materials points to what we might term *culture*, and that emergence of culture allowed humans to exploit resources in regions that at one time would have been beyond the limits of survival for human ancestors. Shifts toward a cold and dry climate, nonetheless, would have pushed back the northern range of habitation and likely separated human populations.

Climate change and *Homo sapiens*

Homo sapiens, meanwhile, emerged in Africa around 200,000 to 300,000 years ago. Scientists long debated the place of origin of *Homo sapiens*. One model suggested multiple sites of origin in Eurasia, but the case for our African origin is now overwhelming, both because of the physical record found through discoveries of fossils and archaeological sites, and also because of genetic analysis. There is much greater genetic diversity among today's human populations in Africa than among all human populations outside of Africa. This pattern of variation is strong evidence for an African origin: genetic diversity has been accumulating in Africa for more than 200,000 years, but non-Africans derive from a subset of Africans who left Africa only some 60,000 years ago. Only in Africa did *Homo erectus* eventually give rise to the core lineages of *Homo sapiens*.

Possibly the most difficult question to answer about the prehistory of *Homo sapiens* is how and why exactly *Homo sapiens* developed modern behavior as demonstrated by more complex art and artifacts. Continued debate surrounds the precise cause of these changes. In one interpretation, the changes took place as early as around 100,000 years ago, but there is also abundant archaeological evidence for an explosion of human creative activity as indicated by artifacts from around 50,000 years ago.

Climate change affected the conditions humans faced during the long prehistory of *Homo sapiens* and during movement out of Africa. The Milankovitch cycles continued to interact to create glacial maxima and interglacial periods. Shorter periods of variability also occurred between peak interglacial periods. During the interglacial period that peaked between 124,000 and 119,000 years ago, warm conditions created high sea levels. Sea-level high-water marks, called highstands, occurred at 124,000 years ago and again at 105,000 and at 82,000 years ago. Overall climate shifted during the substages of the period from 130,000 to 80,000 years ago, but climate cooled after 74,000 years ago.

Within the sequence of glacial and interglacial periods, *Homo sapiens* also experienced and survived several periods of abrupt climate shifts. These included abrupt warming episodes followed by gradual cooling, termed Dansgaard-Oeschger (D-O) events, and rapid cooling episodes known as Heinrich events. These rapid oscillations in climate exposed humans to an abrupt warming—within a decade—that followed a roughly thousand-year cold spell. The warmth that punctuated overall colder climate lasted 200–400y, then gradually gave way to cold again. The release of ice-rafted debris during some of the coldest of these Heinrich events occurs in the sediment record every 7,000 to 12,000y. *Homo sapiens* and any other existing humans therefore experienced significant climate oscillations that probably exceeded any that have occurred during our current Holocene, the period since the definitive end of the last glacial maximum, roughly 20,000y ago.

Changes in thermohaline circulation, or deep ocean circulation, help explain a number of abrupt changes in Earth's climate history, including the Heinrich events and D-O cycles. Oceanic deep water currently forms in polar regions of the Atlantic Ocean where the water is very cold, and the salty Atlantic seawater becomes even saltier due to the formation of sea ice near the poles. These conditions increase the density of the water, causing it to sink to the seabed and flow throughout the ocean depths. Deep water eventually rises again to the surface, creating a circuit that takes roughly 1,000–1,500 years to complete. Pulses of freshwater to the regions of deep-water formation can interrupt this process, which slows the overturning circulation. Because this ocean-wide circuit—the Ocean Conveyor Belt—helps draw warm waters north from the tropics, its weakening can lead to cooling in the North Atlantic, such as seen during Heinrich events.

On a shorter timescale, large volcanic eruptions can create comparatively short and abrupt climate shifts. Large volcanic eruptions reduce temperature temporarily; the degree of cooling depends on the type and location of eruption. To take a recent example, global temperatures dropped by as much as 0.4°C after the Mount Pinatubo eruption in the Philippines in 1991. Temperatures fell by as much as a few degrees Celsius, at least in some regions, after the Tambora eruption of 1815. Using these more recent examples as guides, temperatures could have fallen significantly after the much larger Toba supervolcano of 74,000 years ago, which pushed vastly more ash into the atmosphere. The Campanian Ignimbrite eruption at 40,000 years ago likely led to cooling and occurred in close proximity to a Heinrich event, potentially magnifying the cooling effects.

In addition to these climate shifts, changes in tropical precipitation appear to have had the greatest effect on *Homo sapiens* dispersals. In Northern Africa and the Middle East, changes in Earth's precession drove climate variability in the form of wet and dry periods every 22,000y. This pattern, described as a Pleistocene "pump," expanded habitats available for human dispersal and created paths for dispersal between continents.[10] During wetter periods, hunter-gatherers dispersed northward. These wetter periods in North Africa have been described as the Green Sahara.

Today, the Sahara is by far the largest subtropical desert in the world. Stretching across much of North Africa, it forms an imposing boundary. Travelers across the desert must proceed with caution, carrying stores of water. The bones of many animals that no longer inhabit the region as well as pictures inscribed in rock outcrops in the Sahara demonstrate, however, that the region has not always been so arid.

Just as Milankovitch cycles influence the growth and decay of ice sheets, there is abundant evidence that the precession cycle in particular drives changes in monsoon strength[11] associated with the Green Sahara. An insolation maximum during the subtropical summer causes a stronger summer monsoon, giving rise to the Green Sahara every 22,000 or so years. Evidence for wet periods includes lake levels in Africa, mud deposits in the Mediterranean Sea, and microfossils buried in sediments of the equatorial Atlantic.

Mediterranean mud deposits help us to document the sequence of periods of a Green Sahara. These deposits contain layers that are rich in organic matter, and their presence indicates overlying water with a lack of oxygen. Today's Mediterranean has an overturning circulation that brings oxygen-rich waters from the surface to the sea bottom, aerating the seabed. At times of higher-than-normal river flow, the overturning is reduced, and little to no oxygen reaches these depths. During these low oxygen periods, the biological remains of surface-dwelling plankton are preserved in the form of organic-rich muds. The sequence of these sapropels, as the organic-rich layers are known, provides a record of high river flow that results from strong monsoon rainfall. The intervals between the observed sapropels in the Mediterranean, and thus the Green Sahara periods, are consistent with the 22,000y precessional cycle proposed to drive monsoon strength.[12]

Analysis of microfossils deposited in ocean sediments provides additional evidence for Green Sahara periods. The key microorganism, in this case, is a particular type of freshwater algae observed at 22,000y intervals in marine sediments off the west coast of Africa. As a freshwater species, the algae must have originated from land, thriving at a time of high lake levels. These lakes eventually dried, and winds that blew over the ancient lake beds later carried algal remains out to sea. The 22,000y interval of algal remains, together with the sapropels and studies from other regions, such as cave deposits in China and Brazil, provides support for the hypothesis that precession drives long-term changes in monsoon strength.

As we will see for later periods of human history, severe drought has posed a threat to past human civilizations, and scientists have also pointed to the possible role of aridity, rather than increased precipitation, in helping to propel human dispersal during the long prehistory of *Homo sapiens*. In this model, climate change influenced dispersion more by pushing people out of dry areas than by pulling them into wetter areas. A study of sediment cores from the Gulf of Aden off the Horn of Africa, for example, indicates dry conditions during the main period of *Homo sapiens* dispersal from Africa, though wetter conditions predominated during earlier periods of *Homo sapiens* dispersion.[13]

Changes in Earth's orbit have therefore influenced human dispersal globally, both through the pacing of tropical humidity and through the expansion and retreat of large ice sheets. Abrupt climate shifts may have had a more regional effect. Climate models show a limited effect of D-O events on overall global dispersal of *Homo sapiens*, though these events may have affected the ability of humans to survive in the Levant.[14] As for the possible effect of Toba, some investigations of organic matter show surprisingly little disruption from the eruption. A supervolcano of this size could have created veiled skies and markedly cooler temperatures that could have persisted for more than a year, causing a sharp drop in food supply and famine. In a different scenario, the aerosols released by Toba may have entered the atmosphere in conditions that somehow minimized the cooling effect. According to current findings, Toba could therefore have either produced a harsh volcanic winter or had milder effects. One potential effect of a large volcanic eruption is a significant reduction in population. Genetic analysis indicates that *Homo sapiens* population passed through population bottlenecks in which the number of individuals may have fallen into the thousands. We cannot say precisely what events or factors created such narrow bottlenecks when global human population dropped to levels present today in a small town, but the Toba eruption is one candidate. However current understanding of dispersals indicates resilience by *Homo sapiens* populations in at least some areas to the aftereffects of the eruption. Humans in South Africa, for example, appear to have thrived through the period of the eruption, suggesting their ability to make use of abundant resources.[15]

Much like *Homo erectus*, *Homo sapiens* dispersed out of Africa, and most likely on more than one occasion. The expansion and contraction of savanna and woodland areas periodically opened corridors for *Homo sapiens* dispersal. Indeed, *Homo sapiens* entered the area of the Middle East and either retreated or died out before a later period of dispersal. Such early dispersals have contributed very little genetic material to modern-day non-African populations.[16] Today's non-African populations are descended almost entirely from *Homo sapiens* populations that dispersed out of Africa in the period between 50,000 and 80,000 years ago. *Homo sapiens* dispersed along the southern coast of Asia, but may have dispersed more slowly into Europe.

Even with lower sea levels, *Homo sapiens* took to the sea to reach Sulawesi in what is now Indonesia and Australia. Lower sea levels narrowed the distance between Asia and offshore islands but still required journeys by some kind of seacraft of up to 20 miles or even 60 miles, to reach the continent of Sahul made up of both modern-day Australia and New Guinea. Anatomically modern humans arrived some 50,000 years ago or more in Australia: recent excavations of a rock shelter suggest human presence 65,000 years ago.[17] Genetic analysis confirms that the continent's aborigines descended from a single population that then dispersed widely across Australia. Long dry spells after some 40,000 years ago reduced population density of the hunter-gatherer populations in the region of Lake Mungo.[18]

The arrival of humans and climate change were both possible causes for megafauna extinctions in Australia. Megafauna extinction followed after human arrival, though an earlier date for continuous human presence would mean a longer period of coexistence between humans and megafauna. Remains of megafauna disappear at Lake Mungo after 46,000 years ago, and most megafauna in Australia became extinct by 45,000 years ago. Along with hunting, humans created new pressures for megafauna by remaking the landscape through fires.[19] The threat from humans may have combined with challenges from fluctuations in climate. In northern Australia, for example, megafauna extinctions took place during a dry period with reduced rainfall.[20] Even in small numbers, human entrance into southeastern Australia created additional stress for megafauna during a period of climate change.[21]

A short sea journey was also necessary to reach Japan. *Homo sapiens* established a foothold in Japan by around 38,000 to 35,000 years ago and hunted now-extinct species such as Naumann's elephant.[22] Explaining the cause of extinction depends to a large degree on the dating of the most recent megafauna. Thus older dates, soon after human arrival in Japan would point to a key role for human hunting in killing off megafauna, but later dates during the last glacial maximum (LGM) suggest a key role for climate change.[23]

Neanderthals and *Homo sapiens*

Climate shifts affected all human populations, including Neanderthals. Of all the human extinctions, that of Neanderthals, by far the best documented other human besides *Homo sapiens*, has long posed a puzzle. Neanderthals lived in Europe and western Asia before the arrival of anatomically modern humans, and they coexisted alongside us for at least several thousand years. Neanderthals were shorter and stockier than *Homo sapiens* and had shorter limbs. They had large brains the same size as those of Pleistocene *Homo sapiens*. They made and used sophisticated tools, used fire, and buried their dead, though *Homo sapiens* eventually developed a more complex array of tools. Why, then, are Neanderthals now extinct? Was this extinction the result of competition, climate change, systemic problems particular to Neanderthals, or of some combination of all of these factors?

Both Neanderthals and *Homo sapiens* had survived the previous glacial maxima. Thus, *Homo sapiens* existed in Africa during the penultimate glacial maximum around 130,000y ago, before the most recent glacial maximum 20,000y ago. Cool conditions between 190,000 and 130,000 years ago could have produced a population bottleneck in the small *Homo sapiens* population and pushed humans to coastal areas of southern Africa. Neanderthals, for their part, had also survived the penultimate glacial maximum. Indeed, Neanderthals' shorter limbs are evidence that they were better adapted than *Homo sapiens* to endure cold conditions, though this advantage may only have been modest. Both *Homo sapiens* and Neanderthals would have required clothing to move and thrive in cold regions on the edge of their range during glacial maxima, and glacial maxima likely pushed Neanderthal populations south and caused loss of lineages.

The small Neanderthal population placed the species in an increasingly precarious position. Humans display less genetic diversity than do other mammals, which makes them more vulnerable to environmental change. Modern *Homo sapiens* display far less genetic variation than is found among chimpanzees. Analysis of DNA from Neanderthals suggests even less genetic diversity than for *Homo sapiens*. With a small and isolated total population, modest increases in mortality could have created a glide path toward Neanderthals' extinction.

Competition with *Homo sapiens* may also have threatened Neanderthals. We do not know and may never know all the ways in which *Homo sapiens* and Neanderthals interacted, but as anatomically modern humans moved into the Middle East and Europe, they competed for resources with Neanderthals. We do not know for a fact that *Homo sapiens* always prevailed everywhere in such competition, but *Homo sapiens* population density probably exceeded Neanderthal population density.[24] A reduction in food supplies could have pushed Neanderthals to the brink. In extreme versions of this scenario, the competition could have extended to violent conflict, though we have no evidence of such violence or "warfare." Simply taking over resources would have produced much the same outcome for a small Neanderthal population.

Homo sapiens and Neanderthals interbred, though this was a rare occurrence. New gains in isolating and analyzing Neanderthal DNA show that Eurasians—people who had left Africa—share about 2–3 percent of their genome with Neanderthals. This interbreeding most likely occurred some 50,000–60,000 years ago as indicated by analysis of DNA from the thigh bone of a *Homo sapiens* male from 45,000 years ago found in western Siberia. This individual had the same percentage of the Neanderthal

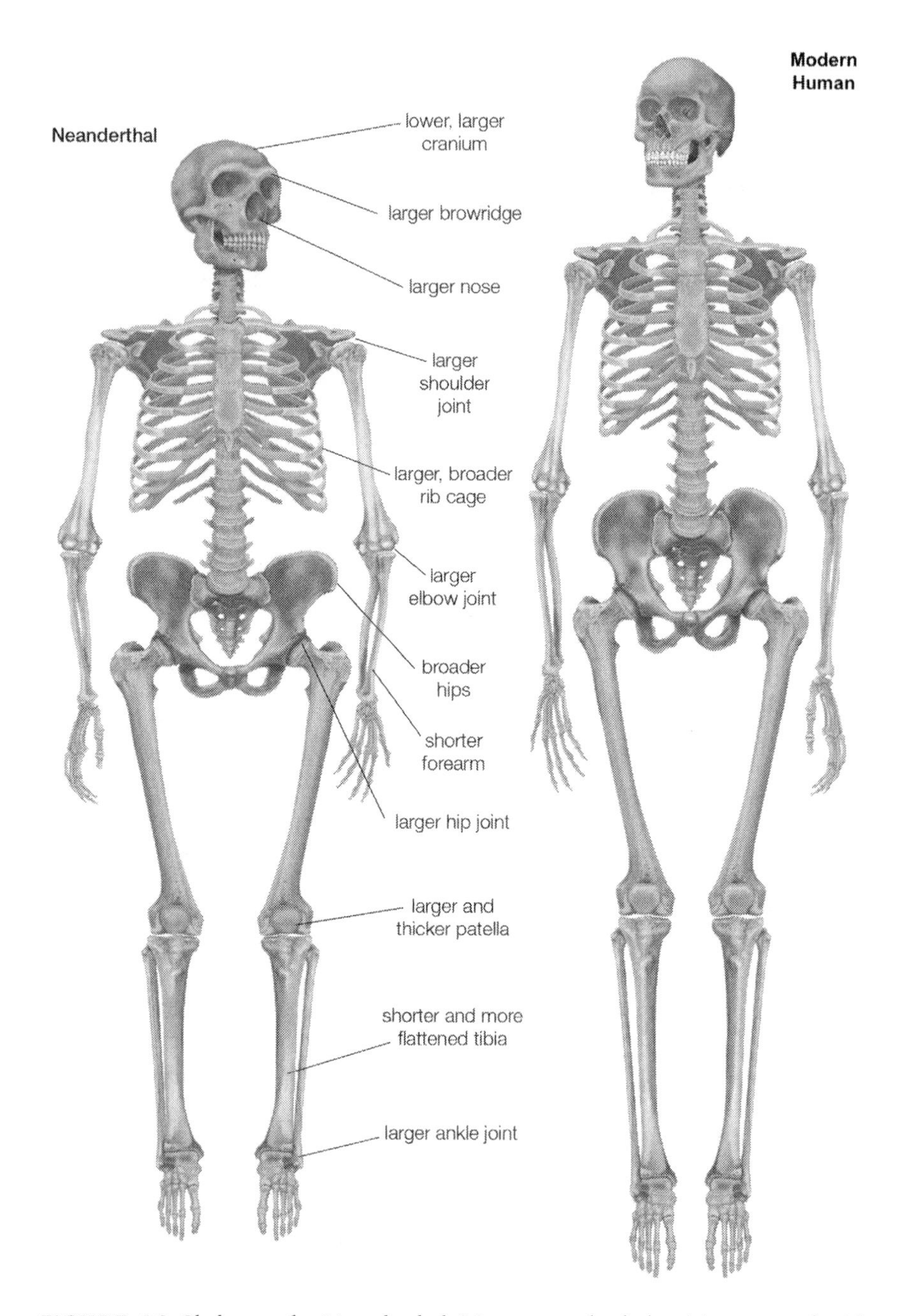

FIGURE 1.3 *Skeleton of a Neanderthal (*Homo neanderthalensis*) compared with a skeleton of a modern human (*Homo sapiens*). Source: Encyclopaedia Britannica/UIG/Bridgeman Images.*

genome as that found in modern Eurasian populations, though the DNA is less fragmented than in modern individuals. Genetic analysis revealing two sets of Neanderthal genes for East Asians and Europeans, and a third set in modern East Asians, suggests that there may have been more than one interbreeding event.[25] In similar fashion, *Homo sapiens* also interbred on occasion with the Denisovans—human populations in Melanesia share some 3–5 percent of their genome with Denisovans.

Debate over the date of the last Neanderthal remains has forced rethinking of the conditions at the time of Neanderthal extinction. The youngest Neanderthal bones at European sites have traditionally been dated by radiocarbon, but the quantity of carbon 14 that remains in samples at ages of 30,000 or 40,000 years ago is so small that even modest background contamination can easily throw off calculations of dates. A new technique for ultrafiltration of collagen, the chief component of bone, has removed small amounts of modern carbon and pushed back the dates for the last Neanderthal sites in much of Europe. Thus Neanderthal bones from sites in Spain once thought to be some 35,000 years old turned out to be some 50,000 years old according to the new techniques. Redating many of these last Neanderthals has changed the picture of Neanderthal extinction in at least two important ways. First, Neanderthals died out well before the LGM: they were not then direct victims of the last "ice age." Second, the period when Neanderthals and *Homo sapiens* coexisted near each other in the same regions was shorter than previously thought.

Even if no single shift in climate was the pivotal cause of extinction for a species that had already lived through a previous glacial maximum, climate change multiplied the challenges for a small population facing competition from *Homo sapiens*. Before the LGM, expanding ice sheets in Europe reduced the range available to both the established Neanderthal population and to the incoming population of *Homo sapiens*. Around 60,000 years ago an ice sheet covered most of the British Isles along with Scandinavia, the Baltic, and parts of northern central Europe. Swings between warming and cooling 50,000 to 40,000 years ago alternately expanded and contracted accessible terrain. Much like other species, Neanderthals would have survived in glacial refuges or refugia when ice sheets expanded, and the effects may have been more damaging for the small Neanderthal population than for *Homo sapiens*. Cooling at around 40,000 years ago, coincident with both a Heinrich event and the Campanian Ignimbrite eruption, would in this scenario have posed a greater threat to any dwindling population of Neanderthals (if they still existed) than to *Homo sapiens*. With the extinction of Neanderthals, *Homo sapiens* were the only surviving humans. The descendants of *Homo erectus* on Flores, once thought to have persisted until 17,000 years ago, may have become extinct by 50,000 years ago.

Human ancestors and humans closely related to *Homo sapiens* displayed resilience in surviving through shifts in climate, but a 2020 study points to a general connection between climate change and extinction of human ancestors and other humans. Analysis of a fossil database indicates that *Homo erectus*, *Homo heidelbergensis*, and Neanderthals all lost much of their climatic niche shortly before their extinction. All of these species dispersed over long distances and their technology and use of fire in principle created significant resilience to climate fluctuations, but the climate niche for the *Homo* species related to modern humans shrank abruptly before extinction. Identifying such climate stress for Neanderthals does not rule out the effects of competition with *Homo sapiens*—thus competition and climate stress could have both amplified each other.[26]

From our present perspective, the possible interaction between habitat fragmentation and climate change foreshadows what is now taking place on a global scale when climate change shifts habitats for a multitude of plant and animals even as encroachment by humans makes many of those same species increasingly vulnerable. If competition for land and food amplified the effects of climate change to help push Neanderthals to the brink, enormous numbers of species of flora and fauna face profound threats today from far greater encroachment by *Homo sapiens* and from the rapid climate change we have set in motion.

The last glacial maximum and *Homo sapiens*

From 33,000 years onward *Homo sapiens* experienced pronounced cooling leading up to the LGM. The LGM revealed both human dependence on climate and adaptability. Cooling and glaciation made some regions uninhabitable for people. Human populations disappeared in regions they had once settled, such as Britain, by around 25,000 years ago. In East Asia, humans persisted in areas of China south of 41° Latitude, though hunter-gatherers may have moved in and out of the region to the north.[27]

Beyond the ice sheets, humans lived in glacial refugia in southern European peninsulas in Spain, southwestern France, Italy, and the Balkans. The Cantabrian reign in northern Atlantic Spain was a refuge for humans and for other animals, including salmon and red deer. Humans in Cantabria hunted animals, such as Ibex, that also thrived in the region.[28] The expansion of ice caused biomes of plants and animals to shift as well. Tundra, today found in the Arctic, retreated south during the LGM, constricting the areas with abundant vegetation and moisture. In Africa, the Sahara was actually even larger during the LGM than today's vast desert, extending some 400 kilometers further to the south.

The LGM encouraged dispersal out of some regions, but humans still managed to gain resources in regions where the cold trend substantially altered the makeup of vegetation and animals. Tundra regions were not altogether emptied of human presence, and people survived in steppes. *Homo sapiens* persisted in Siberia even during the LGM. Except for the northwest, Siberia had few glaciers. Polar desert prevailed in much of the north above belts of tundra and steppes. The climate was harsh: colder, especially in winter, and more arid than in the historic era. To survive in this region during the glacial maximum required extensive use of clothing and fuel. Humans in the region provided for themselves by hunting reindeer, bison, horses, and sheep. Upper Paleolithic human sites from the LGM also show the bones of now-extinct animals such as the wooly rhinoceros and the wooly mammoth. These bones do not always prove hunting: humans may actually have collected mammoth bones to burn as fuel.[29] Upper Paleolithic sites from Germany and northern Switzerland also indicate continued human presence surprisingly close to the ice sheets.

Over thousands of years, life in colder northern climates gave advantage to small, though noticeable, physical traits. The possibility that *Homo sapiens* may have created their own preferences in partners cannot be proved or discounted, but the differences in body type, skin tone, and eye color that humans later categorized as race suggest the influence of climate. A paler skin, for example, could aid in generating vitamin D at northern latitudes.

In their ability to survive in varied regions all the way from savannas of Africa to cold regions near ice sheets, *Homo sapiens* displayed a striking capacity to respond to climate. No human resident of any community above the Arctic Circle can simply venture out unprotected in winter no matter how many generations her or his ancestors may have dwelled in similar conditions. *Homo sapiens* secured shelter and created warm clothing. The evidence for the key innovation of clothing is indirect: no article of clothing from before 34,000 years ago survives. Instead, we find earlier signs of making clothing in tools: sharp blades, awls for making holes, and eyed needles used for stitching together garments.

Hints for how *Homo sapiens* may have clothed themselves in cold climates during the LGM can also be found in the much later example of an individual from the Neolithic dubbed the Ice Man who died in the mountains of Tyrol in the Southern Alps some 5,300 years ago. Covered by ice, his body was mummified and remained intact until hikers discovered his corpse in 1991. He wore a coat and leggings made of animal hides, a bearskin cap, and shoes made of bearskin and deerskin with grass used for insulation. The much earlier *Homo sapiens* of the LGM would not have worn the same clothes—they had no domesticated goats—but the Ice Man's clothing suggests how earlier humans may have clothed themselves to withstand a harsh climate.

Climate change during the LGM both impeded and aided human dispersal. Ice sheets took up so much water that sea levels fell by as much as 140 meters, creating large land bridges. The Bering Strait, which today separates the continents of Asia and North America and the countries of Russia and the United States, was a land bridge during the LGM, now named Beringia. At the same time, ice sheets and extreme cold also slowed movement, especially in contrast to the comparatively rapid pace between

50,000 and 40,000 years ago. As the LGM ended, land bridges narrowed but persisted in the case of Beringia up until 10,000 years ago.

As ice sheets began to recede, humans recolonized areas abandoned during the glacial maximum and moved into entirely new areas. They dispersed north in Europe back into Britain as early as 16,000 years ago, as indicated by the bones of animals butchered for meat. *Homo sapiens* dispersed north from glacial refuges close to the Mediterranean, and movement into Europe also took place from areas in the Near East.[30] The relative contribution of populations from these different regions to the modern European population is the subject of ongoing research and genetic analysis.

In Asia, populations from Southeast Asia and adjacent areas of southern China moved north into East Asia after around 19,000 years ago. In the Transbaikal region of Siberia, human population disappeared between 24,800 and 22,800 years ago.[31] The end of the glacial maximum brought signs of cultural change to China, Korea, and Japan with the adoption of microblade technology. These small sharp blades produced from material such as quartz or obsidian could be attached to wooden stakes to produce spears.[32] This technological shift may show one effect of waves of movement to the north and east.

With the end of the LGM, people also dispersed through northeast Asia into the Americas. Research into the timing of human movement across the Bering Strait and into the Americas has given rise to several possible scenarios. Archaeological and genetic evidence suggests several different pulses of migration rather than a single continuous process. Humans had already reached Japan by 37,000 or 38,000 years ago and arrived in northeastern Siberia by 30,000 or 31,000 years ago. Remarkable human artifacts from this time have been recovered at the Yana River north of the Arctic Circle in Siberia. Findings from the Yana River show that humans made spear handles from rhinoceros horn and mammoth tusk. Ivory recovered from the site contains markings of dots and dashes. It is possible that people who had reached this site could have traveled further east to Beringia and persisted along coastal ice-free areas during the LGM. Beringia was cold, but to a large extent ice-free during the glacial maximum. In this scenario, there was a pause in the human settlement of the Americas.

In an alternate scenario people entered directly into Beringia and continued to move to the south and east into the Americas around 15,000 years ago as ice sheets melted. Another model suggests a shorter pause in migration across Beringia at around 18,000 years ago.[33] Sea-level rise since this period complicates the task of choosing between these and other possible models of human migration into the Americas. Archaeological sites that could help distinguish between these different scenarios of human expansion into Beringia now lie submerged beneath the ocean.

Human populations already accustomed to coastal living dispersed rapidly south along coastal routes. By 14,000 years ago, humans established

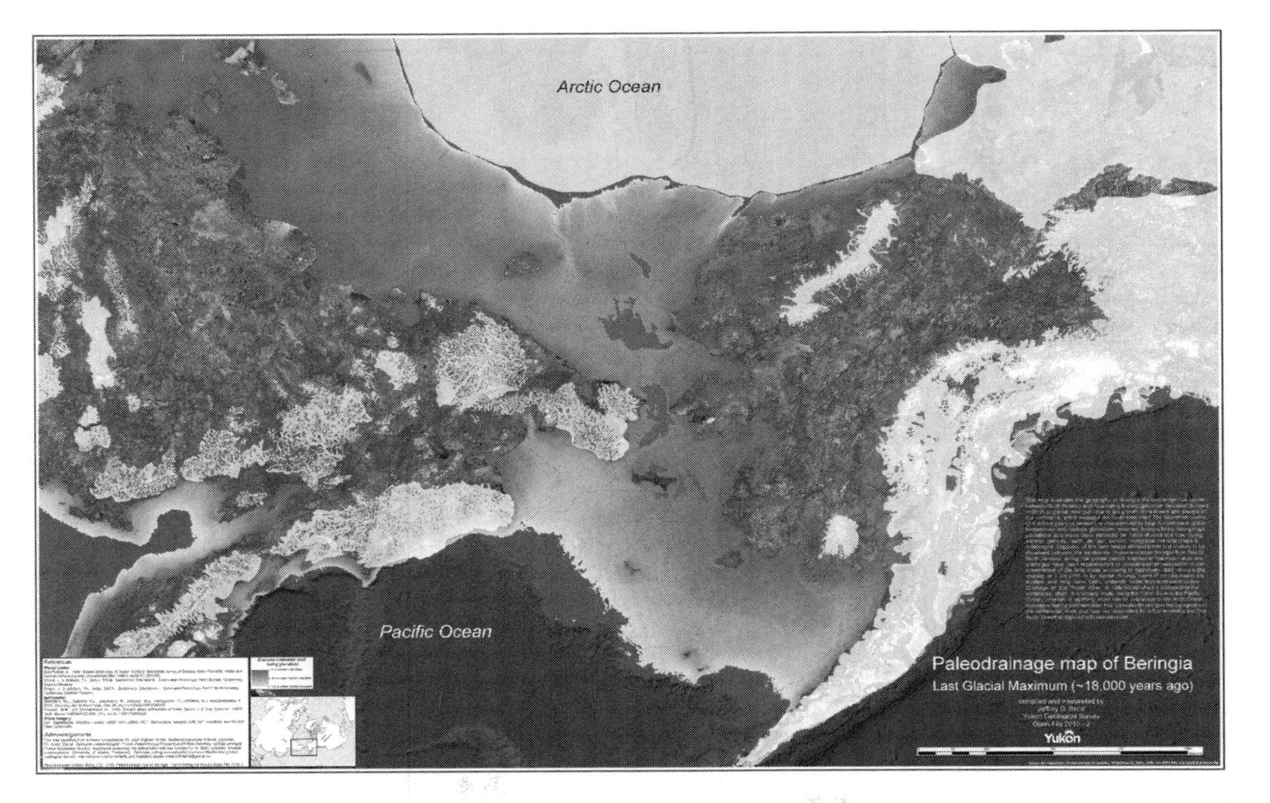

FIGURE 1.4 *Map of Beringia 18,000 years ago. Used with permission from the Yukon Geological Survey. Source:* NOAA - *http://www.ncdc.noaa.gov/paleo/parcs/atlas/beringia/lbridge.html.*

a settlement as far south as Monte Verde in southern Chile, though there is also possible evidence of earlier human presence at Monte Verde and other sites, which, if confirmed, would push back the date for the arrival of *Homo sapiens*. This settlement at Monte Verde was found in a peat bog. Residents lived in huts. Animal remains found at the site indicate that they ate shellfish as well as now-extinct mammals, including gomphotheres, animals related to elephants. They also ate plants and nuts. The residents took seaweed and algae from the ocean, which today lies much closer to Monte Verde.

The same climate shifts that posed a challenge to *Homo sapiens* also affected many other plants and animals. Some survived in glacial refuges where smaller populations created the potential for evolutionary shifts. Others died out, either because of loss of habitat from climate change, from human hunting, or from a combination of these and other factors.

Across much of the world, the fact that humans hunted large animals that no longer exist raises questions about the possible role of humans in the demise of large extinct animals or megafauna. Across continents, large species that existed until comparatively recently are gone: the Megaloceros, a very large deer, the wooly mammoth, and in Australia such large marsupials as the marsupial lion as well as others. There are no longer gomphotheres in the Americas. The pattern was general: North and South America lost most large mammals between 12,000 and 10,000 years ago. The fact that such megafauna had survived the previous interglacial period suggests a decisive part played by human hunters in driving large prey to extinction. However, climate change may also have reduced plant diversity in some regions after the LGM, sharply reducing the supply of a family of protein-rich plants called forbs, an important food source for megafauna.

Extinction of mammals was especially pronounced in North America at the end of the Pleistocene with the loss of 70 percent of megafauna. Taking the frequency of radiocarbon dates of remains as a proxy for population size, one study finds a strong relationship between human population and megafauna extinction for several cases, including for the mammoth, horse, and saber-tooth cat, but the same does not hold for others, including the Shasta ground sloth and mastodon.[34]

On a global scale, however, an analysis of all mammal extinctions since the late Pleistocene suggests a decisive role for humans. Extinction rates spiked after human arrival in multiple regions, with data from islands providing strong evidence as in the case of Madagascar and the Caribbean. Overall, the timing of extinction coincided with human colonization.[35]

Summary

In comparison to the thousands of years of human civilization, the several million years during which ancestral species of humans evolved and the few hundred thousands of years during which *Homo sapiens* emerged made up a comparatively long period of time. The information available from this long era reveals that ancestral species and early humans were both highly dependent on climate and capable of withstanding shifts in climate. Climate change influenced and in several periods helped create the pressures for the natural selection that led to the emergence of modern humans and our ancestors. The cooling and drying that shrank the rain forests of Africa created an advantage for individuals who could eat a wider array of food in more broken forests. In turn, living in and near savannas created advantages for individuals better able to move long distances in search of prey and for individuals better able to communicate and work together. Cooling in African habitats was a major force that helped lead to the emergence of australopiths and to the later rise of *Homo erectus*.

The shifts in climate that accompanies the glacial-interglacial cycles affected *Homo erectus* and the species that emerged from *Homo erectus*, including *Homo sapiens* and our closest human relatives. Glaciation locked up water, creating land bridges that aided dispersal out of Africa, but glacial maxima also reduced the amount of territory in which humans could flourish.

Dependent on climate, humans also demonstrated considerable resilience to shifts and even to abrupt change. Fluctuations between interglacial periods and glacial maxima produced much greater shifts in temperature and sea level than the changes during the much shorter time span of human civilization. Humans also survived through several periods of abrupt climate change, possibly after the eruptions of large supervolcanoes, and also during Heinrich events. Even during the LGM, some human populations managed to exploit resources in northern areas not far from the ice sheets that prevented any permanent habitation. We are then the products of evolution driven at least in large part by climate change, but our ancestors existed during a much wider range of climate regimes than humans have experienced in the recent past. Omnivores who consumed meat resembled some other carnivores in their ability to shift their range of hunting and focus on new prey, and their exploitation of many foodstuffs may have made them more resilient than competing carnivores.

There were, however, critical differences between even the anatomically modern humans who survived through these large climate shifts and us. They lived under profoundly different social conditions. Their total population was miniscule compared to ours, and they apparently passed through population bottlenecks, during which the number of individuals fell into the thousands, creating risk for survival. The dwindling Neanderthal population died out. A total population that plummeted too far could mean extinction, but smaller populations without large fixed settlements also held advantages. The total *Homo sapiens* population required far less food and energy than we do today. Even so, cold spells led to lineage loss, and many human populations went extinct.

CHAPTER TWO

The rise of farming

- After the last glacial maximum
- Younger Dryas
- Cultivation
- The spread of farming
- Abrupt change in the early Holocene
- The end of the Green Sahara
- Complex societies
- Conclusion

After the last glacial maximum

Living in a warming world brought many changes for humans. They experienced shifts in the range of vegetation and of many animals. The warming melted glaciers and unlocked water. Coastlines shifted, land bridges narrowed, and sea levels rose, in some areas inundating previous encampments and routes for travel. Familiar landscapes vanished, and new ones emerged. The pattern of human settlement underwent rapid change. Humans both recolonized regions they had abandoned at the peak of glaciation and dispersed into new areas. They returned to parts of northern Europe previously covered by ice and moved into higher elevations in Europe and Asia. In the Americas, humans traveled up to the margins of ice sheets. Excavations in Wisconsin, not far from the edge of the Laurentide Ice Sheet, which at its peak covered most of Canada and adjacent regions in what is now the United States, revealed tools and mammoth bones with marks that indicate butchering. From their starting point in Africa, modern humans expanded their range of settlement to encompass most of the globe: together, these population movements after the LGM left only some islands, including Hawaii, New Zealand, and the continent of Antarctica as regions without humans.

People exploited a wide array of resources and foodstuffs in the late stages of the Pleistocene. In North America, people of the Clovis culture, named for fluted projectile points first found in Clovis, New Mexico, lived in much of what is now the continental United States as well as in Mexico. People of the Clovis culture hunted large game, including mammoths, elephant-like animals, and bison. A powerful image of the Clovis culture depicts them hunting mammoths throughout the North American landscape, though that dramatic picture overlooks their foraging for plants and consumption of small animals and fish (Figure 2.1).

Humans dispersed throughout South America, adapting to varied environmental conditions from high mountains to rain forests. They dispersed along coasts and created sites at coastal areas such as Huaca Prieta in northern Peru as far back as 14,200 to 13,300 years before present (ybp). Humans also made their way inland. High in the Andes Mountains at an elevation of 4,500 meters, or nearly 15,000 feet, the remains of a camp dating back to 12,400 ybp shows that early hunter-gatherers in the highlands of South America exploited resources in an Alpine environment.[1]

Hunter-gatherers made their way into the Amazon basins. Carbon dates for outcrops at Pedra Furada in a national park in Brazil as well as the origin

FIGURE 2.1 *Clovis spear point, British Museum. Photograph by Mike Peel (www.mikepeel.net)*

of deposits have led to prolonged disputes over the age of settlement, but a site at Caverna da Pedra Pintada (cave of the painted rock) near the town of Monte Alegre in northern Brazil shows that humans established themselves in the region by the late Pleistocene. Scattered sites in Brazil dating back to the period from 15,500 to 12,800 ybp suggest early settlement followed river valleys. More traces of human occupation have been found for the period from 12,800 to 11,400 ybp in multiple regions of Brazil, including the Amazon rainforest, the savanna, and the pampas grasslands in the south.

One of the chief puzzles looking back at this period concerns the fate of the megafauna, the many large animals that had lived through the glacial maximum, but did not survive into or far into the Holocene. The entrance of humans into Patagonia in the far south of South America provides an important test case for explaining the demise of megafauna. A study of mitochondrial DNA from megafauna allows for precise dating of the timing of the extinction of animals such as mammoths and giant sloths. Imagine modern sloths, mammals found in Central and South America, that may weigh some eighteen to twenty pounds—this gives little sense of the enormous sloths that humans entering the Americas encountered. The giant sloth, or Mylodon, weighed upward of 400 pounds and measured some ten feet from snout to tail. Humans arrived in Patagonia around 15,000 to 14,600 ybp, and megafauna and human populations actually coexisted during a reversal to a colder Southern Hemisphere climate between 14,400 and 12,700 ybp termed the Antarctic Cold Reversal (ACR). Rapid warming in the Southern Hemisphere followed the ACR, which coincided with the cooling of the Younger Dryas in the Northern Hemisphere. Extinctions took place around 12,280 ybp, and in only three centuries, 83 percent of Patagonia's large mammal species died out.[2] During this period, pressure from human competition amplified the shock caused by climate stress, leading to a very different outcome from the previous interglacial period when the megafauna survived. In southern Patagonia for example, climate change and human arrival combined to wipe out megafauna, with humans possibly playing the key part in killing off large carnivores, but climate change and ensuing shifts in vegetation were more damaging to large herbivores (Figure 2.2).[3]

In Europe, numerous artifacts and archaeological sites show the growing complexity of thriving human societies after the LGM. Humans created increasingly specialized tools for activities including hunting. Hunter-gatherers built seasonal camps at sites favorable for obtaining food. On the shores of a shallow lake in what is now Belgium, for example, humans gathered plants and hunted game attracted to the waters.[4] In regions including the Rhineland, people increasingly hunted with bow and arrow. The new weapon assisted deer hunting in the boreal forest south of the glaciers. Hunter-gatherers of the late Pleistocene also seasonally collected plants and plant materials in regions including the western edge of the Caucasus in Georgia.

FIGURE 2.2 *Illustration of various megafauna during the last glacial maximum. Getty Images.*

In a swath of Europe running from northern Spain through France into Central Europe, people of the Magdalenian culture, named after a site in the Dordogne River Valley of southwestern France, hunted many species of animals. In a climate still much colder than that of today, they intensively hunted reindeer. Magdalenian hunters employed new tools to hunt their prey in the steppes of Western and Central Europe. With spear throwers, often made with reindeer bone, hunters could strike prey with great force from a long distance. Along lakes and waterways, the Magdalenian hunter-gatherers used harpoons to catch fish.

The spear throwers themselves were often decorated with carved animal heads, reflecting a broader trend toward decorative art most famously preserved in cave paintings. The oldest cave paintings, both in France and elsewhere, predate the Magdalenian culture, which spanned the period from approximately 17,000 to 12,000 ybp, but the Magdalenian produced some of the most startling and remarkable cave paintings, found at numerous sites, including the Lascaux caves in France. Discovered in 1940, the paintings at Lascaux depict animals, including deer, bison, and aurochs. Other caves from the Magdalenian culture show similar subjects. The Magdalenian culture also created portable objects. Engravings of reindeer, frequently absent from cave paintings such as those at Lascaux, have been found on pieces of bone.

Along with hunting, people of the Magdalenian collected plant food—Magdalenian sites contain nuts and stones from fruits. A few sites contain stones that could have been used for grinding wild grains. Humans here and elsewhere made such grain part of their diet well before the advent of farming. Magdalenian peoples may have moved seasonally to establish camps to exploit resources in different parts of their range of settlement.

In Eurasia, as in the Americas, the combination of climate change, human population growth, and advances in human hunting technology placed many large animals in increasing peril. It is unlikely that the warming climate by itself led to mass extinctions of so many large animals. Megafauna had survived previous interglacial periods, but areas into which megafauna had previously retreated likely provided less protection from humans in a warming world. Some of the animals that the people of the Magdalenian culture feasted on, such as reindeer, survived and thrived to the north. However, others went extinct as the world warmed and humans improved hunting technology. The extent to which human activity contributed to megafauna population decline varied by species. Climate change, for instance, appears to have predominated as the main factor in the extinction of the wooly rhinoceros.[5]

Hunter-gatherers in a warming world intensively exploited a wide range of foodstuffs. They were not farmers; they continued to collect food, but there is strong evidence that they gathered grain. In the Near East, as elsewhere, hunter-gatherers already incorporated more grain into their diet before the start of farming. In particular they gathered legumes and wild grasses. Charred seeds and fruits collected from Kebara at Mount Carmel in Israel show the collection of legumes and fruits even before the LGM. Seeds found at a site in the Sea of Galilee from the period of the LGM demonstrate the collection of grasses such as wild barley and wild wheat. Warming after the LGM appears to have encouraged hunter-gatherers to focus more on collecting cereals rather than small-grained grasses.[6]

After the LGM, the Natufians, a culture from the Near East, exemplified the diversification of diet in a warming world. The Natufians hunted animals, including gazelles. They also gathered plants and cereals and created tools, such as sickles and mortars, that could have been used to collect and process food. Before the Natufians, hunter-gatherers in this region employed seasonal strategies of setting up camps at particular sites for the times of the year most useful for collecting food, but the Natufians built sizable permanent villages with structures with stone foundations. Their sites have been found across the Levant, or eastern Mediterranean, in Israel, Palestine, Jordan, Lebanon, and Syria, and late Natufian sites can be found as far north as southern Turkey. They appear to have preferred building their villages in woodland areas.[7] They employed a wide range of tools made from stone and bone and also produced decorative objects such as jewelry made from materials including shells.

In China as well, hunter-gatherers collected wild grasses. They may have turned to tubers and grasses as early as the LGM to supplement their diet during a period of reduced resources.[8] Low sea level during the LGM expanded coastal plains, and the loss of these regions with warming could have deprived hunter-gatherers of lands highly suitable for foraging.[9] At the same time, northern regions became easier to exploit, and genetic analysis suggests northward dispersal.[10]

Younger Dryas

Within the general warming trend after the LGM, humans experienced an abrupt climate shift on a scale that would doubtless shock modern populations. The Younger Dryas, named after an arctic flower (*Dryas octopetala*) that appeared in Europe during this near-glacial cooling episode, interrupted the overall warming trend that followed the LGM. Between 12,900 and 11,600 ybp, temperatures in the Northern Hemisphere dropped abruptly, plunging as much as 10°C within decades. According to the most widely accepted hypothesis, a slowing of deep ocean circulation resulted in an abrupt cooling of the North Atlantic region and initiated the Younger Dryas. As temperatures warmed following the LGM, ice sheets began melting, sending freshwater into the North Atlantic. The cold meltwater flowed by several routes to the ocean, including an initial flow south into the Gulf of Mexico, then later toward the east into the North Atlantic through the St. Lawrence River[11] and north through northwestern Canada and into the Arctic through the MacKenzie River.[12] Fresh meltwater flowing into the ocean causes a decrease in ocean salinity, and therefore its density, effectively slowing the overturning circulation. This slowdown in turn reduced heat transport to the northern polar regions and produced a plunge in temperatures there. An alternative hypothesis posits that one or more extraterrestrial objects struck or exploded over the Laurentide Ice Sheet, triggering the meltwater flow.[13]

The Younger Dryas impact was felt worldwide, with substantial cooling (as much as 10°C)[14] in higher latitudes of the Northern Hemisphere, more modest cooling in Europe (3–4°C), and a slight warming in the Southern Hemisphere. This temperature change, in turn, moved the location of the intertropical convergence zone, or ITCZ, which marks the tropical rain belt. With a cooling Northern Hemisphere, the ITCZ shifted south, which produced a weaker summer monsoon and relatively dry conditions in the Northern Hemisphere, particularly in Africa and Asia.

The Younger Dryas therefore forced humans to respond rapidly to a colder and drier climate. A short comparison to the present day is instructive: how would modern societies react if confronted with a cooling on the order of 3°C (5.5°F)? There is abundant evidence to show that modern societies have on the whole become more resilient to shifts in climate over time, but they have experienced no shifts remotely comparable in severity to the climate change of the Younger Dryas, at least in the most affected regions such as Europe.

Charting the precise human response remains challenging for two reasons independent of the actual temperature record. First, in many regions, the archaeological evidence of prehistoric societies dating back to the Younger Dryas is so fragmentary that we lack a detailed record of material culture just before and during the Younger Dryas. Second, the Younger Dryas was not the only phenomenon that affected societies.

In North America, the Clovis culture that had spread rapidly across much of the continent ended during the Younger Dryas. People of the Clovis culture colonized eastern North America. By around 12,900 ybp, their projectile points were displaying greater diversity, indicating that the uniformity of Clovis technology was starting to break apart as people inhabited particular regions. This diversification became the rule after the onset of the Younger Dryas. Climate stress and megafauna extinctions encouraged more intense exploitation of increasingly varied food supplies, possibly including a greater proportion of plants. The disruption may have been severe enough to cause abandonment of some sites and population decline.[15] In what is now the southeastern United States, human population appears to have risen before the Younger Dryas. As indicated by proxies such as artifacts from the Clovis culture—the Clovis biface projectile points—population rose between 13,200 and 12,800 ybp. Population then dropped from 12,800 to 11,900 years ago before rising again. The sequence in this instance conforms to the pattern we would expect to find if abrupt cooling had hampered the Clovis culture, but other factors could have contributed to these population trends. Small shifts in dating Clovis artifacts would also change our understanding of the relationship between the Clovis culture and the Younger Dryas.[16] More broadly in North America, the Younger Dryas did not end human settlement: small, mobile populations of hunter-gatherers were able to continue to obtain food.

In Europe, the Younger Dryas led to the most significant effects in northern regions. In Britain, remains from humans and from animals associated with human activity show a marked decline. Indeed, no human remains with dates from the Younger Dryas have been found in Britain, though other human artifacts from this period have been identified. Overall, the human population in much of Europe expanded during the warming period that preceded the Younger Dryas.[17]

How did humans respond to the cooling trend of the Younger Dryas? The climate shift created the potential to encourage changes in collecting resources in two directions. Abrupt cooling could have propelled time-tested strategies for hunting and gathering, the methods that humans had employed previously during the LGM. Alternately, humans after a period of population growth could have adopted a different strategy of intensifying the exploitation of a wider array of food such as plants.

The pattern of responses to the Younger Dryas has been most extensively investigated for the Middle East and for the Natufians. Research into the Younger Dryas in the Near East has yielded at least two possible scenarios. Just after a period of population increase, the Natufians had more people to feed in a cold climate. One scenario posits a split response to the Younger Dryas with both hunting and gathering as well as cultivation. Amid colder and drier conditions, some groups intensified nomadic hunting and gathering, leading to a poorer record of archaeological remains.[18] It was also possible to both intensify hunting and gathering and diversify food supply. The abrupt climate shift during a period of rapid social development also gave hunter-gatherers incentive to increase cultivation to sustain the complex communities that they had created. As cooling and drying reduced the supply of some wild plants, such as wild lentils, Natufians collected and cultivated grain in order to keep their settlements and preserve their culture and society. According to scientists who favor the split-response scenario, evidence from Abu Hureyra, a site in the Euphrates River valley in northern Syria, indicates that hunter-gatherers started to cultivate crops because the Younger Dryas caused a decline in wild staple crops.[19] Rye may have been cultivated there as long as 13,000 years ago.

Another scenario sees a less dramatic response to the Younger Dryas and holds that human societies in the Near East, as elsewhere, continued to sustain themselves by hunting and gathering. Research that backs this scenario interprets the seeds not as evidence of early cultivation by Natufians but instead as remnants from animal dung used as fuel. Natufians, in this interpretation, responded to the Younger Dryas cooling by rebalancing their food supply.[20] This scenario raises the question of what would have happened if the Younger Dryas had occurred several thousands of years later. If comparatively large numbers of people depended for their livelihood on cultivation, how would they have responded to an event like the Younger Dryas?

Cultivation

Temperatures quickly rebounded once deep ocean circulation recovered after the Younger Dryas. The ITCZ, along with associated precipitation, again migrated north. With the return to warmer and wetter conditions, agriculture expanded in the Levant. Sites from the period just after the Younger Dryas, an era dubbed the Prepottery Neolithic, yield strong evidence for intensified cultivation, including abundant seeds and granaries for storing grain. There was plentiful wild rye, and with warming in the Holocene, populations made more use of barley (Figure 2.3).[21]

There may be no need to identify either a local response to Younger Dryas cooling or an adaptation to Holocene warming as the single path to farming because farming emerged in more than one locale, even in the Middle East. If the Younger Dryas helped propel hunter-gatherers who had recently experienced population growth to cultivate crops, it also made the prospect of growing grains difficult in colder, higher regions of the interior.[22] A host of sites with early evidence of cultivation have been discovered in the Fertile Crescent, though the exact speed and timing of domestication remain uncertain. Grains, including rye, barley, and einkorn, and legumes have been found across the area. Over time, the number of crops cultivated in each site also increased.[23]

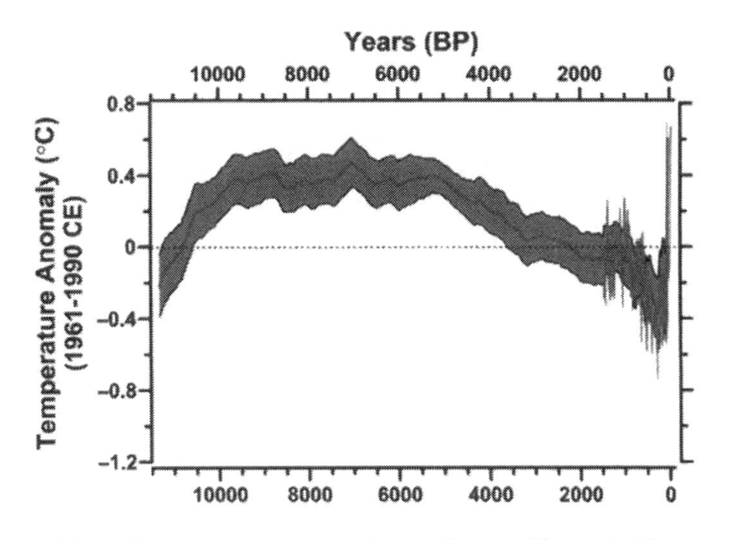

FIGURE 2.3 *Holocene temperature variation. Source: Shaun A. Marcott et al., 2013, "A reconstruction of regional and global temperature for the past 11,300 years," Science 339, 1198–1201.*

Climate change before and during the early Holocene also affected the potential for cultivation and the emergence of agriculture in China. As in the Fertile Crescent, hunter-gatherers in China collected grain before domestication. Ceramics and grinding stones date back all the way to the LGM. In China, late Pleistocene people lived as foragers. They ate animals, including deer, antelope, and wild pigs. Archaeological excavations suggest they occupied some sites only briefly and shifted their base frequently. Thousands of years before the start of farming, they also obtained food from wild grasses and employed grinding stones.[24] The cold climate of the Younger Dryas could have pressed people to get more storable food and therefore to cultivate.[25] In the Holocene, the warmer and more humid climate either stimulated the cultivation of cereal in China or it enhanced cultivation.[26]

After hundreds of thousands of years in which first human ancestors and later humans had survived as hunter-gatherers, the transition to farming was comparatively swift. Humans in region after region took up cultivation of domesticated crops during the Holocene. For millennia, some humans continued to survive as hunter-gatherers, but their proportion of the human population decreased and finally dwindled to the point that only a very small fraction of humans still supported themselves mainly as hunter-gatherers by the early twentieth century.

No change of this magnitude, away from hunting and gathering and toward farming, can be attributed to a single factor, and research has yielded several explanations for the increase of cultivation and domestication of plants and animals. Climate change influenced the conditions for cultivation. Rising population, according to another model, accounted for the shift to farming.[27] In one possible relationship, climate change eased the spread of agriculture, which boosted population. In a second, population growth in the Holocene created the need for acquiring more food. Indeed, population was already rising before the start of the Holocene. Genetic analysis indicates a start to population growth in East Asia, for example, some 13,000 years ago.[28]

Other explanations for intensifying cultivation and the rise of agriculture point to cultural and behavioral change. Humans, before they extensively domesticated plants and animals, had long experience with wild crops and animals that they surely could draw on. Moreover, the favorable climate of the Holocene did not generate a sudden, immediate shift to farming. It could take 1,000 to 2,000 years to develop nonshattering varieties with large grains in which the seeds would not shatter or blow off easily into the wind.[29] Today, we simply assume that grains such as wheat stay largely intact until harvest, but that is not the case for wild varieties: farmers and their customers would certainly suffer if grain dispersed with a puff of wind.

These explanations are not mutually exclusive. Population growth continued during the period of early Holocene warming. The shift toward a warmer and far more stable climate during the Holocene created favorable conditions for a major shift during the Neolithic, or New Stone Age. The New Stone Age, which began about 12,000 years ago in the Fertile Crescent, did not erase the practice of foraging or eliminate hunter-gatherers, but it initiated a decisive shift toward the paramount use of domesticated crops and animals. Climate change, however, did not create some kind of switch that suddenly turned people toward agriculture. Regional conditions shaped the emergence of agriculture in varied locations at different times.

Under any historical approach, southwestern Asia or the Near East was a key site for the independent emergence of agriculture. Over a few thousand years, agriculture spread across the entire region. Early farmers cultivated crops, including barley, emmer (farro), pulses (edible seeds such as peas and beans), and wheat. As early farmers chose to plant some crops, they may also have abandoned or at least reduced their cultivation of some crops such as rye.[30]

Agriculture also emerged and expanded separately in China. Hunter-gatherers in northern China collected wild grasses. The Holocene climate proved favorable to the ancestral forms of millet, increasing yields. Humans, in turn, cultivated and domesticated millet.[31] The precise advent of domesticated rice remains open to question. New Stone Age people in both North and South China collected wild rice by 11,000 years ago.[32] As is the case for grains in the Near East, the speed of domestication remains in dispute, in part because of uncertainty about the extent to which measurement of the size of rice grains can track the pace of domestication.

In a rapid or early domestication scenario, domestication of rice was already under way between 9,000 and 8,400 years ago. Nonshattering rice dates as far back as 8,700 ybp (6,700 BCE) at the Neolithic site of Baligang on a tributary of the middle Yangtze River, but the grain size is still smaller than in domesticated rice.[33] In a slow domestication scenario, hunter-gatherers collected varied wild plants such as water chestnuts and hunted and fished as they gradually domesticated rice.[34] The Holocene generally favored the cultivation of rice, but sea-level rise and associated salinity drove the collection and farming of rice away from the low-lying areas. Rising waters in particular affected the Delta on the lower Yangtze River.

Domestication of plants took place independently during the Holocene in the Americas. The earliest cultivation of moschata squash, which today includes such vegetables as butternut, may have begun as early as 10,000 years ago in far northern South America. Squash was a domesticated crop by almost 9,000 years ago in the Balsas River Valley of southern Mexico.[35] Maize, what Americans call "corn," was domesticated in Central America. Identifying the ancestors of maize eluded scientists for a long time, but there is now overwhelming evidence that corn or maize originated from a wild plant called teosinte, found in parts of Central America and Mexico. Genetic analysis suggests that maize first emerged close to 9,000 years ago in the Balsas River Valley. At the time, residents modified the landscape, cutting down and burning forests and replacing them with plots of land for farming. Stone tools used to grind and mill maize from the region date back nearly that far to 8,700 years ago. The grinding tools still contain the residue from maize.[36] Maize cultivation expanded, and farmers grew maize in the Yucatan Peninsula as early as 7,000 years ago.

There were multiple sites for domestication in the Americas. In South America, the domestication of the potato took place in the Andes Mountains. Remarkably, the early farmers in these steep highlands found ways to make use of a crop that initially contained high levels of toxins. The ancestral forms of domesticated potatoes most likely came from the central Andes, possibly from the region of Lake Titicaca, a large lake between Bolivia and Peru at an elevation of more than 12,000 feet. Dates for domestication range from 7,000 to 10,000, or even more, years ago. The crop spread north and south along the Andes during the pre-Columbian era. Early farming villages in the Zana Valley in Peru ate a diet that included moschata squash by 8,000 years ago.[37]

Along the western coast of South America, complex societies developed from some 5,000 years ago at sites such as Aspero where the Supe River enters the ocean in northern Peru. Another complex site with platform mounds is located inland at Caral. The residents sustained themselves with fish, orchards of fruit, and plants. The question of whether they grew maize gave rise to many years of research. Maize finds from the site of Aspero remain small, but larger concentrations of maize have been found at inland sites in quantities that show deliberate farming of a domesticated crop.[38]

In South America, farming populations along the Andes cultivated terraced fields and collected and distributed water in irrigation systems. Residents of Peru and Bolivia developed strategies for controlling and storing water.[39] In the Peruvian Andes, a system of canals provided drainage for the ceremonial center at Chavin de Huantar (900–200 BCE).[40]

Peoples of eastern North America also began to cultivate several crops. They ate many of the same staples found in Central America. Maize and several varieties of squash entered North America after having been previously farmed in Central America. Other cultivated crops may have had an independent origin in North America. There is evidence that peoples of eastern North America began to cultivate sunflowers and certain varieties of summer squash, before the introduction of corn and other crops from Central America.[41] Chenopod, another crop cultivated in eastern North America before the arrival of Central American foodstuffs, was later abandoned.

Centuries later, European settlers encountered native peoples who cultivated both crops that originated in eastern North America and crops introduced from Central America. English settlers at Jamestown encountered Powhatan Indians who grew squash, maize, beans, and sunflowers. In Massachusetts, the Pilgrims discovered to their benefit that their new neighbors grew corn.

Pastoralism and cultivation also took place in Africa during the Holocene. In the case of Africa, the question of whether native peoples domesticated cattle or adopted cattle first domesticated elsewhere has led to ongoing research. Pastoralists in the Sahel or savanna, the grasslands to the south of the Sahara, first raised cattle as early as 8,000 or even 10,000 years ago, but genetic analysis suggests that the domesticated cattle raised in Africa may have initially come from the Fertile Crescent. Cattle herding spread west and south.

Africa is yet another region with a complex of native crops as well as crops introduced from other regions. As in other regions, hunter-gatherers collected wild grasses. Cultivators in Africa domesticated crops, including millet and sorghum. The process was under way by some 4,000 years ago, and may have started much earlier. An exact date is hard to find because of the lack of a clear early boundary between wild and domesticated crops. Tubers with underground roots or bulbs, including yams, proved an important domesticated crop for people living south of the savanna in the tropics.[42]

Peoples of New Guinea also grew crops. They cultivated taro, a plant with a starchy root, and bananas. Archaeological evidence indicates the cultivation of taro in the highlands of New Guinea as far back as nearly 7,000 years ago, and the transition to such farming may have begun as early as 10,000 years ago.[43] Uncertainty remains about whether people of New Guinea initially cultivated taro or domesticated the plant.

The simple division between hunter-gatherers and farmers does not fully capture the full range of methods employed by hunter-gatherers to remake and shape landscapes. In Australia humans did not farm, but nonetheless set fires to manage use of grazing animals and to control the supply of plants.[44] Natives of Australia also engaged in early aquaculture: at Budj Bim, near the site of Mount Eccles in southern Australia, people set up eel traps some 6,000 years ago.

The scattered timing for the domestication of crops and animals indicates both the importance of climate and the ability of humans to adapt farming and pastoralism to varied regions. Farming in the Near East, China, Central America, the Andes, New Guinea, and elsewhere took place under very different conditions: there was no single environment required for domestication, and humans proved adept at making use of and domesticating a wide range of crops: a potato does not resemble a sheaf of wheat.

In broader perspective, however, the Holocene climate boosted widespread cultivation and domestication. *Homo sapiens*, we believe, had more or less the same intellectual capacities that we do today for many tens of thousands of years, but the expansion of agriculture took place during the Holocene when humans experienced a period of climate stability unlike anything they had witnessed for tens of thousands of years. It is necessary only to think back to the landscape of the LGM: humans had proved versatile in finding resources in varied regions, but the steppes and tundra of the LGM would never have supported intensive farming to the same degree as the far more widespread warm and moist regions of the Holocene.

The spread of farming

Farming and pastoralism spread out during the Holocene from multiple early centers, including the Yangtze and Yellow River Basins, the Levant, Mexico, West Africa, and New Guinea. Either farmers themselves moved and brought farming with them as they established new settlements, or neighboring populations took up farming. In the first scenario, farmers moved, either peacefully or through conquest, into areas previously inhabited by hunter-gatherers. In the second, hunter-gatherers gradually adopted the techniques and methods of farming. Archaeological finds, the investigation of human languages, and more recently genetic analysis have all shed light on this process. Genetic data point to the role of population movement in spreading agriculture by both migration and cultural diffusion.[45]

In the case of Europe, research has identified several possible human migrations during the early Holocene. Genetic analysis indicates a link between early farmers in Central Europe and the Mediterranean and populations from southwestern Asia.[46] In Central and northern Europe, farming and hunter-gatherer populations apparently interacted for a significant period of time. Findings from Sweden suggest the presence of some arrivals from southern Europe as far back as 5,000 years ago.[47] Genetic analysis also suggests the spread of populations that made use of dairy products north and west into Europe. A genetic mutation that allowed adults to digest milk may have given an advantage because the population of lactose-tolerant people could rely on a broader supply of food.[48] The migration north and west of these dairy specialists began some 8,000 years ago.

Farming spread to varied regions in the Americas. In the Amazon, archaeological finds show increasing settlement by 10,500 ybp. During this period, artifacts also indicate growing regional diversity. By 4,000 years ago, humans created villages in the Amazon Basin. Farming also began during this same period with a shift to more deliberate and intentional agriculture.[49]

The spread of farming in the Holocene also led to the rapid growth of human population.[50] Population growth began before the start of the New Stone Age and accelerated during the Holocene for several reasons. The first is most obvious: the production of more food. Food supply affects the frequency with which a mother may become pregnant and give birth. Hunter-gatherer mothers typically breastfeed their children far longer than do mothers in farming societies, and hunter-gatherer mothers wean their children only at age three. This has the effect of spacing out births and therefore of slowing population growth rates. Mothers in a farming society, in contrast, would wean children at a younger age, which has the effect of increasing population growth rates.

Across the world, domestication increased overall population despite some temporary slumps. In Europe population rose quickly during the shift to agriculture. In Africa, population grew rapidly after around 4600 ybp. In Africa, farming may have spread with the migration of Bantu-speaking peoples. A similar pattern between agriculture and population growth holds for Southeast Asia.[51] Overall, between the start of the Neolithic and about 5,000 years ago, world population of humans increased from some 4–6 million to 14 million, though estimates vary.

Once populations shift to farming, it becomes increasingly hard to move back to a hunter-gatherer life. Farmers make massive investments in capital, labor, and time and transform landscapes, by clearing land and raising crops and animals. Population growth creates numbers of people too large to be supported by hunting and gathering. The density of hunter-gatherer populations varies, but population density is much higher for societies that derive most of their calories from agriculture than for hunter-gatherer societies. Farming therefore created the possibility for humans to break through previous demographic constraints. At the same time, farmers became ever more dependent on a climate that could support their way of life. Unlike hunter-gatherers who, in principle, could disperse into new areas so long as population density remained low, farming populations could not leave en masse without suffering massive disruption and danger. An event like the Younger Dryas would cause far more disruption to a large population dependent on farming than to a much smaller and mobile hunter-gatherer population with experience obtaining food in varied ecological niches.

Abrupt change in the early Holocene

Overall population growth during the Holocene did not ensure continuous, sustained growth in all farming regions. Evidence from the period termed the Prepottery Neolithic shows population fluctuations after the emergence of farming in the Levant. During the first era of the Prepottery Neolithic beginning around 10,500 years ago, farming villages increased in size and number in Mesopotamia and the Levant. Such villages also occurred across a much wider area than during the previous Natufian era. However, settlements shrank toward around 8,900 to 8,600 years ago.[52]

This has been attributed to multiple causes, possibly including migration to the north and west, war, the advance of disease in larger communities that provided more hosts, a cooling trend, and more arid conditions as well as to combinations of these factors.[53] Temperatures during this time dropped abruptly. This cooling event has come to be known as the 8.2 k event, or simply, the 8 k event. Similar in some sense to the Younger Dryas, the 8.2 k climate event interrupted the shift toward warming and higher precipitation. In this case the cooling was less than that estimated for the Younger Dryas—ice cores from Greenland indicate a cooling of around 6°C, and cooling averaged around 1°C across Europe.[54] The ice cores show that the cooling lasted for a period of around 150 years, and glaciers expanded on Baffin Island.[55]

Within the relatively stable Holocene, the 8.2 k event stands out as one of the sharpest climate fluctuations since the Younger Dryas. The geographic expression of the 8.2 k event was similar to the Younger Dryas, with the coldest temperatures affecting the northeast Atlantic region, along with drying in Africa and Asia. In contrast to the Younger Dryas, however, there is little evidence for Southern Hemisphere warming. As in the case of the Younger Dryas, a large pulse of freshwater entered the North Atlantic and slowed deep ocean circulation, this time triggered by the draining of glacial Lake Agassiz through Hudson Bay into the Atlantic. This preceded the rise of complex civilizations, so we would find no archaeological evidence of the collapse of large cities. At one extreme, the 8.2 k event could have contributed to the collapse of Neolithic societies or to migration, but a counterinterpretation emphasizes the ability of Neolithic peoples to adapt to this climate shift.[56] At the site of Tell Sabi Ayad in northern Syria, for example, the archaeological record shows numerous changes around this time, including evidence of an increase in textile production and of abandonment of pigs in favor of cattle.[57] This cooling may have countered the further advance of the Neolithic into Europe.[58]

The 8.2 k event, however, did not lead to a permanent collapse of human societies, which continued to develop in the Near East and elsewhere during a new phase, designated the Pottery Neolithic. One interpretation holds that the 8.2 k event led to migration from western Asia into the Balkans.[59] However, farming villages in the Near East proved resilient. At the site of Çatalhöyük in southern Anatolia in Turkey, Neolithic farmers adapted to increased stress on their livestock. At the settlement itself, smaller households emerged before abandonment between 7925 and 7815 BCE.[60] It is certainly possible to find sites abandoned, but they are not situated in the areas likely to have been most affected by the 8.2 k event.[61] The effects may also have varied by region—a study of Scotland that uses evidence of human activity as a proxy for population finds a collapse of population.[62]

The end of the Green Sahara

In much of Africa in lands that today make up the Sahara desert and adjacent regions to the south, fluctuations in climate played a major part in shaping human strategies for obtaining and producing food during the Holocene. In North Africa, the period from the late Pleistocene to the early Holocene brought a period later described as the African Humid Period, the most recent of the Green Sahara. It may seem difficult to believe today when the Sahara is the world's largest warm desert, but from around 12,000 to 5,500 years ago, the Sahara was a region covered with vegetation and lakes. Ancient lake bed sediments show much higher lake levels than at present. Today's Lake Chad, now shrinking because of human-driven climate change, is only a fraction of the size of the ancient paleo-lake that once extended far beyond the shores of the current lake. The abundant supplies of water supported large populations of both people and animals that are either scarce or, in much of the Sahara, altogether missing today. Engravings on rock outcrops in the desert recall this very different past. The engravings depict people swimming, big game hunting, herds of livestock, and wild animals including hippos and crocodiles. The crocodiles of the Green Sahara lived in the lost desert lakes. Today they are mostly gone, though a small remnant population has been found living in caves and near seasonal wetlands in Mauritania in North Africa.

As with other Green Sahara periods, this African Humid Period was caused by an increase in monsoon strength associated with changes in Earth's precession. The Northern Hemisphere summer coincided with Earth's closest approach to the sun around 11,000y ago. The result was about 8 percent greater solar radiation during the summer, which was particularly influential in the lower latitudes that lacked a large ice sheet to moderate the influence of increased insolation. Greater summer insolation increased the temperature difference between large land masses and the surrounding ocean, causing a northward shift in the ITCZ and strengthening the summer monsoon. Monsoonal precipitation increased up to 50 percent during this time.[63]

Human population grew in the Sahara during the African Humid Period. Humans settled throughout the eastern Sahara. Hunter-gatherers and eventually pastoralists both made their home in the region. The rate of population growth was not the same in all parts of the Sahara, but overall population increased rapidly. Human communities formed at lakes such as a paleo-lake (an ancient lake that no longer exists) at Gobero in Niger. Hunter-gatherers who lived at Gobero between 7700 and 6200 BCE created the first known cemetery site in the Sahara. The African Humid Period also improved conditions for hunting in East Africa. Human population of the Sahara fluctuated. People abandoned the site of the paleo-lake at Gobero for about 1,000 years between 6200 and 5200 BCE, but population only declined sharply at the end of the African Humid Period some 5,000 years ago. At Lake Gobero the record of burials comes before 2500 BCE.[64]

The end of the African Humid Period, driven by decreasing solar radiation associated with the precession cycle, created major challenges for human populations. As paleo-lakes dried up, some individuals left.[65] They moved out of the Sahara, either south or east into the Nile Valley. The concentration of population along the Nile immediately preceded the rise of Pharaonic Egypt.

Others intensified use of the most favorable niches. The rapid increase in aridity gave humans reason to more fully exploit novel strategies for obtaining food. Increasing aridity imperiled domestication in rapidly drying areas, but also may have fostered advances in agriculture.[66] With scarce supplies available from hunting and gathering, the benefits of relying on domesticated animals increased. Images from rock art and the remains of dairy fat on pottery indicate the use of domesticated cattle in North Africa in the fifth millennium BCE. Despite their lactose intolerance, people were able to process milk, possibly into a form of butter, cheese, or yogurt, and then consume dairy products.[67] At Wadi Bakht on a plateau in southwestern Egypt people experienced a sharp climate shift around 5,500 years ago. They became nomadic pastoralists until humans abandoned the area with a final episode of drying around 4,500 years ago.[68]

The drying trend appears to have both dispersed and concentrated populations. In West Africa, the overall trend toward drying conditions could have concentrated populations along the Niger and fostered the rise of urban areas with more centralized power.[69]

Shifts in climate may have influenced the routes and timing of the large Bantu migrations that remade the human population map of much of Africa. Today, Bantu-speaking peoples make up the large majority of the population of central and southern Africa. Genetic analysis suggests that Bantu people began to migrate out of West and central Africa some 5,000 to 5,600 ybp.[70] Between 3,000 and 2,000 years ago, a drying trend shrank forests in central Africa. As forests thinned, Bantu farmers were able to move south more easily. New migrants most likely moved at first to the edge of forests and then traveled further into the forest zone. They built villages, used pottery, and appear to have practiced simple agriculture. Findings of domesticated pearl millet in southern Cameroon date back to 400 to 200 BCE.[71] Models based on analysis of large numbers of Bantu languages suggest that Bantu peoples followed emerging savanna areas that broke up the rain forest, both along the periphery around 4,000 years before present and then in some central areas such as the Sangha River by 2,500 years ago. In contrast, rain forest areas slowed migration.[72]

Bantu peoples took up iron working by around 2,500 years before present, and the pursuit of charcoal to smelt iron created another possible influence, along with climate, on the composition of forests. Drying climate created the Dahomey Gap, a savanna that cuts through the rain forest in West Africa. Oil palms spread into and proliferated in the gap. At Lake Ossa in Cameroon, human collection and use of wood for smelting iron could have contributed to reduction in the forest canopy.[73] But the overall decline of Holocene era forests around 2,500 years ago favored the spread of the oil palm in central Africa even without human activity. The regional climate change produced a shift from landscapes of evergreens to savanna in central Africa, but this was also during the period of Bantu migrations, and increased weathering stemmed from human activity as well.[74] In East Africa, as well, it is not always possible to determine the relative influence of precipitation and of human activity on vegetation. Analysis of a core from Lake Masoko in Tanzania shows the effects of burned charcoal, suggesting that Bantu peoples cleared land as they spread agriculture to East Africa.[75]

Complex societies

The expansion of farming during the Holocene helped make possible the emergence of civilization, though the timing for the start of civilization varied greatly by region. The earliest societies commonly identified as civilizations date back to a period when many humans still lived as hunter-gatherers. Neither a warmer and more stable climate nor farming may have required the emergence of civilization according to a particular timetable. However, civilizations repeatedly emerged during the Holocene. Such civilizations shared many qualities. They possessed increasing social complexity and political organization or government. They had larger communities and cities or, in some cases, elaborate ceremonial sites. Their ruling elites and religious leaders erected palaces, temples, and monuments. They possessed denser networks of trade and communication, and in many cases they developed writing.

Civilizations first emerged a little more than 5,000 years ago in Mesopotamia and soon thereafter in Egypt. This period has been described as the Bronze Age, the era when people employed bronze for items such as tools and weapons. Civilizations arose independently in other areas of the world—in China in the second millennium BCE, and in Central and South America in the first millennium BCE. In every case, the stable warm period of the Holocene aided in producing domesticated crops and in raising human population, vital building blocks for civilizations, which rely on extracting a surplus.

In the case of the first human civilization in Mesopotamia, agriculture expanded with the proliferation of farming villages. Farmers generally experienced favorable climate conditions, though sea-level rise in the Persian Gulf submerged coastal areas. Larger settlements as of around 5800 BCE such as the site of Tell Hassuna in northern Mesopotamia could support some 500 people. Specialized buildings dedicated to some kind of religious activity also date back to this era. Increasing social and political complexity is also evident in the site of Tell Zaidan, a community from around 4,000 BCE on the Euphrates River, which contained a temple and signs of an elite culture such as seals or objects used to stamp and perhaps mark possessions.

In the fourth millennium BCE, full-sized cities emerged in Mesopotamia. Uruk, first settled around 4200 BCE, became a center for the Sumerian culture as of around 3500 BCE. Uruk's population increased from some 10,000 people to as many as 50,000. The city possessed a large religious temple complex. Uruk established colonies before its network of colonies collapsed around 3100 BCE. However, despite Uruk's setback, the new urban pattern spread throughout Mesopotamia with the establishment of more than thirty urban centers.

The emergence of civilization in Egypt saw many of the same general phases. By around 5500–5000 BCE there were villages along the Nile River. Towns appeared by the late fourth millennium. Both Egypt and Mesopotamia had large imposing ceremonial religious sites, though Egypt had fewer cities. In contrast to Mesopotamia, where the early cities acted as independent city-states, Egypt was united under the Pharaoh by 3100 BCE.

The connection between climate and civilization was complex. Civilizations both relied heavily on a climate suitable for farming and displayed growing resilience to climate fluctuations. The Holocene climate boosted the ability to collect a surplus upon which cities depended, but researchers have also repeatedly identified stress as a possible cause for social and political change. Thus, drying, within the overall Holocene, may have contributed to state formation in southern Mesopotamia at sites, such as the cities or Ur and Uruk.[76]

Egypt and Mesopotamia both flourished despite a general drying trend in the eastern Mediterranean.[77] Civilizations also emerged in South and in East Asia. In South Asia, the earliest civilization took root along the Indus River in what is now Pakistan. The remains of large cities, Mohenjo Daro and Harappa date back to 2500 BCE. In China, Neolithic settlements spread in several regions, in particular along the Yellow River and the Yangtze River. Early dynasties emerged in the north on the Yellow River, most distinctly in the second millennium BCE. The Shang dynasty ruled farming villages, towns, and cities.

The warmth of the middle Holocene, sometimes referred to as the Holocene climate optimum, boosted emerging civilizations as well as farming areas without complex governments and social organization. Any distinct climate trend could benefit a particular form of life, be it dinosaurs, marsupial lions, or mammoths, but the Holocene climate optimum provided advantages in particular for civilizations built on farming. Extending agriculture sustained a rising population and allowed elites to extract resources to fund the religious and political sites identified with civilization. The ability to collect and store grain on a large scale provided civilizations with resilience in the face of bad harvests. The story of Joseph in the Old Testament provides a striking example. The Pharaoh dreamed of thin, scorched grain and scrawny lean cows that swallowed up healthy grain and cows. He released Joseph from prison to ask him to explain the dream. Joseph told him that famine was coming, and the Pharaoh put Joseph in charge of collecting grain. "Joseph stored up huge quantities of grain, like the sand of the sea; it was so much that he stopped keeping records because it was beyond measure," and Egypt was then able to withstand a severe famine. This account of Joseph's advice and action cannot be historically verified, but it exemplifies the capacity that a farming civilization had to withstand food scarcity.

An analysis of population and urban growth in the Near East suggests that around 2000 BCE settlement patterns became decoupled from climate. This finding does not prove that civilizations had become invulnerable to any climate shock, but indicates increasing resiliency. Favorable conditions could benefit agriculture, but population could grow through dry periods.[78]

Along with advances in technology, art, and architecture and a host of impressive edifices and other building projects, the civilizations that emerged from agricultural societies of the Holocene also created negative effects. Farming produced more food and supported a higher population, but more food did not necessarily equate to better health. Indeed, by many measures the peasants who farmed suffered worse health than hunter-gatherers. Average height, which serves as one proxy for average health across a population, plummeted after the adoption of agriculture. The evidence from archaeology supports the idea that hunter-gatherer societies were surprisingly healthy by this measure. Average height at the end of the LGM in Greece and Turkey was around 5 feet 9 inches, but the figure crashed after the spread of farming and fell to 5 feet 3 inches by 3000 BCE.[79]

Civilizations also served to incubate new diseases. A higher and denser population increased the number of hosts for diseases that previously would have burned out. The same Holocene farming societies that provided the surpluses for building pyramids and ziggurats and the social complexity for creating writing also encouraged the spread of such diseases such as influenza, smallpox, and measles. Higher sedentary populations, for example, provided better conditions for the spread of tuberculosis (TB).

Indeed, one highly dangerous strain of TB first emerged in China around 6,600 years ago at the same time that rice farming was also intensifying in the Yangtze River Valley.[80]

The collection of resources in the Holocene agricultural societies greatly increased opportunities for building highly unequal societies. A hunter-gatherer band can certainly have leaders with greater power and privileges than the average member, but living in a small community that periodically packs up and moves imposes sharp limits on the ability to amass wealth.[81] In contrast, most civilizations that extracted a surplus from agriculture saw the rise of powerful, hereditary elites. The very division of Chinese and Egyptian history into dynasties attests to the might of households that could control and direct a surplus produced by farmers on a massive scale.

Conclusion

The Holocene marked a decisive shift for both climate and human history. Over previous tens of thousands of years, *Homo sapiens* had lived in a world with far greater climate fluctuations. Climate after the Younger Dryas was not static, but fluctuations were far less pronounced. Thousands of years later, we still take this as the norm: a comparatively stable climate.

For human societies, the Holocene also saw an unprecedented change from the past. The descendants of humans who had lived as hunter-gatherers increasingly became farmers. Hunter-gatherers did not instantly vanish, but the proportion of humans living as farmers, and in many cases living in complex societies, increased. More recently, with radical increases in farming efficiency and productivity, the proportion of people employed in farming has fallen, but we still take it as a basic norm that domesticated plants and animals will support a vast human population, far larger than the numbers that ever existed before the Holocene.

CHAPTER THREE

The rise and fall of civilizations

- Mid-Holocene aridity
- Early human climate forcing?
- Late Bronze Age crisis
- Climate optimum: Rome
- Climate optimum: Han China
- Rome and Han China decline and collapse
- Climate and landscape in early medieval Europe
- Summary

On a timescale of several thousands of years, numerous civilizations and human societies emerged, changed, and, in some cases, declined during the late Holocene. Traditionally, climate formed part of the unexamined background to the history of these societies. Thus in studying the history of Holocene societies, historians begin with certain assumptions, including a basic idea of the kind of climate found in the region for any given society. Climate history, however, reveals that even during the late Holocene, many societies confronted significant climate fluctuations.

Integrating the history of climate change with human history takes several possible approaches once we move beyond seeing climate as a mere backdrop to human events. In the most cautious approach, climate contributes to the basic conditions of agriculture and daily life without functioning as a major cause for political or economic change.[1] Societies that thrive during periods of comparative climate stability may, of course, never confront major challenges from climate, but another approach looks at climate as a possible factor that contributes to the rise and fall of civilizations. This takes the most striking form in studies of collapse.[2] An alternative approach, in contrast, focuses on resiliency and on how societies respond to and adapt to shifts in climate along with other external changes.[3]

Civilizations and complex societies of the Holocene in general proved both resilient and vulnerable to climate shifts. If civilizations had emerged at an earlier point, they might well have collapsed in the face of changes in climate. A complex society in Doggerland or Beringia, for example, would have ended up submerged by the sea. Hunter-gatherers who dispersed would leave no Atlantis under the waves. A large-scale civilization in one of the regions most severely affected by the Younger Dryas would likely have suffered great harm. There was no equivalent climate shift during the Holocene, but major civilizations still faced fluctuations in climate. The ability to manipulate the environment and stockpile resources gave civilizations and complex societies resiliency, but even more modest changes could challenge civilizations dependent on stable conditions and a steady supply of water and precipitation.

Mid-Holocene aridity

For climate scientists, the fate of the civilization along the Indus River exemplifies the damage that climate change could inflict on even a highly advanced society. By many measures, the recovered sites of the Indus River civilization continue to impress us. The two largest sites, Mohenjo Daro and Harappa, provide evidence of design and planning with a geometric formation of structures, and foundations of large, imposing buildings. Population estimates for these cities range from 35,000 to 50,000. Mohenjo Daro stands out for its system of wells and drainage systems as well as for a large pool dubbed the Great Bath. Other sites from this lost civilization demonstrate attention to controlling and storing water with drains, wells, channels, and dams. The people of the Indus River civilization also invented a form of writing, though to date the short fragments of script have not been fully deciphered.

In contrast to Mesopotamia, Egypt, and China, the civilization of the Indus River Valley fell into later obscurity. It emerged sometime after Mesopotamia and Egypt, around 2400 BCE. The large Indus River cities preceded the height of Bronze Age settlements in China. However, the Indus River civilization began to decline around 1800 BCE, and the largest settlements of that civilization were abandoned by around 1600 BCE, long before the end of ancient civilizations of Mesopotamia and Egypt (Figure 3.1).

The question of what caused the end of such a complex society has sparked debate. One of the archaeologists who carried out extensive excavations at the site asserted that scattered bodies were victims of a massacre carried out by Aryan invaders who sacked the city and destroyed the civilization. This narrative is dramatic but unconvincing: an assortment

FIGURE 3.1 *The Great Bath of Mohenjo Daro. Source: Wikimedia Commons, https://commons.wikimedia.org/wiki/File:20160806_JYN-03.jpg.*

of skeletons does not prove a massacre carried out by any specific group or the end of a city. Many of the skeletons could also have been buried, albeit in a rudimentary fashion, and there is little archaeological evidence of an act of widespread destruction at the city's end. In place of the invasion hypothesis, other historians suggested possible shifts in the course of the Indus River. Such fluctuations could certainly have led to the decline of some Indus River sites, but would not explain why an advanced society simply did not move to follow a new river course. Finally, newer archaeological studies take issue with the notion of a sudden collapse, by pointing to a shift in settlement patterns to smaller communities to the east. In this scenario some of the peoples of the Indus River survived and adapted to a new way of life. Farming may have moved with migrants in such a pattern, but the Indus River civilization, with its distinct artifacts and script, died out.

Of all possible causes, climate change seems to have been most decisive in undermining the Indus River civilization. Monsoon rains provided indispensable precipitation, and when the monsoon shifted, the civilization faded away. Around 5000 ybp (3050 BCE), a strong summer monsoon in this region fueled intense floods that precluded settlements and formal agriculture. But monsoon strength waned as Earth's precession cycle caused summer insolation to decrease. After 4500 ybp (2550 BCE), this drying trend favored the rise of farming and complex society in the region. Rivers become calmer and floods less extreme, permitting the construction of towns and cities along rivers. However, a continued decrease in precipitation associated with the weakening monsoon threatened agriculture. Geochemical analyses from Bay of Bengal sediments provide evidence for the drying trend. Molecular markers buried in the sediment show that plants adapted to dry conditions became more abundant between 4000 and 1700 ybp (2050 BCE and 250 CE), and predominated after that time. Another indicator of dry conditions, which traces the salinity of the Bay of Bengal, points to the reduced river flow in this region.[4] Humans could have contributed to the problem through overgrazing and deforestation, causing further stress on a reduced water supply. Paleopathology or analysis of human remains suggests that disease increased with climate stress.[5] In one sense, the population proved resilient if some moved east, but such migrants did not preserve their previous civilization.

The increasing evidence in favor of climate change as the prime cause of the end of the Indus River civilization provides important historical lessons. Even an advanced society capable of controlling and manipulating water can meet its demise from aridity. As much as the Indus River people increased their independence from fluctuations in rainfall, they remained vulnerable to a sharp enough climate shock. This example raises important questions: to what point can human societies remain resilient and adapt to climate change, and under what conditions will their strategies fail?

The Indus River civilization was not the only complex society to experience a major challenge from aridity around 4,000 years ago (2050 BCE). Amid the generally favorable Holocene climate, a drying trend at approximately 2000 BCE in eastern Tibet and western China in the western loess plateau, named for the deposits of loess silt, may have damaged Neolithic societies in China. At Sujiawan in northwestern China a forest gave way to a forest steppe and then to a steppe. The drying appears to have affected New Stone Age farming societies, and in some regions residents shifted from farming to pastoralism. Drying also severely affected the Hunshandake regions of Inner Mongolia in northern China and undermined the Hongshan society, a New Stone Age culture that left a trove of jade. A landscape of lakes and rivers gave way to dunes, and for centuries thereafter there were no similar artifacts or evidence of human settlement.[6]

The drying trend around 2000 BCE was not so severe as to undermine all complex societies or civilizations. In Upper Mesopotamia in what is now Syria, some regions of early Bronze Age settlement suffered setbacks. Thus, settlements near Umm-el-Marra in what is now northern Syria east of Aleppo were abandoned, and much of the area in northeastern Syria near the Khabur, a tributary of the Euphrates River, became devoid of settlements, but settlements endured in other regions.[7] Mid-Holocene aridity did not do away with civilization in Mesopotamia, but coincided with the demise of the Akkadian Empire, which the famed King Sargon had established in around 2600 BCE. A sediment core from the Gulf of Oman shows evidence of drought and colder winters with more frequent winter shamals (winds). In this model, drought and severe dust storms may have encouraged migration and weakened the Akkadian Empire as it faced military challenges.[8] A speleothem from a cave in northern Iran further shows evidence of aridity and increased dust, coinciding with the abandonment of settlements in northern Mesopotamia around 2200 BCE. Even as the Akkadian Empire collapsed, however, cities, including Tell Brak and Tell Mozan, showed resilience.[9]

The fall of the Egyptian Old Kingdom in the late third millennium BCE also raises questions about whether the same drying trend that appears to have damaged complex societies in South Asia and in the west of China could also have created similar stress for Egypt. The Old Kingdom saw the rise of a powerful and wealthy state with a rich religious culture. When we imagine the vast structures of ancient Egypt, we are usually thinking of the Old Kingdom: this was the age of the great pyramid building. Later periods of Egyptian history left monuments, but the enormous pyramids at Giza southwest of Cairo as well as the great Sphinx all date back to the Old Kingdom. All required skill in design, planning, and labor, and all rested upon the extraction of an agricultural surplus. This surplus required a steady supply of water for crops. At base, the edifice of the whole advanced Egyptian society rested on the ability of farmers along the Nile River Valley to produce more than they needed to subsist on.

At the end of the Old Kingdom, centralized rule collapsed, and Egypt entered a period termed the First Intermediate Period that lasted from 2160 to 2055 BCE. The building of the vast monuments synonymous with the Old Kingdom came to an end. Internal strife and disorder increased during this time between the Old and the Middle Kingdoms. Texts from the period refer to danger from bandits and frequent deaths.

Historians traditionally attributed the First Intermediate Period to such factors as the long rule by the last pharaoh of the Old Kingdom, Pepi II; power struggles; or invasion, but there is strong evidence that a reduced flood weakened the Egyptian state at this time. The lake at the Faiyum depression in middle Egypt dried up.[10] Numerous studies document a widespread drought in the eastern Mediterranean and West Asia around 4200 ybp (2250 BCE). Geological records from the Nile River Delta in particular highlight low Nile River flow coincident with the collapse of the Old Kingdom.[11] In this scenario, reduction in the Nile floods caused by a drying trend helped propel state collapse.

The First Intermediate Period did not necessarily bring equal hardship to all strata of Egyptian society. In today's world, the most dire effects of climate change often strike those without wealth or power because elites are generally better shielded from the most severe consequences of climate change. However, that was not necessarily true at all times. In Egypt, the central state of the Old Kingdom broke apart, but a decline in the production of luxuries for the elite did not equate to a collapse in broader production of artifacts. In particular, graves for Egyptians below the highest ranks showed a greater abundance of goods, including amulets and beads. By this measure, commoners may not have suffered the same losses as the elite, at least as long as they managed to survive any famine.

Ancient Egyptian civilization also showed resilience to the drying event of approximately 2000 BCE in that the First Intermediate Period did not lead to the collapse of the civilization. The model of kingship survived the Intermediate Period, and the Egyptian state recovered strength during the Middle Kingdom. Drought may have weakened but did not destroy ancient Egyptian civilization; however, the cultural changes that emerged during the First Intermediate Period had lasting effects. The pharaohs of the Middle Kingdom, though immensely powerful, acquired a new image as not simply a ruler but also as a shepherd. A stylized shepherd's crook served as one of the symbols of the pharaoh.[12]

Mid-Holocene aridity may also have created stress for human populations in China. Neolithic societies in China along the Yangtze River came to an end around 2200 BCE, but there are multiple possible causes. Flooding may have led to abandonment of settlements. Climate shifts may also have contributed to collapse. A sedimentary core from the East China Sea indicates a cold period in the Yangtze Delta from 4.4 to 3.8 K BP.[13] However, the pattern of regional climate change was complex: a stalagmite core from a cave in southeast China suggests a decrease in the summer monsoon precipitation in northern China, but an increase in southern China.[14]

Human societies and civilizations flourished in diverse settings during the remainder of the Bronze Age. Holocene agriculture sustained population growth and helped support civilizations in diverse areas of Eurasia. Civilization survived in Mesopotamia and Egypt as well as to the north of Mesopotamia. The Hittite Empire controlled much of Anatolia along with parts of Syria, and in the thirteenth century BCE, the Hittites fought the New Kingdom of Egypt. Bronze Age civilizations also flourished in and around the Aegean Sea on the island of Crete, the site of Minoan civilization, and on the southern Greek mainland at Mycenae. In East Asia, the earliest recorded dynasty, the Shang dynasty, emerged in China around 1600 BCE, though narratives of Chinese history also refer to a previous dynasty, the Xia dynasty, whose existence is difficult to verify. During the Bronze Age, total human population increased from some 14 million people 5,000 years ago to some 50 million by the start of the Iron Age around 3,000 years ago when iron supplanted bronze as the metal of choice for producing tools and weapons.

The intense farming and pastoralization that boosted population growth altered the landscape and local environment in many regions. An empire is not necessary to cut down trees: deforestation extended beyond the areas dominated by elaborate states. Farmers cut trees with flint and later with bronze axes and burned wood. Neolithic farmers in Britain cleared forests.[15] Human activity transformed the landscape in far-flung regions by the Bronze Age.

Early human climate forcing?

On a much smaller scale than in later periods of time, pastoralists and farmers in the Holocene increased emissions of greenhouse gases. The expansion of pastoralism and animal husbandry had at least the potential to create more methane. Clearing land and burning trees, in similar fashion, led to the possibility of increasing CO_2 emissions during the Neolithic and the Bronze Ages. First proposed by William Ruddiman in 2003, the early anthropogenic hypothesis posits that human interference with climate began well before the Industrial Revolution as a result of farming and land use. Evidence for such an early human forcing comes from the observed increase in CO_2 beginning 7,000 years ago, and in methane beginning about 5,000 years ago, a pattern that is not seen at the corresponding point in previous interglacial periods.[16] The timing of CO_2 increase is consistent with the clearing of land for agriculture, and the methane increase coincides with the flooding of land for rice cultivation and expansion of livestock.[17] While the influence of early civilizations on climate is still a matter of discussion, the early anthropogenic hypothesis helps explain records from paleoclimatology and archaeology.[18]

Late Bronze Age crisis

The end of the Bronze Age came with disruption and in some cases catastrophe for complex societies. In Egypt the New Kingdom ended in 1070 BCE when the country was split. In the first millennium BCE, Egypt experienced decentralization, civil wars, and foreign invasions to the point where it fell into the hands of a sequence of foreign empires. In Anatolia, the Hittite Empire collapsed by 1160 BCE, though the Hittite language survived the demise of the state. In coastal Syria, the Ugarit state collapsed in the twelfth century BCE.[19] In Greece civilizations collapsed altogether. The Minoan palace at Knossos in Crete was abandoned in the late Bronze Age, though the site retained some population. Bronze Age civilization did not survive on the Greek mainland. The palaces of the culture of Mycenae were destroyed, and the culture collapsed between 1200 and 1100 BCE.

The change was so profound that the earliest era of Greek civilization vanished into myth. The epics of Homer, composed in the Iron Age, looked back to an earlier Greek civilization, so distant that later readers of an oral tradition put down in words wondered whether there had actually ever been a city of Troy, let alone a Trojan war. The existence of Troy remained uncertain until the amateur archaeologist Heinrich Schliemann, in association with an English consul named Frank Calvert, excavated Troy's ruins in the 1870s.

With the collapse of the Greek Bronze Age, civilization suffered a major disruption. Writing came to an end. The scripts employed by the Minoans and Mycenae, later dubbed Linear A and Linear B, fell into disuse. Linear A has never been deciphered, and Linear B was decoded only in the 1950s. The Greek civilization made famous by cities such as Athens did not begin to emerge until some 300 years after the demise of Bronze Age civilization in Greece.

What caused this series of sharp setbacks suffered by so many complex societies in Southern Europe and the Levant at the end of the Bronze Age? Invasion was one possibility. Contemporary Egyptian texts referred to attacks by Sea Peoples, without providing many details to identify these sea raiders. An inscription from Medinet Habu, a mortuary temple for Pharaoh Ramses III, describes an Egyptian victory over Sea Peoples: "The countries which came from their isles in the midst of the sea, they advanced to Egypt, their hearts relying upon their arms. The net was made ready for them, to ensnare them. Entering stealthily into the harbor-mouth, they fell into it. Caught in their place, they were dispatched and their bodies stripped."[20] In this account Egypt prevailed, but this record does not emphasize the possible costs of victory.

Natural disasters could also have shaken late Bronze Age societies. In the case of the Minoans, archaeologists investigated the damage inflicted by the eruption of Thera (Santorini), a volcano in a small group of islands in the Aegean at around 1600 BCE. This was one of the most powerful volcanic blasts ever witnessed by humans. It left volcanic deposits up to 30 meters thick. It makes sense to consider whether such a massive eruption devastated civilization, but if the volcano and a possible tsunami damaged Crete, it did not immediately wipe out all sites of Minoan culture or erase Bronze Age culture on the Greek mainland (Figure 3.2).

The very transition away from the Bronze Age to the Iron Age was another possible cause of the late Bronze Age crisis. As iron supplanted bronze, a state or civilization without a ready supply of iron would face a military disadvantage. Whether because of this shift to iron, the prowess of outsiders, internal divisions in Egypt, or due to a combination of all these and other factors, Egypt experienced repeated invasions during the first millennium BCE. In the eighth century the Kingdom of Kush in Nubia to the south invaded and held Egypt. Assyria, a powerful military society that rose in the north of Mesopotamia, invaded in the seventh century. Persia took power in the late sixth century. Alexander the Great conquered Egypt in 331 BCE, ending the rule of a last native dynasty, and Rome took Egypt in 30 BCE. The degree to which these external powers exercised direct power over Egypt varied, but the once highly independent civilization that had endured for more than 2,000 years became an imperial province.

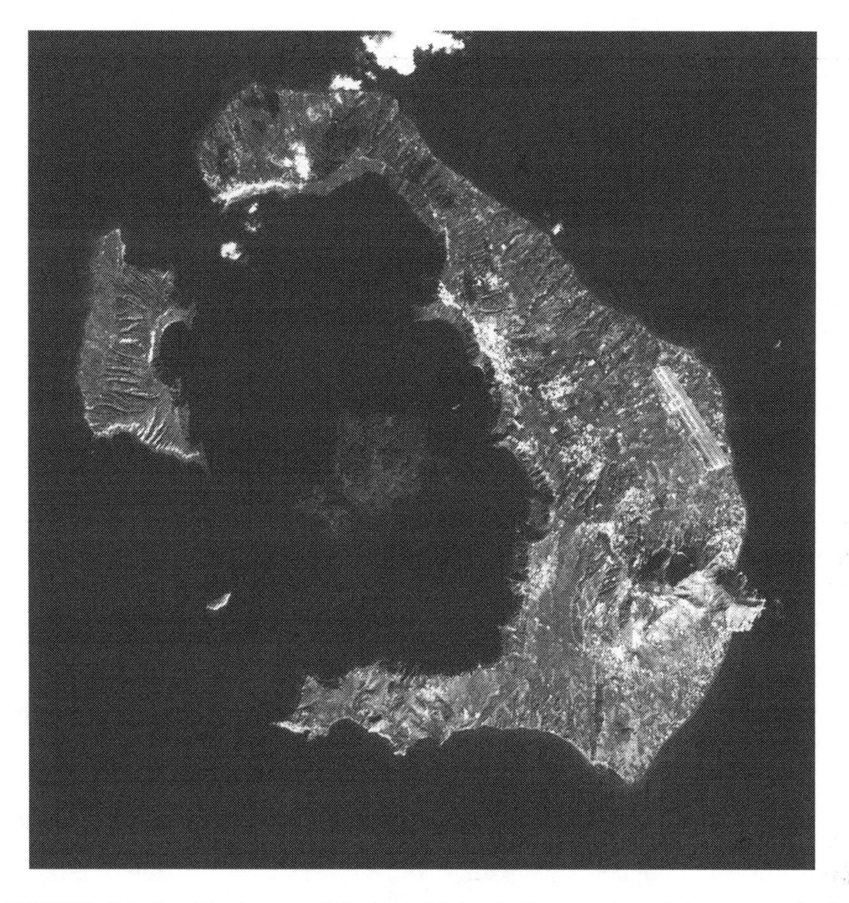

FIGURE 3.2 *Satellite image of Santorini Island, Greece. http://photojournal.jpl.nasa.gov/catalog/PIA02673.Source: NASA.*

Along with singular natural disasters, internal problems, and external attacks, climate change was a possible cause of stress for the late Bronze Age societies of the Mediterranean and the Levant. A sequence of temperature increases in the Northern Hemisphere, followed by temperature decreases and increased aridity during the early Iron Age,[21] resulted in a "hydrological anomaly," or less available water between 1200 and 850 BCE. Records of vegetation from the Nile River Delta point to a series of regional droughts, including the ones around 4200 and 3000 ybp (2250 and 1050 BCE), that would have affected civilizations in this region.[22]

Though showing an overall trend toward arid conditions, such data, however, can be difficult to place in precise timelines with historical outcomes.[23] For the Late Bronze Age, climate variability, rather than simply aridity, may have magnified stress. In the Peloponnese peninsula or southern Greece, the civilization of Mycenae developed elaborate palaces. A dry period in the late Bronze Age coupled with increased climate variability may have both weakened Mycenae and hindered any attempt to restore the palaces, but the rupture was not complete.[24]

Without any single catastrophic event, year after year with an average drop in precipitation would have reduced the supply of food. This was not a sudden climate shock but a cause of growing stress that interacted with other internal and external problems that weakened late Bronze Age societies in the Mediterranean and the Levant. In this scenario, food shortages could also have contributed to the invasions of Sea Peoples, by propelling desperate people to migrate in search of a new way of life.[25] The Sea Peoples, according to this interpretation, were not marauding pirates but environmental refugees. At the same time, a drying trend also encouraged adaptation. In Canaan under Egyptian rule, cultivation of grain intensified during the late Bronze Age. The expansion of dry farming, according to one interpretation, demonstrated the human response to a dry period.[26]

Climate fluctuations in the late Bronze Age did not undermine all civilizations and human societies. Population fell in northwestern Europe before cooling in the early Iron Age.[27] Such timing, if it stands up, does not, however, prove that climate was irrelevant in the fortunes of human societies in areas such as Ireland. If a climate shift did not determine the rate of population decline, a colder period would still have made farming more difficult for societies facing other challenges during the transition from the Bronze Age to the Iron Age.

The abandonment of complex sites in North America in the period after around 1000 BCE has raised the question of whether climate fluctuations during this period generated social stress in the Americas. In particular, a large and elaborate mound complex at Poverty Point in what is now Louisiana includes a series of mounds and ridges built between 3700 and 3100 ybp, but abandoned, according to most interpretations, the sites thereafter. The Early Woodland cultures that followed around 2600 ybp had lower population densities and traded over shorter distances. Along with immigration and technological change, climate change counts among the possible explanations for this cultural and social shift at Poverty Point. Climate change may have increased the likelihood of floods, as recorded in sediments in the Gulf of Mexico.[28] A comparative absence of detailed climate proxies from the region and problems dating and describing the transition to the Early Woodland period as well as the protection from flooding on high ground at Poverty Point itself pose challenges to this climate change hypothesis, but flooding in the region could have disrupted food supplies and trade routes.[29]

Iron Age societies were not immune to the effects of the same climate shifts that created challenges for late Bronze Age societies. The case of Assyria shows both the military power of one of the most successful early Iron Age empires and the possible strain created by arid conditions. From their base of power in northern Mesopotamia, the Assyrian kings waged annual campaigns. Their iron arms and weapons, military engineers, and mastery of sieges led to numerous victories. Assyrian armies conquered Syria, Phoenicia, Israel, Babylon, and Egypt. At times, they employed terror. An inscription declared, "I built a pillar over his city gate and I flayed all the chiefs who had revolted, and I covered the pillar with their skin." Assyrians engaged in mass deportations of some conquered people. Imposing Assyrian stone reliefs depicted the movement of deported peoples, along with other themes, including war and royal lion hunts. Assyrians themselves also gave a religious explanation for conquests. Royal inscriptions referred to expanding the territory of the Assyrians' gods, in particular the god Ashur.

After centuries of asserting its military might, Assyria collapsed comparatively quickly. The Assyrian Empire suffered from civil wars in the late seventh century BCE as well as rebellions, and in 612 BCE forces from Babylon and Medes took the city of Nineveh, the capital city for the Neo-Assyrian Empire. However, the fall of a state, which had caused so much fear among its neighbors for so long, seems surprisingly abrupt. Rebellions and civil war contributed to the demise of the Assyrian Empire, but climate change may also have placed a burden on Assyria. The Near East experienced dry conditions in the seventh century BCE at the same time that Assyria had to provide for a larger population. A contemporary letter written by a court astrologer attests to the difficult conditions: "No harvest was reaped." However, a counterinterpretation disputes the finding that Assyria was overpopulated.[30] A speleothem from northern Iraq indicates a wet climate during the period of the empire's rise, but severe drought that reduced agricultural output in the early and mid-seventh century. The empire's cities benefited from steady rains and a strong agricultural surplus before extreme, prolonged droughts of the empire's last decades.[31] A decline in solar output may also have affected climate during the transition from the late Bronze Age to the early Iron Age. Measurements taken from European peat bogs show a decrease in solar irradiance in this period.[32] According to some models, this decline in solar irradiance led to increased precipitation, which reduced deserts and expanded steppe grasslands in Central Asia and southern Siberia. More plentiful forage would, in turn, have boosted the nomadic population of the region, including a people known as the Scythians. As their population increased, Scythians then migrated westward toward the Caucasus, the Black Sea, and eventually toward Europe.[33] In the fifth century, the Greek historian Herodotus described the Scythians. He believed that they had arrived from Asia. In recounting wars waged by neighboring peoples against the Scythians, Herodotus also provided evidence that could fit into an alternative explanation of Scythian migration as a response to invasion.[34]

Climate optimum: Rome

Within the period of the Holocene, drying trends caused regional stress, but conversely periods with more reliable precipitation benefited civilizations.

The period of comparatively warm and stable climate from roughly 400 BCE to around 200 CE has been described as a Roman climate or climatic optimum. These terms derive from the idea that complex societies and powerful empires of the classical era benefited from the climate of their era. Such optimal conditions, of course, are not fixed: a very different climate, highly inhospitable for such empires, could be optimal for something else. Thus, the much colder climate of the LGM could be described as optimal for reindeer. The term "optimum" itself therefore points to the close relationship between climate change and human history.

Rome and the Han dynasty, two of the largest and most powerful empires of antiquity, flourished during the climate optimum. From a small city-state on the Tiber River in central Italy, Rome expanded to rule all of Italy, the entire Mediterranean Basin, and eventually vast regions of the interior of western and southeastern Europe. Rome's origins were shrouded in legend and in the tale of two brothers, Romulus and Remus, said to have been raised by a wolf. The brothers fought—and the victor, Romulus, named the city after himself. Another legend, made famous by the Roman writer Virgil in his epic poem the *Aeneid*, told of Rome's founding by Trojan refugees. As far as can be determined, eighth century BCE Rome was actually a village or small town. In its early years, Rome had kings, but in the late sixth century BCE Rome abandoned monarchy and became a republic with annually elected consuls and a senate. With its powerful aristocracy, this republic was not a mass democracy.

Precisely because of the power of Rome at its height, Rome provides an instructive example for comparing history with climate change as a mere backdrop to history that integrates climate change. Looking at Rome's conquests, it is easy to imagine the Romans as masters of their own fate, or at least as pitted only against human foes. A standard narrative of Roman expansion features warfare and an expanding military and body of citizens. From its starting point on the Tiber River, Rome steadily grew. This process began long before Rome came under imperial rule. Rome fought a series of wars against neighboring Latins. In the fourth century BCE, Rome defeated and made conquered Latins Roman citizens, thereby strengthening the Roman military. This choice to expand the number of citizens set a model of growth that Rome employed repeatedly over several centuries. To the south, Rome fought against Greek colonies, all the while adopting elements of Greek culture. In the third century BCE Rome struggled against but ultimately defeated King Pyrrhus, who had been invited by Greek city-states in Italy to assist them against Rome.

Rome fought a series of wars, the Punic Wars, against Carthage, Rome's greatest rival for supremacy in the Mediterranean. In the First Punic War from 264 to 241 BCE, Rome took the island of Sicily. The second war, from 218 to 201 BCE, in which Rome fought the masterful Carthaginian general Hannibal, was a hard-fought affair, in which Rome suffered defeats before finally prevailing. The third and final Punic War, from 149 to

146 BCE, was a matter of vengeance. Rome retaliated against rebellion by destroying Carthage, laying the city to waste, and carrying out killings and deportations that foreshadowed genocide. Rome completed conquest of the Mediterranean with a series of further campaigns against Macedonia and entered Syria in the first century BCE. They also moved farther away from the coast: Julius Caesar led the conquest of Gaul.

With Caesar, Rome began the transition to empire. The assassination of Caesar brought a power struggle in which his nephew and adopted son Octavius triumphed. Octavius ruled as princeps, or first man, with imperium or superior powers of command and in effect established the Roman Empire. Expansion continued through the shift in government. The empire pushed further away from the Mediterranean; back into Britain, which Caesar had reached; into Romania; and toward Switzerland and Germany, though German tribes pushed the Romans back to the Rhine (Figure 3.3).

The example of Rome reveals not only the importance but also the limits of integrating climate as a major factor in human history. Vastly expanding on the brief narrative above, the many histories of Roman expansion have not traditionally looked at climate as a major cause for Rome's success. Numerous factors contributed to the sequence of Roman victories and imperial growth. The Romans themselves claimed to fight and win just wars. Whether or not their wars would be seen by others as just, they also benefited from a military elite that won rewards for victory. Many of Rome's generals proved highly capable. Commanders like Caesar came from the aristocratic elite, and when the Roman Empire suffered setbacks in the third century CE, professional career soldiers supplanted commanders from the aristocracy. The long-serving professional soldiers in their legions provided an experienced fighting force.

Rome's ability to incorporate neighbors and conquered peoples into the empire and its military magnified Roman power. The contrast with the famed Greek city-state of Sparta was striking. From the sixth through the fifth centuries BCE, if not earlier, Sparta was the greatest military power on land of all the Greek city-states. In the fifth century BCE Spartans fought in alliance with Athens against the Persian Empire and then fought the decades-long Peloponnesian War with Athens from 431 to 404 BCE. Sparta relied on highly trained soldiers who spent years training, but the number of Spartan soldiers dwindled as did Sparta's military power by the fourth century BCE. Rome, in contrast, built up a very different military model in which men from many cities and regions served in the Roman military.

With so many keys to victory, it is not possible to single out the climate optimum as the single or prime cause of Roman expansion. Other cities, states, and tribal confederacies defeated by Rome experienced much the same climate. Rome did not defeat Carthage or conquer Gaul because of a shift in climate. At the same time, a comparatively stable climate helped sustain the Roman realm and empire over many centuries.

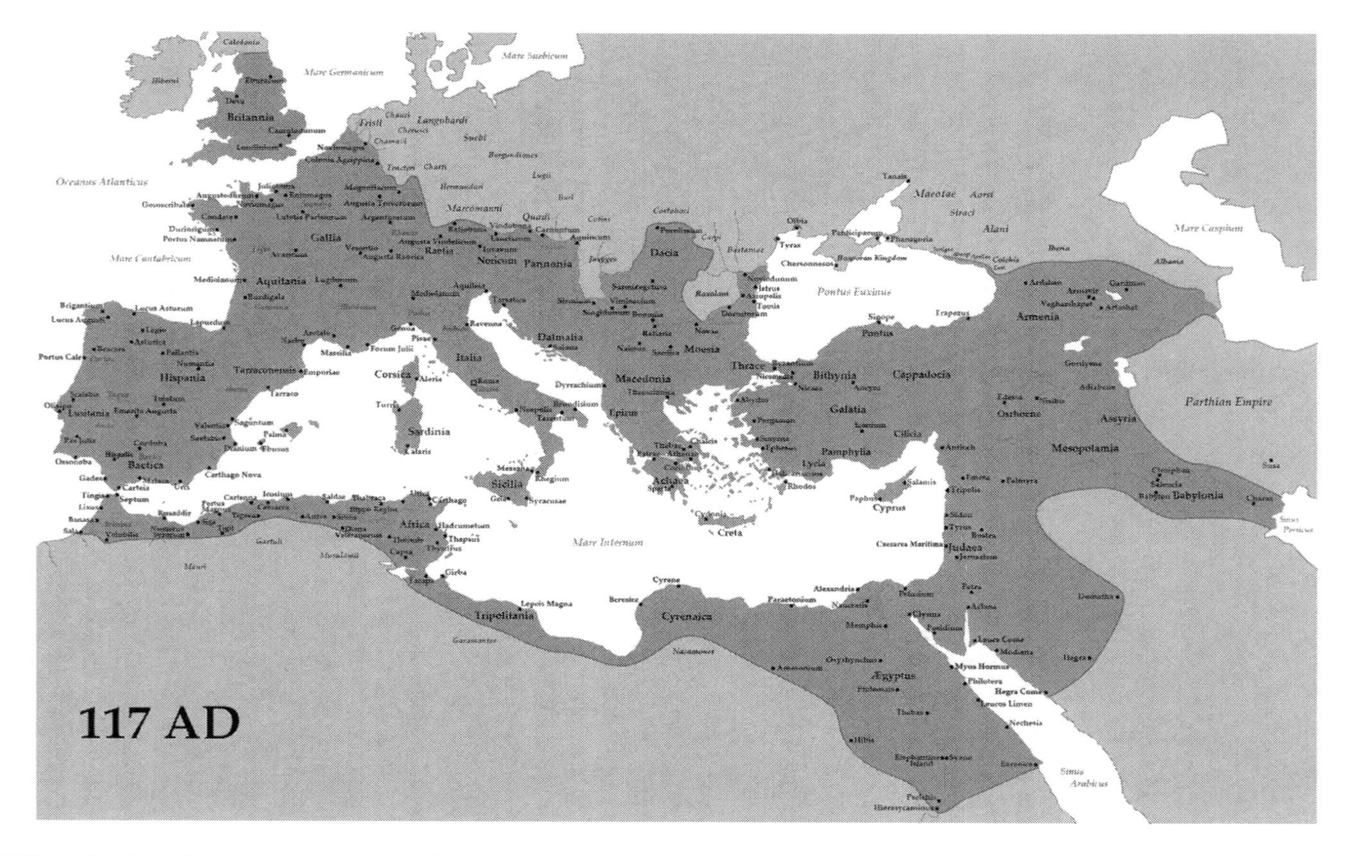

FIGURE 3.3 *Map of the Roman Empire 117 AD. Source: https://commons.wikimedia.org/wiki/File:Roman_Empire_Trajan_117AD.png.*

Rome's population and the area under cultivation grew during the climate optimum. The size of Rome's population is difficult to estimate because there was no single count of all residents. Estimates for maximum population most often range from 50 to 70 million with growth through the second century CE.

The vast and growing population required a steady and reliable supply of food. Roman capacity to feed such a large and far-flung population depended on both human adaptation and on favorable conditions. Rome not only grew but also collected and distributed food. The city of Rome itself took in foodstuffs and grains produced in other parts of the empire. The area under cultivation increased. Already by the first century BCE, Romans engaged in reclaiming land as the Republic expanded.[35] The Roman system of producing and distributing food benefited from the climatic optimum. In Egypt, for example, flooding in the Nile River led to a high ratio of good harvests, boosting agricultural productivity.[36]

In place of feats of generals and emperors, a history of the Roman Optimum could begin by focusing on trees, olives, and grapes. Accounts by Roman authors indicated the shift in range of trees, including beech and chestnut, and of cultivation of olives and grapes during the period of the climate optimum. The author Columella of the first century CE remarked on the belief "that regions which formerly, because of the unremitting severity of winter, could not safeguard any shoot of the vine or the olive planted in them, now that the earlier coldness has abated and the weather is becoming more clement, produce olive harvests and the vintages of Bacchus in the greatest abundance."[37] Olives were being grown in new regions. Indeed, the cultivation of olives in Gaul or France expanded under Roman rule. Roman agriculture also extended the area for growing grapevines to the north.[38] Roman settlement and colonization of conquered provinces created demand for such products as the climate optimum eased the expansion of cultivation.

Climate proxies indicate a general warming trend in the late Republican era. Such warming could have shifted the Mediterranean climate to the north. Evidence collected from the Po River Delta, the Adriatic Sea, and the Alps suggests rising temperatures in Italy. These multiple proxies show a detectable warm period, though one not as warm as that produced in the twentieth and twenty-first centuries. This warm period provided favorable conditions for farmers and reinforced other causes for population growth.[39]

Social and cultural shifts interacted with climate to strengthen these trends. During the climate optimum, Roman innovation and adaptation accounted for much of the success in supplying the empire's population. In far-flung regions of the empire, Romans managed water supplies. To the present day, Roman aqueducts that have endured in some cases for nearly 2,000 years demonstrate ingenuity directing water. The population in arid regions of the empire developed techniques for storing water. Residents of Palmyra and nearby villages in Syria collected rainwater in cisterns and used it for agriculture in a region that today is a desert.[40] In Libya, the remains of Roman fortified farms called *Gsur*, or castles, stand in areas that today receive scant rain and are inhabited by pastoralists who tend to only cultivate grain in valley floors following rain. A wetter climate could explain the ability of Roman farmers to plant in lands that are so dry today, but any increase in moisture was modest. It seems, then, that the farmers of Roman Libya managed to farm dry lands because they excelled at managing water. Farmers stored and channeled water using cisterns, and foggara or quanats, underground channels for water. They channeled water from seasonal rains into fields to grow crops that would otherwise have required more rain: barley, wheat, fruits, herbs, olives, and others.[41]

Within the climate optimum, sudden cooling associated with large volcanic eruptions placed stress on Egypt. The influence of volcanic activity on monsoon rains coincided with revolts against the Ptolemaic state, which, like its predecessors, depended heavily on the annual Nile flood. This last of all Egyptian dynasties was named for the Hellenistic family of Ptolemy, a Macedonian commander, who served Alexander the Great. Ptolemaic rule ended with the death of Cleopatra in 30 BCE as Rome, led by Octavian who soon became Augustus Caesar, the first Roman emperor, made Egypt a province. Historians have ascribed revolts in the Ptolemaic period to multiple causes, but revolts frequently followed volcanic eruptions that would have disrupted the monsoon rains. A model shows that aerosols released by volcanoes shifted the ITCZ south, thereby weakening the monsoon rains in their northern range.[42] A volcanic eruption coincided the end of the Ptolemaic state in Egypt and the shift from the Roman Republic to the Roman Empire under Augustus Caesar. Again, the traditional history of these events finds many causes other than climate, including social tension and the power struggles between leading generals, but an eruption of the Okmok volcano in Alaska in 43 BCE shortly after the assassination of Julius Caesar in 44 BCE led to a brief, but intense, cold period.[43]

As Rome expanded, climate influenced the empire's dimensions. Roman power thrived in the Mediterranean Basin and in adjacent lands. At its height, Rome pushed its boundaries farther north, capturing all of Gaul, the Balkans, and much of Britain as well as Alpine lands. However, Roman power faltered as Roman forces pushed further into Central Europe. After winning so many victories around the Mediterranean and in Gaul and Spain, Rome suffered a decisive defeat in 9 CE at the Teutoburg Forest, where German tribes ambushed Roman legions in dense woods.

The ability to thrive in a harsh and often cold climate became part of the German image for Romans. The Roman historian Tacitus emphasized such qualities. He distinguished Germany's climate from that of Italy or other Mediterranean regions. "Who would relinquish Asia, or Africa, or Italy, to repair to Germany, a region hideous and rude, under a rigorous climate, dismal to behold or to manure," Tacitus asked, "unless the same were his native country?" Their landscape and climate made the Germans tough: "To bear hunger and cold they are hardened by their climate and soil."[44]

In Britain, climate also shaped Roman perceptions of inhabitants and the setting for the Roman boundary. Cassius Dio, a Roman consul and historian, stressed the hardiness of British tribes that withstood cold in the north: "They can endure hunger and cold and any kind of hardship; for they plunge into the swamps and exist there for many days with only their heads above water, and in the forests they support themselves upon bark and roots."[45] Roman legions at times entered Scotland, but Rome stayed close to the present southern boundary of Scotland along Hadrian's Wall in the 120s and 130s CE. Hadrian's successors tried to push the boundary north, before falling back to the line along the wall. The Roman/Byzantine historian Procopius, writing in the sixth century, described the wall as marking a sharp division in climate:

> Now in this island of Britain the men of ancient times built a long wall, cutting off a large part of it; and the climate and the soil and everything else is not alike on the two sides of it. For to the east of the wall there is a salubrious air, changing with the seasons, being moderately warm in summer and cool in winter . . . But on the west side everything is the reverse of this, so that it is actually impossible for a man to survive there even a half-hour.[46]

Climate optimum: Han China

In China, the Sui and Han dynasties similarly created a vast empire sustained mainly by agriculture during the climatic optimum, and, as in the case of Rome, standard narratives of Chinese history treat climate mainly as a backdrop. In 1046 BCE, the Zhou dynasty overthrew and replaced the Shang dynasty. From its core in the Yellow River Valley, the Zhou expanded to the west and the south, but began to fragment by the eighth century BCE. As the Iron Age emerged, there was no strong single state in China. The power struggles between competing states formed the backdrop for the thinking of the philosopher Confucius. Living during a period of conflict, he praised an imagined past of order and called for hierarchy in which people followed their proper role in society. One of his analects, reportedly compiled by his students, describes Confucius as having stated, "Let the ruler be a ruler and the subject a subject; let the father be a father and the son a son."[47]

A powerful, centralized, Chinese dynasty emerged only during the short-lived Qin dynasty of 221–206 BCE. The Qin emperor conquered the competing Chinese states. He imposed order by building a strong bureaucracy and by standardizing weights, measures, and writing. He supported the philosophy of legalism, which called for strict punishments to ensure good behavior under the state. The goals of legalist writings at least converged with his aims of increasing state authority. The Qin Empire collapsed soon after the death of the founding emperor, but this did not return China to the prolonged conflicts of the warring states period. Instead, the Han dynasty took rule over China in 206 BCE.

Under the Han dynasty, China became a large and stable empire. The Han dynasty blended Confucianism with legalism and allied the emperors with Confucian scholars. The empire ruled over the bulk of areas held by later dynasties and established a strong presence along the western and northern frontiers. At its height of military power, the Han dynasty pushed west during the reign of the Emperor Wudi, who ruled from 141 to 87 BCE. Chinese armies ventured to Central Asia in support of trade and colonization. The empire erected cities with walls built out of pounded earth to offer protection from steppe nomads. Emperor Wudi settled colonists on the Ordos Plateau of what is now Inner Mongolia in northern China. Armies and colonists also pushed south. As in nearly all periods of Chinese history, seminomadic peoples on the frontier proved difficult to control and subjugate, so Han emperors relied on diplomacy and trade as well as on military campaigns to maintain a presence in the west and north.

With a focus on the rise and fall of dynasties and methods for rule and integration, prevailing narratives do not emphasize the role of climate, but as in the case of Rome, the stability of the climate optimum made it easier to expand agricultural output. Han dynasty farmers employed a wide array of tools and techniques to raise yields, and the state sponsored irrigation projects. Agricultural experts wrote texts outlining models for improved farming. The population increased from some 20 million to close to 60 million by the midpoint of the dynasty, though it fell sharply in the early first century CE during a period of severe floods.

Climate, as in the case of the Roman Empire, also helped mark the Han dynasty's boundaries. Ambitious rulers such as Wudi pushed China westward, but only at an enormous cost. The Han dynasty dominated the agricultural lands of China, but struggled to assert power over arid and cold regions to the west and north. China built fortified towns in the northern zone.[48] Along the northern frontier, the Han dynasty focused in particular on regulating a nomadic people known as the Xiongnu. Chinese perceptions of the Xiongnu and of their ability to thrive in a difficult climate resembled Roman views of Germans, though Mongolia was very different from Germany. Simi Qian, the historian who wrote during the reign of Wu and was harshly punished by the emperor, described the lands of the Xiongnu as "submissive wastes."[49]

As in the Roman Empire, the large population also depended on successful management of water. Indeed, China exploited the Yellow River so effectively that the very scale of engineering came to pose a risk. Flood-control projects on the Yellow River expanded to hundreds of miles of levees. The intense farming that sustained the Han dynasty's large and growing population increased erosion and in turn led to further construction of levees. In CE 14–17 the levee system broke down in a series of massive floods that caused many deaths and crisis in the empire.[50] Rebellions interrupted imperial rule until restoration of the Han in a new capital to the east, Luoyang.

The apogee of the Roman Empire and the Han dynasty marked a new high mark for overall human exploitation of the natural environment. Never before, in all of human history and prehistory, had humans tilled so many fields, raised such large quantities of cereal, or raised so many animals. With the clearing of land for agriculture, deforestation increased, with a concomitant rise in CO_2. Flooding of fields for rice cultivation can account for much of the observed increase in methane since 2000 ybp, with the raising of livestock contributing to the methane emissions as well.[51]

Rome and Han China decline and collapse

Over several centuries, the Roman Empire and the Han dynasty came to an end or collapsed. That very word "collapse" has met with frequent critique from historians who have argued that narratives of political collapse overlook elements of continuity. Roman collapse did not mean that the Roman Empire eroded at the same pace in all locations. The end of imperial rule in the western empire occurred in 476 CE, but the eastern empire survived for many centuries afterward. That eastern empire became known as the Byzantine Empire, but in its day, it was still termed the Roman Empire. In the centuries immediately after the demise of the empire in the west, the culture of the eastern empire endured, before the Byzantine Empire lost much of its territory and underwent cultural, social, and religious changes that took it very far from the society of antiquity. The remnant of the Byzantine Empire endured until the Ottoman Turks conquered the city of Constantinople in 1453, putting a definitive end to the Roman Empire.

Despite the persistence of Rome in the east, it is useful to speak of the collapse of the Roman Empire so long as we understand what that means. Collapse did not immediately extinguish the culture and society of antiquity in all places at the same time—it is true that life continued on in the east without such disruptions longer than in the west. However, collapse had many real dimensions. Urban living all but vanished in the west of Europe and retreated in the east. Such change came far more slowly and less evenly to the eastern regions of the Roman Empire, but even Constantinople became a shadow of its former self, with large sections of the city in disrepair, and the population in the late Byzantine era fell to a fraction of the level of earlier centuries. In the former western empire, collapse erased the complex state and bureaucracy in wide sections of the empire: power became highly decentralized and fragmented. Population plummeted. The amount of cultivated land decreased and many areas became waste or forest.

In terms of culture, literacy dropped, and reading and writing became the province of a cadre of religious specialists. Even Charlemagne, king of the Franks from 768 to 814 and the most powerful ruler in Western and Central Europe during the first half of the Middle Ages, could not read or write, and failed at the task when he tried to learn late in life. As his biographer Einhard recounted, "He also tried to write, and used to keep tablets and blanks in bed under his pillow, that at leisure hours he might accustom his hand to form the letters; however, as he did not begin his efforts in due season, but late in life, they met with ill success."[52]

In China, the Eastern Han dynasty emerged after the crisis of the first century CE, but the renewed Han dynasty in its new eastern capital faced persistent challenges. Imperial power eroded, and the dynasty suffered from internal struggles. Settlements and colonization in the west also burdened China with the costs of maintaining frontiers alongside restive seminomadic peoples. Within China, the empire confronted a series of rebellions by Daoists and others. Factions, including eunuchs and Confucian scholars, engaged in protracted power struggles at the imperial capital. All the while, the provincial commanders entrusted with propping up the dynasty became increasingly powerful until the eastern Han dynasty came to an end as China split apart into multiple kingdoms in 220 CE.

The imperial model did not end in China as subsequent dynasties would prove, but by several measures China experienced decline during and after the period of Han imperial breakup. Population fell, and the area under cultivation dropped, especially in northern China. Nomadic groups repeatedly moved into the former imperial realm.

If a stable climate contributed to the success and expansion of the Roman Empire and the Han dynasty, what role, if any, did climate play in collapse and decline? Multiple factors contributed to the rise and fall of Chinese dynasties, including the actions of rulers, relations between ruling elites, relations with neighbors, and cycles of rebellion, but dynasties also typically flourished during eras with greater precipitation. Thus, the area of desert decreased during the period of the western Han dynasty from 206 BC to 24 CE.[53] In general, the earlier western Han dynasty enjoyed better grain harvests. During drier periods, the state was also more likely to suffer from fiscal problems.[54]

In the case of Rome, the causes of collapse have long fascinated historians. Going back centuries, historians pointed to multiple events and phenomena as possible causes of Rome's decline. They stressed both external and internal causes. From the frontiers of the empire, barbarians, Germanic tribes, and others attacked the Roman Empire until they overran it. Urban living faded, weakening the empire from within. Inside imperial borders, the rise of local magnates and military strong men eroded the influence of the imperial government at the center. That government itself grew more corrupt and less effective. Economic imbalance grew between the eastern and western empires: long after the western empire had begun to decline, the east and Constantinople remained comparatively wealthy. Rome's defenses also weakened: the military lost its edge, relying increasingly on barbarian recruits and eroding the distinction between friend and foe. As Rome's religion changed, the rise of Christianity struck some authors as a cause of decline, a charge rejected by St. Augustine in his work the *City of God*.

Historians have identified all of these as well as other possible explanations for the collapse of Rome, however they have defined collapse. A complete listing of all hypotheses advanced for the demise of Rome would be encyclopedic, but even though some explanations might not convince, the sheer number of plausible arguments indicates that Roman collapse and decline stemmed from multiple interacting causes. Seldom, however, did the historians of Rome's decline and collapse stress the effects of climate change. The standard historical interpretations of decline did not advance any kind of deterministic argument based on climate, though the eighteenth-century author Edward Gibbon, who coined the phrase "decline and fall," discussed the possible influences of climate on people.

With so many possible causes of decline, it is difficult to conclude that climate change determined the fate of either the Roman Empire or of the Han dynasty. In the case of Rome, the capacity to manage water and distribute food made Rome resilient to fluctuations in weather, but Roman population growth also made the empire more vulnerable to more decisive climate disruption.[55] Even after the third-century crisis, the Roman Empire showed resilience. Other factors, including declining capacity to respond to harvest shocks, may have predominated over climate change as a factor in changing settlement patterns in the western Roman Empire.[56]

Climate instability during the late Roman era affected agricultural production. The Roman climate optimum with higher humidity gradually ended in the late second century.[57] Drying in the third century came at the same time that Rome almost collapsed. Good harvests became more infrequent.[58] The climate fluctuations in late antiquity took a different form in the eastern empire. Higher humidity and precipitation improved conditions for farming in parts of the eastern Mediterranean and Anatolia.[59] A more favorable climate was one of several factors that helped classical civilization endure significantly longer in the eastern than in the former western regions of the Roman Empire. Settlements in the eastern empire flourished with the cultivation of grain, olives, walnuts, and fruit in regions including the interior of southwestern Anatolia before a shift at the end of the sixth century CE when farmers abandoned fields and orchards, and pastoralism increased. South-central Anatolia also experienced population growth in late antiquity before the shift to drying when cultivated cereals and walnuts gave way to steppes and forests of pines and cedar. Palestine similarly benefited from a moist climate in areas such as the Golan Heights and faced a contraction of farming in the seventh century.[60]

Far from Rome's imperial core, climate fluctuations affected the empire's fortunes. The Nile flood in the late imperial era became less reliable for farming, and a series of large droughts in Central Asia may have added to the pressures exerted against both Chinese imperial rule and the Roman Empire. The massive droughts appear to have propelled migration: residents, including the Huns and Avars, moved west from Central Asia to the frontiers of Rome where they posed a major military threat. The historian Kyle Harper writes that "the entire Asian steppe, for the first time in its history, shifted its weight hurling its most advanced state formations against the west" and describes the Huns as "armed climate refugees on horseback."[61] Tree ring records from north-central China show that Central Asia experienced three multi-decadal droughts centered around 360, 460, and 550 CE, each of which is roughly coincident with a period of invasions.

Climate fluctuations stemming from the El Niño-Southern Oscillation (ENSO) may have caused these Central Asian "megadroughts." ENSO conditions fluctuate between the El Niño (warm) phase, expressed as warm ocean waters in the eastern equatorial Pacific Ocean, and the La Niña (cold) phase, with cooler waters. During El Niño (warm) phases, a weakening of the trade winds allows warm waters to migrate eastward along the equatorial Pacific. The warmer surface temperatures fuel stronger convection cells in the atmosphere. This alters atmospheric and oceanic circulation patterns in such a way as to produce a positive feedback: even more warm water piles up on the eastern side of the Pacific and upwelling of cold deep waters along Peru's coast weakens. The opposite conditions—strong trade winds, increased upwelling, and colder surface waters in the eastern Pacific—characterize La Niña (cold) phases. Shifts between these two phases currently occur every two to seven years.

El Niño and La Niña conditions affect weather and climate worldwide. ENSO cycles shift the jet stream, and thus storm tracks, in midlatitude regions. The shifting of the jet stream during El Niño winters increases precipitation in southern California, for example, while a more northerly track of the jet stream during La Niña years produces drier conditions across the southern United States and wetter winters in the Pacific Northwest. Indonesia, northern Australia, parts of South Africa, and northern Brazil experience wetter conditions during La Niña, while equatorial Africa and the southeast coast of South America are drier. Central Asia experiences dry conditions during La Niña and wetter conditions during El Niño. Tree ring records indicate prevailing La Niña conditions during the times of the Central Asian megadroughts,[62] coinciding with the three droughts that occurred during the Hun-Avar migrations (Figure 3.4a&b).

Without any written records describing the Huns or the Avars before they traveled west, it is not possible to say what other internal factors might have induced them to move out of the steppes. The Huns, as the historian Procopius stressed, lacked their own writing. He described them as "absolutely unacquainted with writing and unskilled in it to the present."[63] Much the same lack of information surrounds the Avars, who also came from the steppes and reached the Caucasus by the mid-sixth century.

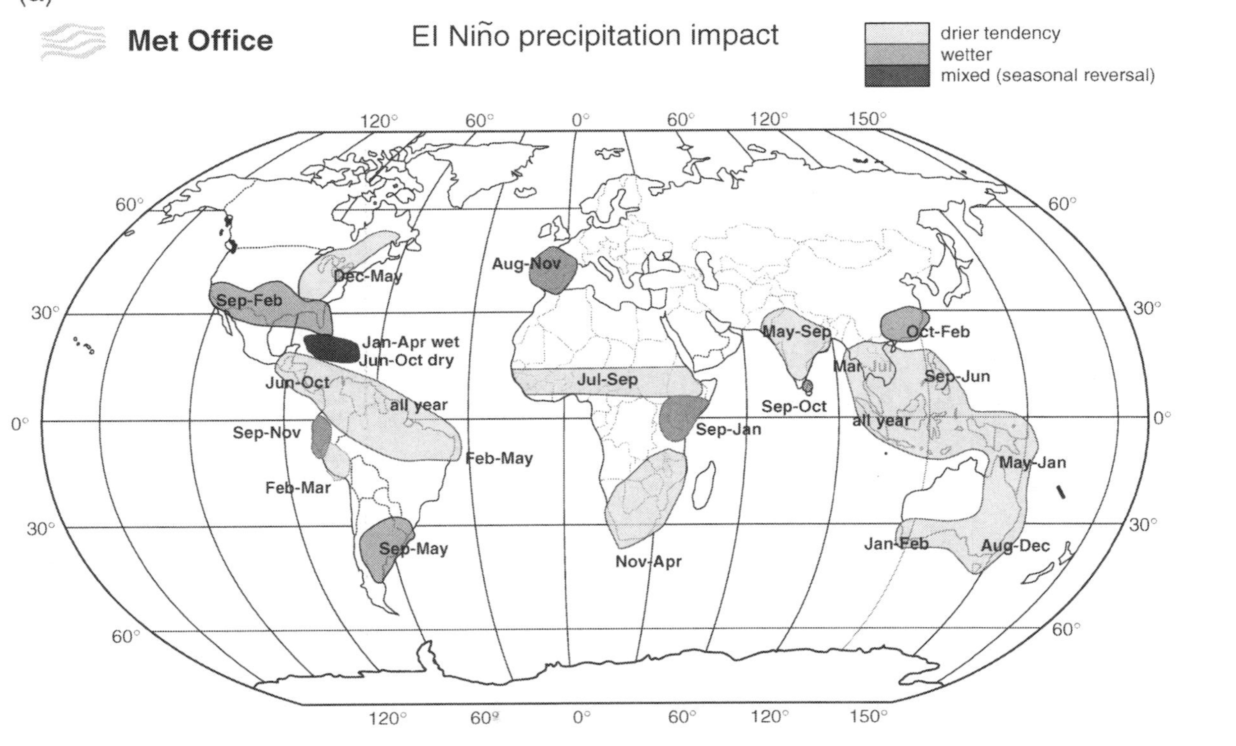

FIGURE 3.4 *(a) Map of El Niño precipitation effects. (b) Map of La Niña precipitation effects. Source:* NOAA, *http://www.metoffice.gov.uk/research/climate/seasonal-to-decadal/ gpc-outlooks/el-nino-la-nina/enso-impacts.*

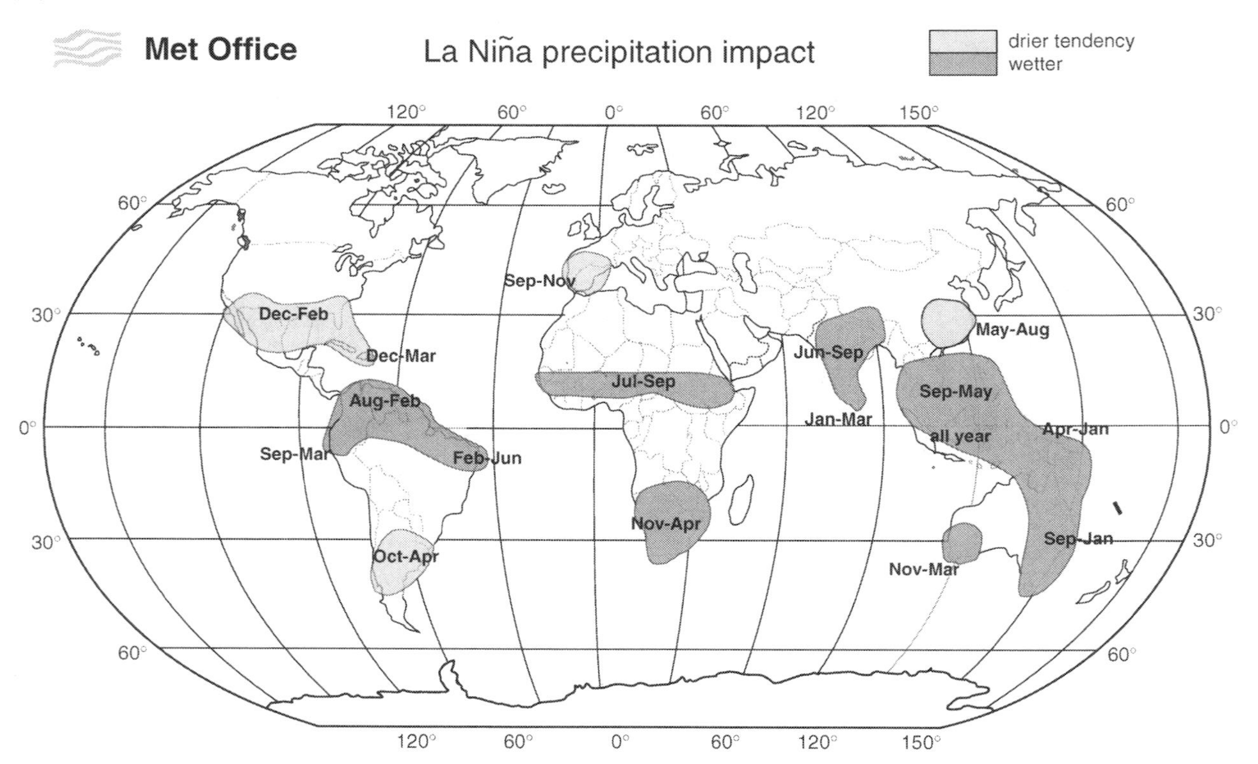

FIGURE 3.4 *(continued)*

Both the Huns and the Avars, like so many other seminomadic peoples of the steppes, were skilled at war. Mounted on horseback, the Huns advanced rapidly and assailed their foes with bows, lassoes, and lances. The Avars similarly took advantage of their formidable cavalry and made use of a composite bow. They also employed siege engines and may have introduced the iron stirrup to Europe. Both Hun and Avar archers could fire their bows rapidly while riding at full speed.

The migration westward of Huns and Avars presented a new potential threat to the Roman Empire. The Romans were long acquainted with Germanic tribes who had pushed Rome out of most of Germany in the reign of Augustus Caesar. After the debacle at Teutoburg Forest, Rome established a stable frontier along the Rhine and Danube Rivers. The empire faced a new round of barbarian attacks across the Rhine in the third century CE just as it also suffered defeat at the hands of Persia in the east, but Roman power revived in the late third and the early fourth century under new emperors with a military background: Diocletian and Constantinople.

Drought in Central Asia helped promote the westward "barbarian" migrations or invasions toward the end of the western Roman Empire. As Huns and Avars migrated west out of Central Asia, they, in turn, came into contact with Germanic tribes. Mere contact between Central Asia nomads and Germans did not invariably lead either to war or to victory for Huns and Avars, but the increasing competition from the east helped propel more Germans into the boundaries of the Roman Empire. Rome integrated many such Germans into its military, but the influx also destabilized Rome. In 376, Fritigern, leader of the Tervingi/Visigoths, sought to flee competition with Huns by entering the Roman Empire. Valens, who at the time was away on the Persian frontier, accepted Goths on the condition that they submit to Roman rule and taxes and supply troops. Tens of thousands of Goths then crossed the Danube River in numbers far greater than Rome had anticipated, and local order soon broke down on the Roman side of the Danube. More Goths, Greuthungi, soon thereafter followed across the Danube to escape the Huns.

Valens might not have been able to stop the influx even if he had tried, but the Goths he had allowed to cross the Danube surrounded and destroyed his army outside the city of Adrianople on August 9, 378. The Roman soldiers fought bravely for hours as a soldier present that day narrated: "When the barbarians poured down in great hordes, trampling horses and men, and when in the tight press of ranks there was no room for retreat and no opening."[64] The emperor, Valens, himself was killed, though the Goths were unable to follow up on their victory by taking major cities or fortresses. The new emperor, Theodosius, reached accommodation with the Goths as subjects who still retained their own autonomy. The defeat at Adrianople did not lead directly to the fall of Rome, but Rome could not control powerful Gothic leaders inside the empire. In 410, Alaric, the leader of the Visigoths, sacked the city of Rome after the emperor Honorius refused to accept Alaric's demands.

The Huns themselves soon followed German tribes into Roman lands. The Huns built up a powerful confederation in the early fifth century and engaged in diplomacy as well as war. Asking for hostages served as one way to keep peace. Thus, Flavius Aetius, a Roman commander of the fifth century who was later dubbed last of the Romans by the historian Procopius, lived among the Huns in his youth and later at times commanded Huns. In the 440s, however, Attila launched invasions of Rome, causing massive damage in the Balkans, and in 451 he attacked Gaul. Aetius, along with the Visigoths, defeated Attila at the battle of Chalons in 452. The Huns by themselves did not end the western Roman Empire, but the constant attention to the Huns made it all the harder for Rome to retain pieces of its fragmenting empire. Despite his success in holding Gaul, Aetius could do little to stop the loss of Spain.[65]

After the fall of the western Roman Empire, westward migration continued to pressure the surviving eastern empire. Drought on the empire's periphery continued to contribute to migrations westward.[66] As in the case of the Huns, drought in the steppes was a major possible cause of the westward movement of the Avars toward the eastern Roman or Byzantine Empire. Byzantine rulers alternately fought with, allied with, or granted tribute to the Avars. Preoccupied with exhausting wars with Persia, the Byzantine Empire could not afford to divert major resources to sustain a full-scale war against the Avars. Byzantine forces defeated Avars in 601 toward the end of the reign of Emperor Maurice. After a rebellion by Phocas, a Byzantine officer, that ended with the assassination of Maurice in 602, the Neo-Sassanian Persian Empire renewed war against the Byzantine Empire. Emperor Heraclius, who overthrew Phocas in 610, barely managed to avoid total defeat. As Persian forces occupied Byzantine provinces, Anatolia and Syria, Avars renewed attacks in the Balkans and in 626 also reached the walls of Constantinople, though they never took the city.

Though no single barbarian incursion led to Roman collapse or defeat, the cumulative burden of protracted raids, wars, and movements of refugees weakened the western empire and burdened the eastern empire. We often think of barbarians as marauders, invading rich lands to ride off with treasure or, in a different image, as back-country folks dazzled by the wonders of civilization and interested in imitating and adopting some of the customs and culture of those they also threatened. Climate change provided another starting point for this long series of barbarian incursions. The appeal of civilization still helps account for movement into old imperial centers, but severe droughts also generated population shifts. The ensuing series of migrations and incursions helped destabilize an empire that already faced many other threats.

The Byzantine Empire and other regions may also have experienced a phase of abrupt cooling after a possible volcanic eruption in 536. Contemporary authors described a veil of dust. Tree ring records confirm the cooling,[67] and volcanic eruptions followed in 540 and 547. The sequence, according to analysis of tree rings, produced a pronounced cooling trend, "a late Antique

Little Ice Age," just as the Byzantine Empire was starting to expand west into areas of the old western Roman Empire.[68]

Historical attention has recently turned to the possible effects of the Late Antique Little Ice Age (LALIA) on the Eastern Roman Empire and on the region of the former Western Empire. The onset of the LALIA coincided closely with a devastating outbreak of plague, named Justinian's plague for the reigning Roman or Byzantine emperor. The plague arrived in Alexandria in 541, struck Constantinople, and reached Rome in 543. Pronounced cooling and heavy precipitation with flooding in Italy and in Anatolian damaged societies already experiencing shock from the plague. According to one model, the convergence of disease and the Late Antique Little Ice Age helped to bring the society of antiquity to an end in the west of the empire and severely weakened the Byzantine Empire in the east.[69] Critiques of this model point to regional variation in climate, other factors, and the complexity of linking cause with effect. A study of the region of Petra in what is now southern Jordan, for example, found that changes in the settlement of marginal zones did not track with shifts in precipitation.[70]

Other regional studies, however, show change during the period of the LALIA. Elusa in the Negev desert in the south of the Byzantine Empire was a flourishing city with public buildings, including a theater, gymnasium, public baths, churches, and management of trash from the fourth to sixth centuries, but this came to an end in the middle of the sixth century as the plague and the Little Antique Little Ice Age struck.[71] Climate change interacted with other factors, but there was a broad pattern evidence of deterioration, at least for the Byzantine Empire in Anatolia with a cold period in the second half of the sixth century and less intensive exploitation of land in the seventh century with the expansion of natural vegetation and less pollen from fruit trees. Both Sicily and the southern Levant took on importance for the Byzantine Empire and the Abbasid Caliphate respectively, but experienced a shift to drier conditions that damaged local agriculture.[72]

The history of Rome and of the Han dynasty points to the general benefits of a favorable climate. The complexity of the climate record and regional variation make it difficult to attribute any single event in the history of either empire to climate. More broadly, however, empires thrived and populations grew, with interruption in the Chinese case by first century CE floods, during a period of comparative climate stability. Amid many causes for a complex process of collapse, decline, and transition, imperial power eroded as that period of comparative climate stability came to an end.

Climate and landscape in early medieval Europe

With the end of the western Roman Empire, the classical human landscape also vanished in much of Europe. Archaeologists have found that large numbers of the sites in regions of Italy and France occupied during antiquity or the classical era were deserted by the sixth and seventh centuries. In the Rhone Valley of southern France the number of sites in the fifth century fell to one-third the number of sites of the second century CE. In the northeast of Gaul, many Roman villas and farmhouses were abandoned. Field plans changed as well in the north of what is now France. Beyond human dwellings the area of cultivated land shrank and forests grew, and farm animals were also smaller in size than in Roman times.[73]

The continuing trend of thinning settlement for centuries after the eclipse of Roman power in Western Europe has most often been attributed to the ongoing barbarian migrations that helped weaken the empire in the first place, but there are few signs of violent disruption. A period of greater climate extremes with more intense cold and heavy precipitation may have contributed to the ongoing change in the early medieval landscape. In particular there is evidence of heavier flooding of the Rhone River and of expansion of Alpine glaciers in the sixth and seventh centuries.[74]

Beyond Rome's former German border along the Oder River, population sank sharply during the period of the late western Roman Empire and afterward. Rome's weakness and chances for plunder provided incentive to move west into former Roman territory. At the same time, a deteriorating climate gave added reason to migrate away from the region of the Oder. As population dropped, fields became scrubland, further reducing agricultural productivity.[75]

Summary

From the Bronze Age into the Iron Age, the Holocene climate continued to provide generally favorable conditions for agriculture and for the emergence of complex societies and civilizations. The scale and dimensions of complex societies increased over several thousand years from the emergence of farming villages up to the Roman Empire and the Han dynasty. The general pattern of urban living and complex states supported by an agricultural surplus spread to more regions, at least until the sharp contraction in much of the former western Roman Empire. Political and social history reveals ruptures and discontinuities, but the persistence of the basic pattern or model of civilization through many political transitions indicates resiliency.

Among Holocene climate fluctuations, shifts toward aridity posed significant challenges. In the most extreme cases, movement of the monsoon belt was a major factor in undermining urban societies of the Indus River. The very long history of urban civilization in the Fertile Crescent demonstrated ability over time to adapt to climate fluctuations, but climate appears to have contributed to widespread setbacks at the end of the Bronze Age. Drier, and in some regions cooler, conditions also gave incentive to nomadic or seminomadic people to migrate west toward the Roman Empire.

Climate change and human activity interacted to shape human landscapes. Records and accounts of crops and agriculture from the Roman Empire thus provide possible proxies for climate history, but economics and cultural preferences also drove planting. With the expansion of population, the human role in making landscapes increased through antiquity.[76]

CHAPTER FOUR

Climate and civilizations of the Middle Ages

- Climate of the Middle Ages
- North Atlantic: The viking era
- Warmth in Europe
- Internal colonization
- Asian monsoon and drought
- Drought and the Tang and Song dynasties
- Mongols and climate
- Southeast Asian expansion
- American floods and droughts: The Maya
- The Mississippi and Ohio Rivers
- Chaco Canyon
- Precipitation in South America
- Summary

As the case of Rome suggests, regional climate trends could either benefit or create challenges for civilizations and complex societies. Climate fluctuations did not determine a particular fate for Rome. Administrative capacity and a diverse economy made Rome resilient, but shifts in climate still interacted with other factors to magnify the problems facing the late Roman Empire. The same pattern was generally true for complex societies after the period of antiquity. Periods with favorable climate conditions benefited agriculture and the expansion of trade, but climate fluctuations also in some cases contributed to crises for particular complex societies, especially when regional climate shifts led to prolonged and severe drought or to major shifts in precipitation.

Historians have often referred to the period after the Roman Empire as the Middle Ages or the medieval era. In much of Europe, people lived alongside ruins of structures that they had not built and no longer knew how to construct. Indeed, Roman stones provided a good source of building material in the former imperial provinces. Massive structures, such as aqueducts, provided a kind of quarry. In England, construction took place in and around pieces of London's former Roman Wall. As far away as Wales, medieval builders both built upon the remnants of Roman walls and repurposed Roman materials. In the east, the Empire survived until the Ottoman Turks finally conquered the city of Constantinople in 1453, but there too, it became obvious by the seventh and eighth centuries that people were living in a fundamentally different time period. The Hagia Sophia, the enormous basilica that dominated the city's skyline, incorporated expertise and skills lost by the later Byzantine period.

The Middle Ages saw profound political shifts away from centralized power. The empire itself crumbled. Where it survived in the east, it shrank into a tiny shell. The barbarian kingdoms that followed Rome were often short-lived. The idea of kingship endured, but it was clear to all that there was no longer a Roman emperor. The decentralization and fragmentation that had started during the late Roman Empire accelerated as Europe entered a period of feudalism with a multiplicity of local magnates, and lords who could not easily be managed or controlled by anyone who claimed to rule as monarch. In the east, the staying power of the Byzantine Empire did not stave off deep change. In the early period after the demise of the Roman west, it seemed possible that a revived east might expand to the west, but an outbreak of plague in 541–42 CE caused population loss and weakened the empire. In the late sixth and early seventh centuries, the Byzantines confronted attacks from Germans and from Avars, and struggled for decades to survive massive wars against the Neo-Sassanian Persian Empire. The emperor Heraclius managed to defeat Persia, but almost immediately afterward, the Byzantine Empire faced invasion from an entirely new source—Arab armies spreading out of Arabia after the formation of Islam. These drove the Byzantines from Egypt and Syria, among other regions.

Given the scale of change, it made sense for historians to speak of a new time period—the Middle Ages—even though historians would later stress the slow transition to the Middle Ages in the eastern Mediterranean and Levant. Because world history as a discipline began in large part as an expansion of a Western-centered history, that periodization survived even though the historical reasons for speaking of a Middle Ages were weaker in other regions. In China the end of the Han dynasty coincided very roughly with Roman chronology, but it would not make sense to speak of a Middle Ages at all in some regions. Research and analysis of climate change in human history sometimes make use of the term "Middle Ages" that corresponds roughly to the era between approximately 500 and 1300, but even for regions where historians might use other terms to identify historical periods, we can still track the relationship between climate and history during this part of the Holocene.

Climate of the Middle Ages

Similar to the imperfect terminology adopted by historians, the climate of the Middle Ages is often referred to as the "Medieval Warm Period," a term first introduced (as the "Medieval Warm Epoch") by Hubert Lamb[1] in 1965 when describing the apparent warmth of Europe that lasted a few centuries around 1000–1200 CE. Although his original use of this term came with the acknowledgment that regions such as Asia did not appear to be warmer during the Middle Ages, its adoption into the climate literature has led to the misleading impression that temperatures were higher globally compared to the modern day. Recent climate research has revealed substantial variation in both the timing and the geographic extent of warming during this period. North America, Europe, and Asia exhibited warmer temperatures earlier than 1000—between approximately 830 and 1100 CE—while a multicentury warm period was observed later in South America and Australia (1160–1370 CE).[2] There is also evidence for cooling in some regions, such as the tropical Pacific. The term "Medieval Climate Anomaly" (MCA)[3] instead conveys that warming was not synchronous or consistent across the globe, and encompasses changes in precipitation patterns, which may have had a greater impact on civilizations during the Holocene. During the Middle Ages, several regions of the globe experienced sustained drought, particularly the western United States, northern Mexico, southern Europe, equatorial Africa, and the Middle East. In contrast, regions such as northern Europe and the eastern South Africa were wetter during the MCA. The precipitation in Asia during the Middle Ages showed regional variability, with some areas experiencing more arid conditions, while others appeared to be more moist.[4]

Processes that drove climate change and variability during the MCA and the subsequent Little Ice Age (LIA) included external factors such as solar radiation and volcanic activity as well as internal climate feedbacks. A history of solar activity can be reconstructed based on the measurement of radioactive isotopes, such as Beryllium-10 and Carbon-14, which are produced in Earth's atmosphere and subsequently incorporated into geological records. Because their initial formation is dependent on cosmic rays, these isotopes serve as a proxy of the sun's strength through time. Be-10 levels in ice cores and C-14 abundance in tree rings both indicate an uptick in solar activity during the MCA.[5] It was also a relatively quiescent time for volcanic activity. Because volcanic eruptions release aerosols that reflect sunlight, they generally lead to temporary (1–2y) cooling. This is particularly true for explosive eruptions in the tropics because the aerosols are injected higher in the atmosphere and transported farther by global winds. Deposition of sulfate aerosols preserved in ice cores provides a record of past volcanic activity. The low abundance of sulfate aerosols in ice cores around the time of the MCA suggests a minimum in volcanic activity, which may help explain the relative warmth of this period.

The increase in solar radiation may have triggered ocean-atmosphere interactions, such as ENSO and the North Atlantic Oscillation (NAO), that explain the climate patterns of the MCA.[6] Several regions of the globe experienced dry conditions during the MCA, including parts of North America, East Africa, and southern Europe. The lower precipitation in East Africa reduced Nile floods. At the same time, wetter conditions prevailed in areas such as the Sahel and South Africa. We find much the same precipitation patterns during modern La Niña,[7] which suggests that a prolonged La Niña state existed during the MCA, consistent with the cooler temperatures that have been reconstructed for the tropical Pacific during this time.

Along with ENSO, a second climate oscillation, the NAO, may have played a key role in medieval climate shifts. Regional climate patterns that persisted during this period are broadly consistent with ones that develop during a positive phase of the NAO today. As with ENSO, atmospheric pressure changes govern the NAO. A greater pressure difference between the subpolar and subtropical Atlantic produces stronger westerlies during the positive mode of the NAO. As a result, regions such as northern Europe and the US Atlantic coast experience warmer and wetter winters, while the Mediterranean, Greenland, and northern Canada tend to be cold and dry. The effect of the NAO on the Nile River basin is more complex, but a positive NAO during the MCA may help explain the lower floods during this time.[8]

North Atlantic: The viking era

In Europe and the North Atlantic, the relative warmth of the MCA influenced migration patterns, the expansion of governments or states, and farming. In an era of political, social, and cultural disruption, some of the trends that had shaped late Roman history extended into the post-Roman era. The waves of migrations that remade Rome's population and ultimately contributed to the demise of Roman power continued. Germanic peoples—Angles, Saxons, and Jutes—migrated into Britain. Lombards moved west into northern Italy. Migration was not simply Germanic—Slavic tribes moved into much of Eastern and Central Europe during the sixth and seventh centuries. Some historians date the end of the era of migrations to around 700 CE, but starting in the late eighth century, Vikings began to travel far from Scandinavia.

The inhabitants of Scandinavia, or Vikings, migrated and traveled farther and more rapidly than any other peoples of this era of migration. These Norsemen engaged in far-flung raids. It is difficult to date the absolute start of Viking raids, but the first to attract wide attention struck Lindisfarne, a small island off the coast of Northumbria in northeastern England. Lindisfarne was a holy isle, a center of Celtic Christianity, where the relics of St. Cuthbert, who had taken a major part in Christianizing northern England, were held. In 793, Vikings attacked Lindisfarne, killing monks and taking treasure. The assault shocked Christians such as Alcuin, a leading scholar born in Northumbria who had become a resident at the court of Charlemagne. "The church of St. Cuthbert is splattered with the blood of the priests of God, despoiled of all of its ornaments; a place more venerable than all in Britain is given as prey to pagan peoples," he lamented.[9]

Lindisfarne was no isolated incident but a harbinger of waves of raids and attacks carried out by Vikings. In 795 Vikings attacked Iona, a holy island with a monastery in the Inner Hebrides off the west coast of Scotland, and Vikings attacked Iona again in 802. Traveling by sea in their longships, Vikings landed along the coasts of Europe. The shallow draft of their boats also enabled them to travel along rivers. They raided Normandy in the north of France and traveled upriver along the Seine River to raid Paris in 845. In the same year Vikings plundered Hamburg in northern Germany.

Vikings expanded on their pattern of raiding to occupy new lands. In the late ninth century Vikings held large parts of England, which became subject to the Danelaw, or the law of the Danes. They established a kingdom in Ireland around Dublin, and for centuries the Norse or Norse-Gaels there were known by the name Ostmen, or men of the east. Along the north coast they settled islands off the coast of Scotland and to the north and west of the Faroe Islands. South of the British Isles, Viking rulers established themselves in Normandy, where they became Christian. Normans traveled into the Mediterranean and for a time held the island of Sicily. In 1066, William Duke of Normandy invaded England in the Norman Conquest and defeated the last Anglo-Saxon king, Harold, at the Battle of Hastings. An attack only weeks earlier by a Norwegian army had depleted Harold's forces at the Battle of Stamford Bridge.

Vikings voyaged still farther, crossing long stretches of water in the North Atlantic to the west. They reached Iceland in the ninth century, and an influx of settlers rapidly divided up the island. Genetic analysis of Iceland's modern-day population now indicates that Viking men took with them Celtic women. From Iceland, Vikings traveled west to Greenland and founded settlements in the late tenth century. Erik the Red is most often identified with this settlement, though he was most likely one of several to promote settlement. He is also credited with the name Greenland, perhaps employed to persuade prospective settlers to make the long journey to a land mostly covered by ice.

The Norse settlers in Greenland built a settlement on the southwest coast (the western settlement) and another on the southern tip (the eastern settlement). The population of some 5,000 kept sheep farms and traded with Norway. They also built churches and a cathedral, Gardar, along with a palace for the bishop, who became Greenland's largest landholder.

Still farther west Viking journeys reached all the way to North America. Vikings reached Newfoundland in what is now Canada around the year 1000. They established a settlement at L'Anse aux Meadows on the northern tip of Newfoundland with several wood-framed turf houses and workshops, including a smithy. Other sites for Viking landings and settlements have been suggested along the coast of North America, some based on flimsy evidence and speculation, but it is very possible that Vikings established other brief camps along the coast of Canada. However, the Vikings did not stay for long in North America (Figure 4.1).

What led the Vikings to travel so far from Scandinavia in so many directions? The stunned accounts of their looting of monasteries and treasures suggest that they wanted loot: in this interpretation, Vikings were much like pirates. This view shares similarities with one interpretation for the migration of seminomadic tribes in which peoples moved across the steppes in search of what they could seize from complex societies.

It would be ahistorical to discount the Vikings' interest in precious items, but they did not travel simply to steal. The most widespread popular image of the Viking probably envisions a warrior on a ship or at war, but many of the Norsemen and women were farmers who cultivated fields and raised

FIGURE 4.1 *L'Anse aux Meadows. Source: Isabel Lieberman.*

animals. Some farmed their own land, while others worked in the fields of chieftains. They were also involved in commerce and trade. In England, for example, Vikings set up a settlement called Jorvik, or York, where they engaged in extensive trade.

Alternately, the Norse traveled far because they could do so and wanted to seek out new homelands. The Vikings were masters of the longboat and gained the ability to travel great distances. At the same time, conditions in Scandinavia could have also given them motive to leave. In one interpretation, population growth in homelands with limited arable land and short growing seasons could have propelled emigration. However, Vikings continued to farm in Scandinavia. They exploited new lands both within and outside of Scandinavia. Indeed, settlements expanded in Viking homelands until the fourteenth century.

No single factor by itself "determined" Viking migration. Indeed, historians have long known that few complex trends stem from a single cause. Following this logic, Vikings did not set out from Scandinavia simply because of climate, but regional climate change during the medieval period eased and aided Viking expansion in several respects. The warmth in northern high latitudes during this period sustained population growth, aided ocean travel, and improved conditions for Viking colonists. A longer growing season and shorter winter contributed to the population growth in Scandinavia that encouraged Vikings to leave for new lands. Once they left Scandinavia and traveled to the west, a reduction in sea ice made long-distance ocean travel comparatively easier, though a Viking voyage of any sort would likely strike today's sailor as highly risky. As they traveled across the North Atlantic, a warm period would have created better conditions for colonization. Iceland was a hard country to colonize with a short growing season, large glaciers, and active volcanoes, but Norse Iceland managed to gain a population of some 80,000. Overgrazing, deforestation, a very slow rate of soil replacement, and volcanic eruptions made it hard to sustain that population.[10]

Greenland provides a complex case for analyzing the possible influence of the MCA on Viking expansion. Vikings making their way across the North Atlantic would have benefited from the relative absence of ice floes. Once they had settled on Greenland, the isolated Norse population living at the very end of European trade networks had to support themselves with resources from Greenland. They managed to do so during the warmth of the medieval period, but faced much greater challenges during the cooling phase that followed. A temperature reconstruction from west Greenland lakes indicates a generally warmer climate ~850–1100 CE, during the period of Viking migration, with a subsequent cooling of approximately 4°C within eighty years.[11]

Recent research challenges the notion of a warm Greenland during the MCA and questions the role of climate change in Viking settlement and later abandonment. Alpine moraine deposits, used to reconstruct the extent of glaciers during the past millennium, indicate that glaciers in western Greenland were at their LIA extent by the onset of the MCA, soon after the Norse settled there.[12] This glacial evidence suggests overall cooler summers, though it does not rule out occasional years of warmth. Temperature estimates based on air sampled from a Greenland ice core also indicate a cooler Greenland during the MCA.[13] On the other side of the Atlantic, however, temperatures were relatively warm during the MCA. This temperature contrast—a warm eastern North Atlantic and a cool western North Atlantic—usually occurs during a positive phase of the NAO, and is therefore consistent with the NAO in driving medieval climate patterns.

The case of the Thule, the population that came into contact with Norse colonists, has also raised questions about the possible interaction between climate and human migration. Viking sagas, composed from earlier traditions, told how Norse who ventured to the far west encountered people described in the sagas as Skraelings, or savages. These were the Thule, ancestors of today's Inuit. The Thule were not long-established populations in the far northeast of North America. They migrated to the east at roughly the same time that the Vikings moved west. The Viking withdrawal from North America did not end contact, because the Thule people moved to Greenland, settling in the north. Vikings therefore reached Greenland sometime before the settlement by the Thule or Inuit.

Today's Inuit, or Thule, displaced earlier inhabitants of the North American Arctic and sub-Arctic. These earlier people, now termed the Paleo-Eskimos, entered North America between 4,000 and 6,000 years ago, after earlier migrations across the Bering Strait. They adopted advances in the bow and arrow from Eurasia and the harpoon from hunters in the Pacific and Bering area and moved across the north as hunters. They lived in regions such as Baffin Island, Hudson Bay, Labrador, Newfoundland, and Greenland, but their range of settlement shifted at times, affected by climate, changes in the tree line, and competition with Indians to the south. By 1200 to 1300 CE, however, the Paleo-Eskimo culture had vanished. Until recently it was thought that some Arctic populations were descendants of the Paleo-Eskimos, but genetic analysis shows that all modern Inuit are descended from the Thule. It is not possible to say what exactly led to the demise of the Paleo-Eskimos: disease, competition from the Thule, violence, or some combination of these and possibly other factors.

Starting about 1,000 years ago, the Thule began to migrate east. The Thule, like the Vikings, were skilled travelers. They employed dog sleds to traverse snowbound landscapes at speed. By water, they traveled by boats sided with walrus skin. As was true for the Vikings, multiple factors contributed to the expansion of the Thule. Thus, Thule hunted whales for blubber, and they may have sought raw materials, such as iron that they had previously acquired through trade with Siberia. As with the Vikings, it is possible that comparative warmth may have eased the movement of the Thule, but there is also evidence of regional cooling in the High Arctic, which would show the resiliency of the Thule and their ability to respond to regional climate shifts.[14]

Warmth in Europe

The evidence of a warm MCA is strong for Europe, and the same medieval warmth that likely benefited Vikings in the eastern North Atlantic also prevailed during a period of slow but real revival of states and state power in Europe. The Middle Ages began in much of Europe with a sharp fall in population, abandonment of cultivated land, and fragmentation of political power. After the early post-Roman period, state power very slowly began to revive, though to extremely modest levels. A patchwork of kingdoms, for example, dotted post-Roman Britain, and Germanic migrations by Angles, Saxons, and Jutes added to the complex mix of peoples. Monarchs ruled realms such as Mercia, Northumbria, Wessex, Sussex, Kent, and other—some no larger than a modern county. Some, such as Alfred the Great of Wessex, amassed power to the point where they titled themselves king of the Anglo-Saxons. The Anglo-Saxon monarchs suffered losses from Viking invasions, but by the end of the first millennium the tradition of a king of England had been firmly established.

In former Roman Gaul, Frankish warlords established themselves as the chief political leaders after the end of imperial rule. Clovis I consolidated a Merovingian dynasty in the fifth century CE. Over several centuries, the Merovingian royal house grew weak until its mayors, or chief servants, the Carolingians took over real power and finally made themselves kings. Charlemagne, the greatest of all the Carolingians, created an empire that stretched into former Roman lands in Italy as well as into regions that had remained independent of Rome across the Rhine and the Elbe Rivers in Germany. Charlemagne was also crowned emperor on Christmas Day 800 in Rome. With this title, he sought to link his royal house to the old imperial title and also proclaimed himself a peer of the Byzantine or Roman emperor in the east in Constantinople.

The reemergence of royal authority was halting and partial. Charlemagne's empire was divided among his heirs. Monarchs in England, France, and elsewhere, for centuries, sought to navigate a feudal society in which their greatest supporters were also potentially their greatest threats. Central authority in the form of royal law or justice gave way to a multiplicity of feudal authorities, and many towns held independent powers in regions such as Germany and the Baltic.

State power, as the example of the Carolingians shows, did not emerge in a linear fashion after the demise of the Roman Empire, but by the eleventh and twelfth and thirteenth centuries, a period termed the High Middle Ages, European states were substantially more organized. Royal dynasties had become more firmly established, and in many cases had greater staying power than short-lived barbarian kingdoms of the late Roman and immediate post-Roman era. Monarchs and feudal lords still presided over a mainly rural landscape, but towns experienced significant growth, especially compared to the early Middle Ages.

The MCA in the North Atlantic may have provided favorable conditions for the growth of state power and expansion of trade and agriculture in Europe by the High Middle Ages. Far too many factors affected the fortunes of any single dynasty to attribute the rise of a royal household to climate, but the climate shift from the very early medieval period enabled royal households as a group to collect greater resources. The expansion of agriculture supported by the warm period provided a greater surplus for states to extract and contributed to the growth of trade and towns. Royal houses therefore could draw on more luxury goods and specialists as well as on greater agricultural wealth.

Integrating climate into our understanding of this period does not remove a human role in shaping outcomes. The ingenuity of farmers and their adoption of new techniques raised yields. European farmers widely adopted the moldboard plow, pulled by oxen and well suited for cutting through heavy soils. They also employed a variety of plows to cultivate other types of soil. The new plowing techniques helped in the cutting and sowing of fields in dense soils near riverbeds. The invention of a new collar for horses also made it possible for farmers without teams of oxen to employ these plows. Modifying crop rotation and gains in knowledge about how to obtain and apply fertilizers further boosted yields.

A climate favorable for farming helped multiply the effect of these changes. The warm period allowed farmers to grow crops at higher elevations and at higher latitudes. Grapes have sometimes been used as a proxy for climate because the classic winemaking grapes fare badly at extreme cold temperatures. It is therefore striking that English vineyards prospered during the period between 1100 and 1300 CE.[15] The regional warming could have aided grape cultivation, though growers' expertise and shifts in taste and demand could also explain this pattern. The extension of farming upslope to higher elevations provides even stronger evidence of improved conditions for growing. Settlements also reached farther north in cold areas, such as along the coast of Norway. In Sweden, settlements extended into areas previously inhabited by the Sami, seminomadic reindeer herders of northern Scandinavia, often termed Lapps in English.[16]

The extension of cultivation supported a marked boost in European population. The increased production did not end the threat of famine, but the overall result was striking. After sharply falling in late Roman times, Europe's population more than doubled from some 30 million in 1000 to 70–80 million in 1340. This growth did not stem solely from climate change: social and cultural trends, including changes in the average age at marriage, influence the rate of increase or decrease of population. Such records on average age at marriage are patchy and scant for the Middle Ages, but the average age of marriage may have changed. If so, cultural factors and climate combined to lead to population growth.

Internal colonization

Europeans farmed new regions on their own frontiers during the MCA. Well before Europeans established colonies around the globe, they colonized border zones in Europe. In Britain, colonization expanded cultivation into wetlands, including fens, or low-lying country in the east, the north in Yorkshire, and in Wales. Colonization of Wales dated back to the Anglo-Saxon era, but the new Anglo-Norman elite carried out far more extensive colonization. Some of the colonists were actually Flemings, or Flemish, but the majority were English.[17] In some cases the newcomers drove out natives, but the colonists did not simply take over land: they also pushed farming into previously uncultivated areas.[18] New Anglicized place names and deeds leave a record of the process.

In the twelfth century the Anglo-Normans began to colonize Ireland. A papal bull issued by Pope Adrian IV in 1155 supported the takeover of Ireland by King Henry II of England. Anglo-Norman forces entered Ireland in 1169 to support one of the rivals in a civil war, and in 1171 Henry II himself arrived in Ireland. Normans invaded and settled in Ireland. Prince John landed in Waterford in 1185, and Anglo-Norman lords established themselves in Ireland, though much of the population remained Gaelic. The new Anglo-Norman settlements attracted English colonists in the late twelfth and early thirteenth centuries. Gaelic power rebounded in the later Middle Ages, before renewed phases of English conquest and colonization from Britain in the early modern era.

The expansion of cultivation took place in many regions of Europe during the medieval period. An epoch of draining and claiming land took place in the low countries, today's Netherlands, and much of Belgium, and in coastal areas and swamps in regions including England, France, Germany, and Italy.[19] In Holland, reclamation of peat marshes began with draining the bogs, but the loss of volume of land made reclaimed land vulnerable to flooding and required further construction of canals and dikes. Later generations of Dutch farmers in the early modern era faced the choice of abandoning the reclaimed peat lands or engaging in far more elaborate and ambitious projects. With their expertise in reclaiming land, the Dutch also carried out similar projects in the internal colonization of medieval Central and Eastern Europe. They took part in eastward migration into regions such as Prussian-Holland in Prussia in what is now Poland.

Large waves of migration took place in Europe's east. During the Roman Empire and after the fall of the western Roman Empire, Germans had generally moved west, but in the High Middle Ages, German migration shifted toward the east. The Magdeburg Charter, an appeal to colonists from 1107 to 1108, described these "heathens," the pagans of the east, as "very bad," but also stressed the promise of riches for settlers in a land "rich in meat, honey, grain, birds." Colonists would gain both spiritual and worldly rewards: "You can both save your souls and, if it so pleases you, acquire the best land to live on."[20] Under the leadership of military religious orders, most famously the Teutonic Order, the colonization campaign advanced all the way to the lands that today make up the Baltic states. Along Baltic shores and in forests in the interior Germans fought wars against pagan natives and established themselves in settlements, including Riga, today the capital of Latvia.

The German migration east left Germans in many lands. Germans created colonies and enclaves in Poland, Bohemia, and Hungary. The number of German towns increased by an order of magnitude in the thirteenth century, and the pace of migration did not begin to slow until the fourteenth century.[21] Settlers from the Netherlands and Flanders also moved east. In some cases, local rulers and lords invited settlers for their expertise in particular occupations. Germans settled, for example, in Transylvania, initially at the invitation of King Geza II. German speakers today still live in small numbers in some of these regions, including towns in Transylvania, now within Romania's borders, but it is easy to overlook the extensive German colonization of eastern Europe during the Middle Ages because most Germans either fled or were expelled and transferred from these lands at the end of the Second World War in the largest single episode of forced migration in modern European history.

European migration extended all the way to the Near East. In 1095, Pope Urban II preached for the First Crusade to aid the Byzantine Empire and to take Jerusalem. Knights took up arms as military pilgrims and made their way east to Syria and to Palestine. There the fractious forces of Latin Christendom assaulted Jerusalem in 1099. Repelled on their first assault, they circled the city in penance, launched a renewed attack, and took and sacked Jerusalem. The crusaders established a series of kingdoms, and in 1145 the pope announced a crusade, the Second Crusade, to attempt to reverse the eroding Latin position after the fall of the crusader state of Edessa in 1144. After Saladin took Jerusalem in 1187 the Third Crusade marked an unsuccessful attempt to retake the city.

Religious motives initiated the crusading era in the Near East, but climate affected the crusades as well. The general expansion of agriculture and population during the MCA created greater potential for expansion by the Latin Christian population of Western and Central Europe, though years immediately preceding the First Crusade may have seen a burst of poor weather. Crusading armies, once they reached the Near East, also confronted a climate they were ill-equipped to deal with.[22]

Asian monsoon and drought

The climate of the period corresponding to the High Middle Ages was not uniformly favorable for expanding cultivation and states everywhere. Just before the First Crusade, the Byzantine emperor Alexios I confronted attacks from Turks and Turkish peoples in the late eleventh century during a period of severely cold winters. Accounts and chronicles dating back to the tenth century described extremes and harsh conditions in the eastern Mediterranean, Egypt, Anatolia, and Iran. A chronicle from the city of Isfahan in Iran described snows so heavy in 942– 43 "that people were not able to move around."[23] As is still true today, any single report or even a series of reports on bad weather may not signify any shift in climate. Thus, a series of reports about heavy snows in eastern North America from February 2015 did not give an accurate sense of overall climate trends—the winter was globally warm. The preponderance of contemporary reports over time can reinforce the thesis of a shift beyond normal variation, but climate proxies provide a more complex picture. Records from the Nile show low floods during this time. Climate proxies, including tree rings, point to cold in the eleventh century in the Central Asian steppes, a "big chill" in the words of one historian, along with dry conditions in Iran and Eastern Anatolia.[24] The record for western Anatolia and the southern Balkans is more mixed without pronounced cooling and with average humidity, but the eastern Mediterranean grew drier from the late twelfth century onward.[25]

Droughts were exceptional between 300 and 900 CE, but the Nile flood became less reliable by the middle of the tenth century. Between 950 and 1072, drought became ten times as frequent compared to the prior centuries, and over a period of 125 years, there were 27 years with low Nile floods.[26] When the Nile remained low, water did not reach the height necessary to enter canals and irrigate fields. As was the case for medieval precipitation patterns in general, variations in ENSO and the NAO influenced the fluctuations in the Nile River flow. The Nile flood is primarily controlled by monsoonal precipitation in the Ethiopian highlands, driven by the seasonal migration of the ITCZ. Persistent La Niña conditions and a positive mode of the NAO likely reduced precipitation in the Nile River basin, creating low floods and resulting famines.

The failure of the Nile floods damaged not only Egypt but also nearby societies that depended on the ability of Egypt to produce a surplus of grain.[27] Famine, in this interpretation, contributed to disorder, rebellion, and political breakdown. In the case of Egypt itself, a brief revival of Abbasid rule came to an end, and new rulers, the Fatimids, took over. The Byzantine Empire took advantage of the crisis in Egypt by briefly retaking lands not held under Byzantine rule for centuries. In 1024–5, Fatimid rulers responded to renewed drought by seizing grain transports and opening granaries. Egypt suffered prolonged famine from 1065 to 1072. The ostensible Fatimid rulers struggled meanwhile to control their Turkish soldiers. By the late eleventh century, the Byzantine Empire faced multiple threats. Its grain surplus shrank. Meanwhile, the empire experienced attacks by seminomadic pastoralists, including the Pechenegs, and the Byzantine Empire suffered defeats in 1049–50.[28] A devaluation of the currency weakened the once formidable army, which was defeated by the Seljuk Turks at the Battle of Manzikert in 1071.

The cold spell in Central Asia created major challenges from the Byzantine Empire to Persia and regions still further east. The cold climate in Central Asia propelled pastoralists to migrate elsewhere. Pechenegs, and Oghuz and Seljuk Turks, moved westward. In Iran where, according to one interpretation, cotton cultivation had created much prosperity, the cold shift struck as cotton production declined in northern areas and as seminomadic peoples moved into Iran.[29] A chronicle described the effects of disorder, famine, and cold in 1040: "Nishapur was not the city I knew from the past: it now lay in ruins . . . A great number of people . . . had died from hunger . . . the weather was bitterly cold and life was becoming hard to bear." Disorder in Iran helped set off a diaspora that increased the influence of Persian culture in South Asia.[30] Turkish power increased as well in Iraq and Anatolia. The Buyid dynasty collapsed in Baghdad, and after a period of civil war Seljuk Turks took Baghdad in 1060.[31]

This pattern of climate shifts in Central Asia that dislocated pastoralists shared general similarities with cycles of drought and migrations during the late Roman Empire. In both cases, it is possible to find many factors for migration and for the defeats suffered by the rulers of urban areas and complex societies, but in both the late Roman era and in the eleventh century, cold and dry conditions may have given nomadic and seminomadic peoples reasons to move west in the first place. A second interpretation points to the possible effects of cold on the one-humped female camel that Turks bred with two-humped male camels, producing a camel well suited for Silk Road trade. The one-humped camel did not tolerate cold well, so cooling forced camel breeders to move south.

Drought and the Tang and Song dynasties

In East Asia, climate fluctuations influenced China during a new cycle of dynastic revival and collapse. In contrast to the collapse of the western Roman Empire, the demise of the Han dynasty in China did not end the cycle of empire in China. Another strong dynasty emerged with the Sui dynasty in 581. Yang Jian, a commander from the north, gained power over northern China and took the south during the 580s. The Sui dynasty was short-lived, collapsing in 618, but the Tang dynasty followed immediately. Tang China built an immensely wealthy state for its time with a capital at Chang'an in northern central China. A political and economic center and a city of numerous religious shrines and temples, Chang'an's population reached some 1 million by the eighth century.

Tang China benefited from extensive regional and long-distance trade networks. Trade and communication flourished along the Silk Road that crossed deserts to the west into Central Asia. Buddhists, Christians, and Jews traveled along the Silk Road, and Buddhism became a major religion in China during the Tang dynasty, before suffering persecution under the Emperor Wuzong of the late Tang era.

Both the Sui and the Tang dynasty emerged during a period of warmth that spanned 551–760 CE.[32] In the middle of the eighth century, the Tang dynasty began to weaken. The immediate cause of the crisis stemmed from the emperor Xuanzong/Minghuang's misjudgment. The emperor relied on Turkish commanders to guard the frontiers and expand China's powers. One of these non-Chinese commanders, An Lushan in his Chinese name, gained increasing power. In 755, An Lushan rebelled, and the emperor fled to Sichuan. The Tang dynasty survived the rebellion, but internal conflict marked the late Tang era. The Emperor Wuzong, who ruled from 840 to 846, moved to crush Buddhism, closing most temples and seizing their property. Meanwhile, regional warlords proved increasingly difficult to control. The Tang dynasty endured repeated rebellions until it finally collapsed in 907.

Much as for the Roman Empire or the Han dynasty, a political and military narrative of Tang decline can treat climate simply as a backdrop for a narrative of the feats and mistakes of emperors and their rivals. Climate, at the same time, intensified the problems confronting the late Tang dynasty as desertification increased in the ninth century.[33] A stalagmite record from China indicates a strong summer monsoon from 190 to 530 CE followed by a steady weakening through 850 CE. The monsoon remained weak, with several sharp minima, until 940 CE.[34] These monsoon changes, which likely contributed to both the Mayan and the Tang decline, appear to be related to a southward shift of the ITCZ.[35] Climate was only one of multiple factors influencing warfare between Chinese states and nomads to the north, but on the whole nomads were more likely to achieve victory during cold dry periods such as the late Tang era.[36] Nomadic groups expanded south into central plains during such periods.

Seminomadic peoples on China's frontiers themselves showed both vulnerability and resilience to fluctuations in climate. To China's west, the Eastern Turkish Khanate Empire came to an end in 630 CE. Internal factors contributed to this demise, but Chinese visitors remarked on the effects of drought and cold from Dzud events or unseasonable frosts following the eruption of a volcano in 626 CE.[37] The Uyghur Empire, in contrast, that emerged in the eighth century thrived despite drought, possibly because of growing trade in horses with China and the emergence of Silk Road trade.

China was fragmented and divided after the Tang dynasty, but this interlude was much shorter than the similar period after the Han dynasty.

In 960, a commander in one of the northern Chinese states reunited China under a new dynasty, the Song dynasty. The Song did not try to reestablish Chinese central power over the northern frontier, where states with roots in seminomadic peoples remained strong. Instead of pushing to the north or west, the Song established a capital at Kaifeng on the Yellow River and the Grand Canal. This period of Song rule from Kaifeng was known as the Northern Song dynasty.

Song China stood out during its era as an extraordinarily prosperous state. Population grew, both in the countryside and in cities. The population of Song China reached some 100 million by 1100 CE, a total that vastly exceeded that of any other state in the world at the time. Multiple cities had populations of 100,000 or more, and the populations of the cities of Kaifeng and of Hangzhou each reached 1 million. Along with other factors including a stable political system, trade, and technological advances, climate contributed to this record of growth. The reliable precipitation was one important condition for the sharp population growth. In particular, a strong monsoon benefited Song China.[38] China also made intensive use of rice as the population tripled during the Northern Song era.

In the twelfth century the Song dynasty lost control of its lands in the north of China to a Jurchen invasion. The Jurchens spoke a Tungusic language, distinct from both Mongol and Turkish, found in regions of eastern Siberia and in Manchuria to the northeast of China. The Song dynasty did not fall but moved to the south and established a new capital at Hangzhou. Despite losing territory, the Southern Song dynasty thrived economically. Song China manufactured porcelain or china and innovated in many areas, such as printing and the production of gunpowder. Song military engineers developed new weapons such as rockets and bombs, but Mongol advances in the late thirteenth century ended the Song dynasty. Kublai Khan, a grandson of Genghis Khan, invaded the Southern Song. The Song employed their new weapons against the invaders, but the Mongols adopted Song technology and also gained some Chinese support. In 1276, Kublai Khan took the Southern Song capital Hangzhou, ending the Southern Song dynasty and consolidating the power of the Mongols' Yuan dynasty in China.

Mongols and climate

As for the history of so many other empires, traditional historical narratives treat climate as a backdrop, in this case attributing the Mongols' many victories to their military prowess and effective leaders. The Mongols possessed great skill in war. From an early age, boys practiced riding, raiding, and hunting. Indeed, they started learning to ride and shoot bows almost as soon as they could walk. Mongol warriors were expert marksmen, even as mounted archers, who fired composite bows to deadly effect. On horse, they traveled long distances and carried out highly organized attacks. These formidable warriors from the steppes became a far more dangerous military force when Genghis, or Chinggis, Khan united the Mongol clans under his command in 1206. Genghis Khan held a great hunt, called the Nerge, to practice large-scale military maneuvers. The Mongol hunters encircled their prey as they would surround their enemies at war. They also developed the practice of feigning retreats to lure out and then attack their enemies.

In Genghis Khan's day, the Song dynasty of China had already retreated to the south of China, replaced in the north by the Jin dynasty founded by Jurchen tribes. In 1211, Genghis Khan invaded the lands of the Jin dynasty. The Mongols also pushed west along the Silk Road, taking ancient trading centers, including Samarkand and Bukhara. After Genghis Khan's death in 1227, his heirs continued to expand Mongol control to the south. By 1279 they conquered the remaining territory of the Southern Song dynasty in southern China. Mongol warlords also pressed far to the west. Batu, one of Genghis Khan's grandsons, sacked Kiev. Mongols established themselves in Russia as the Golden Horde and collected tribute until the late fourteenth century. Further west, Mongol armies reached all the way to Hungary in 1241. In the Near East, Hulagu, also a grandson of Genghis Khan, sacked Baghdad in 1258, bringing an end to the Abbasid Caliphate.

As accounts of Mongol victories make clear, climate was far from the only reason for Mongol success, but climate fluctuations nonetheless may have assisted the Mongols in their expansion. If drought could drive nomadic migration, so too could periods with reliable precipitation. The Mongols benefited from favorable climate conditions in their homeland just as they expanded far out of Mongolia and established a vast realm in the early thirteenth century. Analysis of tree ring data from central Mongolia indicates that several severe droughts occurred during the MCA, particularly 900–1064, 1115–39, and 1180–90. This last period, during Genghis Khan's early years, coincided with instability in the political realm of the Mongols and may have contributed to his rise to power. The cool period in Central Asia came to an end with warming that peaked in the early 1200s,[39] and an extended period of higher rainfall occurred between 1211 and 1225. This record indicates that the Mongols expanded during a time that was, overall, wetter and warmer compared to any other time during 1112 years.[40]

In many cases, whether in Rome, Han China, or medieval Europe, climate contributed to state expansion because of benefits for agriculture, but for the Mongols, it was horses that really mattered. The Mongol horses were renowned for their hardiness, but carrying out military campaigns across thousands of miles required large numbers of horses, and each Mongol warrior had several. The favorable spell of climate in the early thirteenth century boosted Mongol power because the wet and warm period provided good conditions for raising and feeding horses.

The Mongols also had to procure food for their horses as their army traveled. Indeed, the Mongols were not nearly as formidable in regions where they could not easily feed large numbers of horses. Such difficulties provisioning horses helped blunt the power of Mongols and other seminomadic warriors as they moved south in South Asia and toward Southeast Asia. Even as they defeated the Song dynasty and other once-mighty states, climate therefore set limits on ultimate Mongol expansion on the frontiers of the Mongol realms.

Southeast Asian expansion

South of China, climate in the period that corresponded to Europe's Middle Ages aided the expansion of major states and complex societies. A stronger monsoon predominated in Southeast Asia from around 950 to around 1250, and favorable conditions for agriculture helped support the development of several states in Southeast Asia. These included the Khmer State with its capital at Angkor, the Pagan Kingdom in Burma, and the Dai Viet state. These three states have been described as "charter states" because they marked the first large, powerful, indigenous states in these regions.[41]

Between the ninth and the fifteenth centuries, the Khmer Empire was a major power in Southeast Asia. At its peak, the Khmer Empire extended far beyond the borders of today's state of Cambodia, though the Khmer also suffered defeats at the hands of their rivals such as the Cham, who lived along the coast of Vietnam, before the Khmer expanded in the late twelfth and early thirteenth centuries under Jayavarman VII.

The temple complex of Angkor Wat built during the Khmer Empire remains an imposing archaeological site to the present day. Angkor Wat was constructed in the first half of the twelfth century when Suryavarman started the building of the temple complex. The "Temple City" was a political and religious site first centered on Hinduism but soon converted into a Buddhist site. It grew to encompass hundreds of temples. The central tower, dedicated to the Hindu God Vishnu, rose to a height of more than 200 feet. A moat encircled the temple complex. As capital of the Khmer Empire, Angkor was also an enormous city for its time with a population in the hundreds of thousands—as high as 750,000 (Figure 4.2).

The Pagan state in what is now Myanmar rose to prominence at roughly the same time as the Khmer Empire in Cambodia. Closely allied with Buddhism, the Pagan state established thousands of Buddhist temples, monasteries, and stupas. To this day, some 2,000–3,000 sites remain, though many are ruins. The capital city of Pagan, with thousands of Buddhist monks, was a main center of Buddhism and attracted visitors from South Asia and from other parts of Southeast Asia. Pagan had a population of more than 50,000, and hundreds of thousands of peasants lived in nearby regions. As population grew, the entire Pagan state expanded. Buddhist temples and monasteries marked frontiers of religion and cultivation.[42]

FIGURE 4.2 *Angkor Wat Temple, Cambodia. © Dirk Radzinski/Bridgeman Images*

The expansion of state power during this era can also be seen in the history of Vietnam. First independent from China in the tenth century, the Dai Viet, or Great Viet, state emerged as a strong power in Vietnam in the eleventh century. Its capital lay at Thang Long on the Red River Delta in what is now Hanoi. The Dai Viet state competed both with China to the north and with a more Hindu region to the south under the rule of the Cham, who today are an ethnic minority in Vietnam. Like its neighbors, the Dai Viet state experienced strong population growth, with an increase from 1.6 to 3 million people between 1000 and 1300.[43]

Varied factors contributed to state building, expansion of agriculture, and population growth in these different states in Cambodia, Burma or Myanmar, and Vietnam. In the case of Dai Viet, relations with Chinese states to the north as well as the adoption of technology influenced the pace of growth. Increased trade may have benefited all three states. It is possible that more consistent communication and contacts also reduced mortality to certain diseases, by making diseases that had once caused epidemics instead endemic.[44]

Along with these varied possible causes of expansion, climate trends favored the emergence and growth of states and societies in Southeast Asia. There were extended periods with high rainfall in the twelfth century and in the late thirteenth century. Reliable, strong precipitation boosted agriculture, which in turn aided expansion and construction in Pagan.[45] Similarly, strong monsoons supported Khmer expansion and colonization. The relationship between climate and growth is more complex for Dai Viet, which was located in a wetter region than either the Khmer or the Pagan state. Higher precipitation may have raised populations in uplands, leading to migration down to the Delta.

American floods and droughts: The Maya

As in other parts of the world, complex societies in the Americas derived large surpluses during the Holocene. The climate in this region between 500 and 1300 CE was characterized by variability: multi-decadal droughts interrupted warm and wet periods. The long history of complex societies in the Americas shows resiliency to shifts in climate during the Holocene, but several civilizations and states also experienced more abrupt transitions and even sharp declines as measured by the abandonment of cities. Prolonged drought, in particular, posed a major challenge to complex societies of the Americas.

Several climate shocks had already altered and in some cases toppled some of the most complex societies in North America before Europeans entered the Americas. The remains of a large abandoned settlement stand to this day in the northwestern corner of New Mexico at Chaco Canyon, but the entire site was abandoned by the twelfth century. In numerous sites along the Mississippi River and the Ohio River as well as areas of the southeastern United States, pre-Columbian peoples established sites later described as mounds. Cahokia, near the modern city of Saint Louis, contained more than 100 such mounds. Intensive farming in the eleventh and the early twelfth centuries supported a large population that reached into the thousands, but Cahokia was abandoned before Europeans ever reached the region. Farther south, Spanish adventurers and conquerors met Maya, but the Mayan world had already experienced severe disruption. The very fact that the desertion of large imposing structures stands out can skew our understanding of the development of American societies. Drought in several instances had severe effects in particular regions, but there was no general collapse of indigenous cultures across the Americas.

The Maya created the longest continuous society and culture in Central America. Emerging in the farming areas of Central America and Southern Mexico, the first Mayan settlements dated back to about 800 BCE. Within some 500 years, Maya started to build more elaborate ceremonial centers for religious purposes. The Mayan society continued to grow until it reached its height of complexity during what we now call the Classic period of 200–900 CE. The Maya built ceremonial sites, most notably pyramids, and cities in a belt that extended from Guatemala and Belize through the Yucatan Peninsula and to the west into what is now Chiapas. Maya employed multiple kinds of cultivation to support a relatively large population during the Classic period. They created orchard gardens, built terraced fields, and exploited wetlands. To enrich fields, they engaged in burning and flooding. They also managed forests to obtain fuel. [46]

There was no single Mayan empire but rather a series of city-states. During the Classic era, the population of the largest city-states exceeded 50,000 people. Tikal and Calakmul were two of the largest city-states of this period. The ruins of Tikal in northern Guatemala contain five large pyramids: the tallest reaches a height of more than 200 feet. Mayan hieroglyphs at Tikal tell of the city's wars against neighboring city-states and of both defeats and victories, but the inscriptions on the dated stelae or carved stone slabs or pillars end at 869.

Calakmul at Campeche in Mexico predated the Classic period and grew to become the largest city of the Classic period with a population of some 60,000 people and pyramids standing to a height of close to 150 feet. More Stelae still stand at Calakmul than any other Mayan site. The inscriptions tell of complex relations with neighboring Mayan city-states, both tributaries and at times rivals of Calakmul, and of wars between Calakmul and Tikal. In the seventh century, Calakmul won victories against Tikal, but the city suffered defeat by Tikal in 695—the victors killed captives as human sacrifice. The inscriptions end by the early tenth century, and by the time the Spanish reached Calakmul, the population amounted to only a very small fraction of the total of the Classic era.

Many of the Mayan sites were already in ruins by the time Spanish conquistadors came to the Americas, and centuries later, Mayan sites are still being found. In 1570 Diego Garcia de Palacio found the ruins of the city of Copan in what is now Honduras. The ruins at Copan included mammoth plazas, a 10-meter-wide stairway, and nearly 2,000 glyphs. The population reached some 25,000 people, but the royal dynasty came to an end in the ninth century.

As in the case of Rome, the question of why the classic Mayan civilization collapsed has generated a host of answers as well as debate over the meaning of collapse. Certainly, Maya survived past the end of the Classic period. The Spanish met Maya, and there are still Mayan populations today. At the same time, the sheer number of abandoned sites points to a major and real disruption in Mayan society at the end of the Classic period. We now possess numerous hypotheses: earthquakes, disease, rejection of female children, decadence, peasant revolts, invasion, or forced migration carried out by invaders. For most of these explanations, there is little supporting evidence. Warfare between Mayan city-states provides another possible cause. The deciphering of Mayan inscriptions overturned the notion that Mayan elites were peaceful. The great city-states of Tikal and Calakmul, along with their allies, fought long wars. However, if warfare contributed to the end of the Classic period, it did so only very slowly, because major city-states fought wars over many generations.

The gap between higher populations in Mayan times and lower populations today suggests that the Mayan city-states placed great pressure on available resources: food, fuel, and water. At Tikal and Calakmul, builders gave up the wood they had long used for making beams, Sapodilla

(*Manilkara zapot*), and turned to substitutes, suggesting that they were exhausting timber supplies. They also stopped using lime as a material for plaster. The Maya had proven their resilience over many centuries, but urbanization and population growth over many years caused shortages of materials that they depended on.[47]

Drought caused by a shift in climate interacted with several other factors to bring about the end of the Mayan Classic period. An arid period with several intense droughts weakened a society that had already pushed the limits of its environment. Maya adapted to their landscape over many centuries, but their large population and deforestation made them less resilient to drought. Depletion of resources in a dry period, in turn, contributed to the potential for warfare between Mayan city-states that already had a long tradition of fighting with each other. Deteriorating conditions in collapsing urban centers also gave Maya incentives to move away. A shift from land to coastal routes could have reinforced such migration.

Although Mayan collapse likely stemmed from several factors, episodes of drought most likely destabilized Mayan civilization of the Classic era. Multiple studies point to the strong effects of climate shifts. Increased monsoonal precipitation filled lakes in the Yucatan during the early Holocene. As summer insolation waned, the region started to become drier beginning around 3,000 y ago, and the very peak in aridity around 800–1000 CE coincided with the collapse of the Maya. The Mayan cities of the Yucatan and Central America proved highly sensitive to reductions in rainfall of some 40 percent.[48] Drought was most intense in the regions that displayed the strongest evidence of collapse. Mayan societies had previously shown resilience to dry periods, but greater social complexity and more severe drought overwhelmed their capacity to adapt.[49]

The shifting of the intertropical convergence zone may have brought about the drought associated with Mayan collapse. We can read these climate changes from layers of marine sediment from the Cariaco Basin to the north of Venezuela, which reveal migration patterns of the ITCZ. Today, the annual shift in the ITCZ leaves distinct light and dark banding, and sediments in the basin preserve that pattern throughout the Holocene, allowing scientists to track past changes. During the winter and spring dry season, the ITCZ is positioned to the south, leading to strong trade winds that strengthen upwelling, which in turn initiates greater algal growth that is then preserved in the basin as lighter-colored sediments. The annual northward shift of the ITCZ in the Northern Hemisphere summer marks the rainy season in Venezuela, when more dark-colored sediments from land are delivered to the Cariaco Basin. Titanium (Ti) concentrations in the land-derived sediments document the ITCZ migration in this region. The Maya lived north and west of the Cariaco region, but the same ITCZ shifts recorded in the Cariaco sediments also influenced the Mayan cities in the Yucatan. The terminal collapse of the Maya between 750 and 900 CE coincided with low sedimentary levels of Ti. In particular, Ti minima corresponded to dry events around 760, 810, 860, and 910, dates consistent with a three-phase model of Mayan collapse.[50] A Yucatan stalagmite record[51] and sediments from nearby Lake Chichancanab[52] provide additional evidence about the influence of drought on the Maya. Debate nonetheless continues regarding the degree to which drought played a role in the collapse of this and other Mesoamerican civilizations.[53]

The Maya's history demonstrates both the resilience and the dependence of complex societies on climate. The Maya adapted to their landscape and climate. They engineered water, land, and forests to produce enough food to support dense populations, and they did so over long periods of time. Even their collapse was not total—Mayan culture and society did not reach the terminal point. At the same time, the very ability to adapt and innovate also meant that Maya were able to place ever-greater stress on their landscape and supply of soil, wood, and water. They therefore became more vulnerable to the aridity and large droughts that struck in the late Classic period.

If Mayan culture did not come to an end, it nonetheless changed. Climate change, in this case, may have brought the greatest shock for elites. The Maya whom the Spanish first met had not abandoned social hierarchy. When Hernán Cortés and his companions first defeated Mayan warriors on the Yucatan coast, he asked for a delegation of the leaders and chiefs. He received gifts of gold ornaments and young women, including Dona Marina, or Malinche, who became both Cortés's mistress and his invaluable translator during his subsequent expeditions in Central Mexico. However, the dynasties, whose wars and feuds figure prominently in the Mayan hieroglyphs from the Classic era, appear to have come to an end. Others who closely served the royal houses or provided them with luxury goods would not have been able to continue their previous way of life. With less ability to concentrate a massive surplus, societies with less elaborate hierarchies, such as those on Hispaniola, would have been better able to adapt to the drying trend.[54] The shift in precipitation brought similar stress in southern Central America.[55]

Shifts in precipitation also affected human societies in Central America beyond the Mayan heartland. East of Mexico City, the city of Cantona reached a population of some 90,000 by 700 CE and supplied obsidian to sites on the Gulf of Mexico. A long drying period between 500 and 1150 did not immediately undermine Cantona: indeed, migrants from other arid regions may have moved to Cantona. However, the city's population fell between 900 and 1050 CE to only 5,000 people during one of the driest periods in nearly 4,000 years.[56]

The Mississippi and Ohio Rivers

Major changes in precipitation posed a challenge to other complex societies in pre-Columbian North America such as the culture of the Mississippi and Ohio River Valleys identified with ceremonial mounds. The Spanish adventurers came across some of these mounds. In 1539, Hernando de Soto embarked on an entrada, or expedition, north of territory already conquered by Spain into lands that are now in the United States. In search of land and gold, he landed in Florida and traveled north into the Appalachians and then generally west, reaching the Mississippi River in 1541. De Soto died in 1542, but some of his men made it back to Mexico City. The Spanish historian Oviedo y Valdes compiled an account of de Soto's journey based on the diary of de Soto's secretary, Rodrigo Ranjel. From this account we know that de Soto came across some remnants of the Mississippian culture that had erected large mounds and earthworks. In what is now Georgia, de Soto and his men entered a village with a mound upon which they set up a cross. At a place called Talimeco, possibly near present-day Camden, South Carolina, the Spanish found a "village holding extensive sway; and this house of worship was on a high mound and much revered." After crossing the Mississippi, "the Christians planted the cross on a mound."[57]

In clashes and battles along their route de Soto and his men killed many natives and also suffered many losses themselves. They also found a native population severely depleted and weakened by disease carried to the Americas by Spain, which had already apparently spread further than the Spaniards themselves. De Soto and his men on the ground carried Eurasian infectious diseases with them, setting off new trains of transmission. As in Mexico and elsewhere, native peoples with no resistance died in extraordinary numbers. In one interpretation of this period, the many deaths severely weakened native societies of the late phase of Mississippian culture. Raised temple mounds ceased to function as key political and religious sites.

Epidemics unleashed by the Columbian exchange between the Old and New Worlds would cause misery, sickness, and death for many natives of the Americas, but the contraction of complex societies cannot be attributed solely to diseases carried by the conquistadors. The Mississippian culture that de Soto and his band of Spanish adventurers encountered had suffered setbacks before the first direct and indirect contacts with Europeans. Warmth during the MCA may have contributed to the expansion of the Mississippi culture and of sites in what is now the southeast of the United States.[58] After the peak population of Cahokia and nearby regions in the early twelfth century, population density fell. By the fourteenth century, peoples of the Mississippian culture had left Cahokia as well as other sites in the Ohio and Mississippi Valleys. The settlement at the mounds near Kincaid in southern Illinois reached its peak in the early 1200s, but mound building ended around 1300 and the settlement was abandoned by around 1450. That same pattern of abandonment was true for many sites in the Mississippi and Ohio River Valleys.[59]

The paucity of sources makes it difficult to reconstruct the full history of the cultures of the Mississippi, but research on climate suggests that shifts in precipitation may have placed mound builders under stress, either through greater aridity or through greater flooding. In one interpretation, long droughts between the mid-twelfth and the early thirteenth centuries weakened the societies of the mound builder. The cycle of drought undercut intensive agriculture in the Cahokia region. The water table fell, and reduced precipitation threatened the intensive maize cultivation necessary to sustain the comparatively dense population. Drying out would also have lowered the stock of fish. Such trends did not make collapse inevitable for all Mississippian sites, but shook the largest population sites most dependent on the best conditions for agriculture.[60] A second interpretation, in contrast, sees damage from greater flooding at the end of the arid period. A study of sediment cores indicates that Cahokia emerged during a period with fewer floods, but declined with the return of more severe flooding.[61]

The effects of any such shifts in precipitation varied by type of society and by region. Small sedentary villages in the Monongahela River Valley in Pennsylvania appear to have formed and reformed, independent of any large-scale long-term trends.[62] Abandonment of large sites such as Cahokia struck at the most centralized manifestation of Mississippian culture. Religious and political elites who benefited from larger-scale extraction of a surplus may have found themselves occupying less exalted positions. However, such an outcome did not by itself equate to complete collapse or the extinction of culture.

Chaco Canyon

Shifts in aridity and precipitation had the greatest effects in the most vulnerable regions of human settlement such as the American Southwest. A striking example comes from the history of a complex society that flourished well before contact with Europeans in what is now New Mexico. Between the 800s and approximately 1150 to 1200 CE, a civilization built a large and elaborate settlement at Chaco Canyon in the northwestern corner of New Mexico (Figure 4.3).

Builders at Chaco Canyon designed and erected large multistory stone houses. Several thousand people may have lived at the central site, though population estimates remain uncertain. The people of this culture constructed many other great houses along roads. The remains of the roads, no longer standing, are not visible to the naked eye but can be traced by satellite images. The material culture of Chaco Canyon made abundant use of turquoise. Excavations have revealed some 200,000 pieces of turquoise. Some turquoise came from local sources, but Chaco Canyon also traded turquoise along trade networks that extended into what are now Colorado, California, and Nevada.[63]

FIGURE 4.3 *Chaco Canyon. Source: Isabel Lieberman.*

We have no equivalents to the Mayan script that tell us about the elites, but burial patterns suggests a strong hierarchy at Chaco Canyon.[64] Items buried with the dead, or grave goods, from the largest of the Great Houses, named Pueblo Bonito, included tens of thousands of turquoise beads. One small room contained no less than 25,000 turquoise items. There were also vessels containing traces of cacao that could only have been acquired via long-distance trade.

To support their complex society, the people of Chaco Canyon employed several methods to obtain and store water. The very location in a canyon placed farmers closer to the water table. In addition, Chaco Canyon and nearby communities diverted and stored water using dams and canals. Small dams intercepted runoff and steered water into canals. There were also several large dams, including a 130-foot-long masonry dam. The water supported the cultivation of beans, corn, and squash.[65]

In the twelfth century, the history of Chaco Canyon as a major population center came to an end. Any surviving population departed, and the great buildings were left abandoned. In this case, as in virtually all others, the choice of the term "collapse" has caused controversy. By any normal understanding of the term, Chaco Canyon collapsed. We would probably say the same about London, New York, or Shanghai if their populations abandoned those vastly larger cities, leaving structures that eventually fell into ruin. The end of Chaco Canyon, however, did not necessarily mean the extinction of all its peoples who may have migrated to other southwestern communities. By the late thirteenth century, Anasazi also abandoned recently constructed cliff dwellings and moved to the south and east. Thus the elaborate cliff dwellings at Mesa Verde in southern Colorado were abandoned in around 1300.

Why did ancestral Pueblo peoples abandon Chaco Canyon in the late twelfth century and prominent cliff dwellings around 1300? Their departure has been attributed to war, but this explanation lacks evidence despite some abandoned Anasazi sites that contain physical remains suggesting death by violence and even cannibalism.[66] Alternately, the culture could have suffered economic losses from a shift in the turquoise trade, though it is not clear why that would have led to complete abandonment of imposing sites. The hypothesis of an ideological collapse from some dispute over religious ritual is virtually impossible to test in the absence of any records.

As the most densely populated site in a dry region, the people of Chaco Canyon may have placed too great a strain on limited resources, including wood and water. Because the builders at Chaco Canyon employed wood beams, it is logical to wonder whether they could have deforested the surrounding area and run out of timber. In one interpretation, Chaco Canyon's people used so much wood that they deforested the lands they depended on, but the source of wood and the rate of deforestation remain in dispute.[67] Whether or not Chaco Canyon's people cut down too many trees in their region, they confronted a series of intense and long droughts. Such fluctuations in climate endangered the food supply in an already dry region.

It is not possible to run a test on the abandonment of Chaco Canyon, but climate change was an important cause for the demise of the center. Peoples in the region had proven resilient in enduring long droughts centuries previously, but a series of intense droughts would have posed a severe challenge to a large population center.

Precipitation in South America

Throughout the pre-Columbian era, societies along the west coast of South America built on long traditions of storing and steering water. Furrows that remain in the ground near Lake Titicaca, a vast mountain lake high in the Andes, survive from a system of canals of the Tiwanaku empire, which reached its height of power between around 500 and 900 CE. On the North Coast of Peru, Chimu, a major cultural site from around 1100 until the late 1400s, adapted to environmental variability, including both floods and drought. Overflow weirs limited the damage from high waters.[68] In the later Chimu era, people built structures away from areas most susceptible to flooding. Drought likely posed an even greater danger. Farmers of the Chimu era employed aqueducts and other irrigation networks. Along with these kinds of hydraulic engineering, communities in areas such as northern Chile responded to the risk created by environmental shifts by changing crops, moving fields, and increasing trade.[69]

The wide persistence of human settlements and activity in western South America shows that climate fluctuations by themselves did not lead to doom, but complex societies faced shifts between drier and wetter periods.[70] People of the Nazca, on the southern coast of Peru, for example, adapted to a dry setting, but they still benefited from more rain. The area received a boost in precipitation between approximately 800 BCE and 650 CE. From 650 to around 1150 CE more moisture was available to the east in the area around Lake Titicaca on the borders of what are now Bolivia and Peru. Precipitation increased again in the Nazca from 1150 to 1450.[71]

The possible effects of climate shifts on complex societies in western South America have given rise to debate. In northern Peru, sand covered over the irrigation system of Moche, a complex society on the North Coast of Peru. The capital was abandoned in the sixth century CE, and the culture moved east toward the interior to higher terrain with more water. A shift in climate to a dry period provides a possible cause for such migration.[72] But another interpretation points to social change as the most likely key factor.[73] In another possible example for the influence of climate change in South America, the society of Tiwanaku thrived for centuries on the shores of Lake Titicaca before a long dry period started in the eleventh century. The capital of an empire, Tiwanaku contained numerous ceremonial sites, including temples and pyramids. Terraced fields surrounded the capital. This large and complex society declined and came to an end in the eleventh and twelfth centuries. In one scenario, the state saw its power erode and ultimately declined because of a decrease in precipitation.[74] Another interpretation, in contrast, sees any changes in agriculture as separate from the fortunes of the Tiwanaku state.[75]

Correlation does not prove that climate shocks led directly to political and social change in western South America, but genetic evidence shows that periods of migration corresponded to climate change. DNA samples from populations in southern Peru between 840 BCE and 1450 CE indicate two major phases of migration. Toward the late period of the Nazca culture, migration into the central Andes increased from coastal valleys. Later, the end of the Wari and Tiwanaku empires around 1200 CE appears to have encouraged migration back toward the coast.[76] In the case of the Wari empire, in Peru's highlands, a long drought between 900 and 1350 CE coincided with the empire's end around 1100 CE. Internal conflict was the most immediate cause of a crisis that saw rising rates of violent injury as well as a decline in diet, but drought exacerbated the challenges of the late Wari period. [77]

Over a long period, multiple complex societies had emerged in western South America, and shortly before the Spanish conquest of much of the Americas, the Inca state rose to power along the Andes. Multiple factors contributed to the Incas' ability to subdue and incorporate many of their neighbors into an empire that stretched across more than 2,000 miles. Drawing on their military capacity and on diplomacy, the Incas also expanded during a period of warming in the region, which would have enabled them to exploit a broader array of high land areas.[78]

Summary

Investigating climate and human history for the period after the end of the most powerful empires of the classical era or axial age in Eurasia demonstrates the potential influence of regional climate shifts. Strong evidence for a warming period in the High Middle Ages of Europe points to the interaction between a favorable climate and European expansion both within Europe and in the North Atlantic. Regional shifts in other regions, including Central Asia, Central America, and areas in the interior and southwest of what is now the United States, also show that climate shifts and, in particular, changes in precipitation placed constraints on complex societies. Human societies developed significant resilience to climate variations during the Holocene, but complex states reached a point where their ability to store resources failed to meet the demands imposed by long droughts.

The MCA has frequently surfaced in recent discussions of climate change. In particular, attacks against either the science of climate change or the findings that human activity has become the main forcing agent of climate change frequently point to the Medieval Warm Period. In one version, the period was so warm that current global temperatures and temperature trends are not striking. These claims draw heavily on the most impressionistic evidence, in particular references to grapevines in England and Vinland in North America. Cultivated wine grapes, like many other crops, can serve as one of many proxies for climate shifts, but their cultivation also depends on other factors. Farmers may plant grapes or replace them with other crops because of changes in taste or competition from other grape-growing regions. Similarly Viking references to Vinland (which may not even refer to grapevines) or the naming of Greenland as green provide no exact record of past temperature.

Another problem with using the MCA to attempt to discount the key role of human forcing of climate since the Industrial Revolution is that the warmth was most likely regional. Some regions may have been as warm as today, but overall the warming was not global and was not synchronous across regions. The regional expression is consistent with the internal climate variations (ENSO, NAO), which may have been caused by a slight increase in solar irradiance.

Finally, the Medieval Warm Period cannot serve as logical evidence against human forcing of climate change because the causes of climate shifts during the Middle Ages were different from those today. Humans in the Middle Ages engaged in farming and in many economic enterprises, some of which involved using peat or coal as fuel, but there was no Industrial Revolution, no internal combustion engine, and no logarithmic rate of growth in extraction and combustion of fossil fuels.

CHAPTER FIVE

The Little Ice Age

- Climate of the Little Ice Age
- The Little Ice Age in the North Atlantic
- The Little Ice Age in Europe
- Little Ice Age in East Asia
- Little Ice Age in the tropics
- Seventeenth-century crisis
- Seventeenth-century crisis in China and Japan
- North American colonization in the Little Ice Age
- Cultural and social effects
- Adaptation
- Abrupt climate change in the Little Ice Age
- Summary

A complex cooling trend punctuated in several regions by markedly cold phases influenced human history over the centuries following the MCA. This period became known in historical and scientific literature as the Little Ice Age. Though the precise timing and severity of cooling varied, the fluctuations in climate affected human societies in widespread regions, particularly in Europe and in those surrounding the North Atlantic, but also in Asia and North America.

During the periods of most pronounced cooling, the LIA created challenges for societies and states in many regions of the world. At the same time, societies varied greatly in their responses to the LIA. By the early modern era, some of the more prosperous societies demonstrated great resilience.

Others experienced periods of crisis, but continued to adapt and develop until the cold spells became a distant memory. Others, especially in locations that already made them more vulnerable to cold, faced profound threats.

As the case studies in this chapter show, the LIA had strong effects in the North Atlantic and in Europe. Indeed, much of the early literature on the topic focused on these regions. In North America, European colonization began during the LIA. Study of the LIA and its interaction with human history has also extended to major societies across Eurasia from the Ottoman Empire to China.

Climate of the Little Ice Age

By Little Ice Age, we should think of a period of marked cooling by today's standards, but not of a single event that shaped all parts of the globe at the same time. Historical documentation of glacial advance in Europe, which was particularly extensive during the late 1600s–1700s, provided some of the first evidence for a colder period that spanned the latter half of the second millennium. Although there were generally cold conditions in the Northern Hemisphere between 1300 and 1850, this "little ice age" was not globally synchronous, nor was it a continuously cold period. Records indicate that the coldest time period in Europe was the 1600s, but parts of North America did not see their coldest temperatures until the 1800s, while East Asia experienced consistently cooler temperatures during this time. Temperatures also fluctuated within the LIA. A particularly cold period struck Europe during the 1430s. Contemporary authors described damage to grain, vineyards, herbs, and livestock. Food production fell, and food prices rose.[1] The Little Ice Age, however, was not a period of hundreds of years of uninterrupted cold weather. Thus much of Europe experienced heat and drought in the early 1470s, especially in 1473–4, but the 1480s and 1490s saw wet cold years.[2] Nevertheless, the temperatures appeared to be colder than the centuries prior, and certainly colder than today. The complexity of the temperature record has led to the suggestion that cold snaps did not amount to any clear cooling trend, and even prompted skepticism about the very idea of a LIA. However, as the historian Sam White responds, "there is not at present any substantial debate about whether global climate cooled between c.1300 and c.1850, or whether that cooling had significant human impacts."[3]

The evocative term "Little Ice Age" should also not lead to the idea that this period amounted to another glacial maximum. The LIA cooling that is evident in many records amounts to 1°C or 2°C. In contrast, the LGM saw temperatures around 10°C colder on average.

To determine the causes of the LIA, climate scientists consider external factors, including solar output and volcanic activity. On a timescale of decades to centuries, solar variability stems mainly from sunspots, magnetic storms on the sun's surface that increase solar output. Variations in the number of sunspots currently follow an 11y cycle, but historical sunspot records, which began in the early 1600s, reveal prolonged periods in the past with very few sunspots. Several of these sunspot minima occurred during the LIA, notably the Maunder Minimum that spanned from 1645 to 1715, as well as the Spörer Minimum (1460–1550) and the Dalton Minimum (1790–1830). Such sunspot minima generally coincide with minima in temperature, but whether or how sunspot minima might lead to decreased temperature particular to the LIA remains a topic of debate and ongoing research (Figure 5.1).

Active volcanism in the thirteenth century likely contributed to the cooling of the LIA. At first glance, this connection seems surprising. Volcanic eruptions typically lead to short-term cooling, so how could volcanic activity by itself lead to the centuries-long cooling of the LIA? A possible answer lies in climate feedbacks: several large eruptions in quick succession could have initiated sufficient cooling to trigger sea ice growth. The increased reflection of sunlight by ice and snow, in turn, would create additional cooling. This is known as the ice-albedo feedback, and is one of several feedbacks that amplify climate changes. The timing of increased volcanic activity appears to be consistent with the idea of a volcanic forcing for the LIA: cooling would have had to begin in the 1300s for glaciers to have reached their maximum extent by around 1600.[4] The degree of cooling associated with volcanic activity is difficult to reconstruct, but there is evidence for numerous eruptions in the 1200s and another around 1450, indicating that eruptions provided some cooling effect

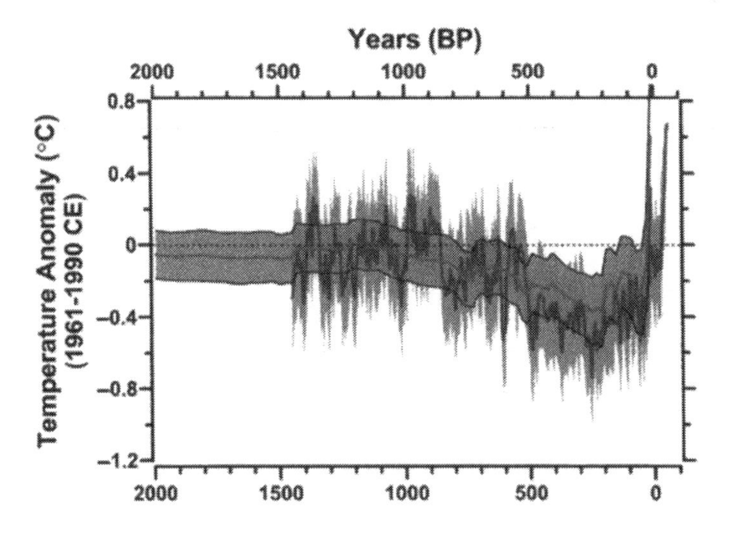

FIGURE 5.1 *Two thousand years of temperature variation. Source: Shaun A. Marcott et al., 2013, "A Reconstruction of Regional and Global Temperature for the Past 11,300 Years," Science 339, 1198–201.*

for the LIA. Volcanic eruptions likely intensified one of the coldest periods of the LIA, the Grindelwald Fluctuation, between 1560 and 1630[5].

In addition to solar variability and volcanic eruptions, changes in atmospheric CO_2 concentration may have played a role in LIA cooling. Concentrations of CO_2 dropped from 284 ppm around 1200 CE to 272 ppm by 1610,[6] after Northern Hemisphere cooling. Greater solubility of gases in the ocean as temperatures cool can in principle reduce atmospheric CO_2 concentrations, but not sufficiently to account for the full observed decrease in CO_2 during the LIA. CO_2 may also have decreased as forest reclaimed agricultural lands when pandemics caused sharp population decline. The black death that struck between 1347 and 1352 reduced Europe's population by 25 million. Population fell by 50 million between 1492 and 1700 when Europeans introduced new diseases to the Native Americans. China suffered population decline in the eleventh and twelfth centuries, and again in the seventeenth and eighteenth centuries. The steep regional decline in population led to the reforestation of abandoned farm land; new tree growth reduced CO_2 in the atmosphere, leading to cooling.

These external forcings influence climate on yearly to centennial timescales, and all likely contributed to the cooling of the LIA to some degree. Internal climate dynamics also become important on this timescale. For example, a slowdown in deep ocean circulation would provide another possible explanation for the cooling of the LIA.[7] As we saw in the case of the Younger Dryas, meltwater slowed deep ocean circulation and plunged the North Atlantic region into near-glacial conditions. On a much smaller scale, warmth during the MCA could have increased melting and slowed circulation. This could explain the cooling patterns observed in this time period. Other climate phenomena may have contributed to the geographic patterns of temperature as well as precipitation anomalies during the LIA. In a reversal of the prolonged La Niña and positive NAO conditions of the MCA, the opposite modes (El Niño and a negative NAO) prevailed during the LIA.

The Little Ice Age in the North Atlantic

Regardless of the ultimate cause of the LIA, the cooling trend that characterized the North Atlantic had severe consequences for human settlements established at the margins for farming and pastoralism. Cooling caused the greatest difficulty in farms and pastures of the north or at high elevation. One of the chief historical debates over the effects of the LIA in the North Atlantic concerns the fate of the far-flung colonies established by Vikings in the North Atlantic during the previous MCA. The Norse settlers established a mixed economy in Greenland with two main settlements, an eastern settlement in the far south of Greenland and a larger western settlement along the west coast. Altogether the population reached some

5,000 people. They traded products, mainly to Norway. Europeans could acquire the leather and woolen cloth that the Greenlanders sold from other sources, but exports of walrus tusks and polar bears from Greenland stood out as distinctive goods in the European market.

To support themselves, settlers created farms, about 250 in total, in the western and eastern settlements. At first they replicated their previous practice of animal husbandry by raising cows and pigs, but they soon stopped raising pigs, possibly because they wreaked so much damage on Greenland's soil.[8] Sheep fared better than cows in Greenland because they could be kept outdoors grazing far longer than cows, which had to be housed indoors for approximately nine months each year. The number of goats, animals well suited to eating scrub, also increased over time. This shift away from pigs and cows provides strong evidence that the Norse adapted to living on Greenland. Despite their previous experience, they made the necessary adjustments to turn to animals better suited for life on Greenland. Simply moving to a region with a different landscape, resources, and climate did not by itself determine their fate.

The Norse also hunted and adapted to life on Greenland in their choice of prey. The island's large herds of caribou provided a key source for meat, and so, too, did seals. The Norse learned when and where seals migrated to the vast island, and in spring the Norse set up camps to hunt seal. Over time the Norse depended ever more on hunting and on seals to obtain food. Surprisingly, given the abundant ocean waters, the Norse do not appear to have engaged in fishing, as measured by the very small numbers of fish bones recovered at their sites. Jared Diamond suggested that this absence of fish may show that the Norse developed a taboo of some kind against eating fish.[9]

Though Norse settlers adapted their way of life to Greenland, their presence placed a strain on the island's resources. They quickly cut down the meager supply of trees, and the slow regeneration of soil, especially with grazing animals, meant that they could not easily obtain new sources of timber. Other than the occasional driftwood or timber import from Norway, they could obtain new timber by undertaking dangerous trips to log along the coast of Labrador. In turn, the absence of charcoal deprived the Greenlanders of the fuel source necessary to smelt bog iron.

The Greenlanders survived for centuries before they confronted the LIA in an already marginal region. Temperatures dropped sharply in Greenland within just a few decades in the fourteenth century (4°C in ~80y).[10] A microfossil record of a sediment core from Disko Bay further shows cooling in western Greenland around 1350.[11] Pronounced cooling created more difficult conditions for pastoralism. A shorter growing season reduced the meager harvest of hay, placing the raising of sheep and goats at risk. Longer harsher winters and shorter colder summers cut the margin for error in obtaining and storing food. Under these circumstances, it made sense for the Greenlanders to obtain a rising proportion of their diet from seals, but this must have been an adaptation made only under great pressure. With limited resources to begin with, the Norse of Greenland had no reason to voluntarily cut their consumption of any source of food.

Icier waters also made ocean travel harder and more risky. At the best of times it had never been easy for the Greenlanders to travel in their small vessels to go north to hunt for walrus or to seek timber, but these trips would have become increasingly dangerous, and the severe shortage of wood meant that the Greenlanders could not easily build new boats. Longer periods of sea ice also obstructed the vital exchange of resources between Greenlanders and other Norse outposts and Europe. Communication with Norway suffered. Only sporadic ships traveled between Norway and Greenland by the late fourteenth century, and communication stopped altogether by the early fifteenth century. In 1492, well after the last recorded voyage from Norway to Greenland, Pope Alexander VI wrote, "Shipping to that country is very infrequent because of the extensive freezing of the waters— no ship having put into shore, it is believed, for eighty years."[12] The pope overlooked the distant voyages of fishermen, but his basic point was correct: Greenland had fallen off the European map.

Other factors combined with climate change of the LIA to threaten the prosperity of Greenlanders. Even as the cooling and ice in the North Atlantic gave sailors reason to avoid the dangerous and long journey to Greenland, other commercial changes simultaneously cut into trade with Greenland. The Norse settlers on the island depended greatly on walrus tusk exports. Communities organized the walrus hunt. Indeed, the hunt was so ubiquitous that the site of nearly every Norse farm in Greenland contains fragments of walrus bone.[13] But trading patterns in Europe shifted away from Greenland as the Hanseatic cities gained power. These North Sea and Baltic ports favored a growing trade in items such as fish and fish oil rather than a more limited luxury trade for items such as walrus tusk, and for those Europeans who wanted to purchase ivory, other sources of supply became available to Europe's east.[14]

Increasingly peripheral within extended European trade networks, the Norse of Greenland also faced potential competition from the other migrants to Greenland, the Inuit, or Thule, who had moved east across northern North America and replaced the Paleo-Eskimos. The two lived apart: there is no genetic evidence of any intermingling between the Norse and the Inuit. The Norse provided only scattered references to their contacts with Inuit but enough to show that their relationship was primarily hostile. Accounts from the fourteenth century refer to attacks by people the Norse called Skraelings, including the killing in 1379 of eighteen men and the enslavement of two boys. Along with the loss of life, these deaths would have likely inflicted severe economic and psychological damage on a small population.

In some interpretations, the Greenlanders also failed to learn from their Inuit competitors. There is debate over the degree to which the Greenlanders supplemented seals with fish, but they do not appear to have engaged in fishing to remotely the same degree as the Thule. In addition, the Thule hunted more varieties of seal than did the Norse, such as the ringed seal, which can be hunted in winter. If the Norse had cooperated with or even learned from the Thule, they could have acquired more food in winter. They could have hunted seal packs during annual seal migrations in the spring and the ringed and bearded seals in winter that surface at blowholes. Adopting Inuit hunting techniques and watercraft would have been invaluable for broadening the Greenlanders' limited food supply.

From these multiple variables, scholars have produced divergent explanations of the end of the long-established Norse settlements in Greenland. In a catastrophic explanation climate change combined with growing isolation and bad relations with Inuit to wipe out the Norse Greenlanders. An alternate explanation, in contrast, sees the end of the Norse settlements as a matter of choice. Greenlanders by the fourteenth century shifted their diet to hunting seals—a point of consensus with the more catastrophic interpretation. They acquired more protein from the sea. They even fed seal and fish scraps to their pigs, before abandoning the raising of pigs altogether. Unpleasant living conditions, limited diet, and few chances for contact with the world beyond Greenland gave the island's youngest residents cause to leave Greenland for Iceland and points west. Population decline caused by the fourteenth-century plague in Iceland and Scandinavia could also have made land cheaper for Greenlanders who decided to emigrate.[15]

The alternate scenario stresses human choice more than the effects of climate change, but in any scenario, the Greenlanders had made choices and shown significant ability to adapt. They obviously chose to go to Greenland and to stay there in the first place, and though the warmth of the Middle Ages made their trip easier than it would have been otherwise, they still chose to remain in a region that almost immediately proved significantly different from Iceland, let alone from Norway. They managed to survive in Greenland as long as they did only by, first, adopting pastoralism to the island and, second, by increasing their consumption of seals and food from the sea. In other words, the catastrophic scenario does not overlook the role of human choice or deny that the Greenlanders showed significant capacity to adapt.

This alternate interpretation stresses human decision-making, but does not dispense with the influence of climate. In one interpretation, climate change during the LIA so depleted the Greenland population in their isolated and exposed colonies that they died out. In the second interpretation, climate change brought them to abandon many of the economic activities that they had carried out and that made life so unsatisfactory that the dwindling population left. Changing consumer tastes may have reduced demand for walrus tusks, but stormy seas also made it increasingly difficult to undertake the walrus hunt.[16] This less dramatic scenario is possible, but there is only scant historical evidence of migration out of Greenland. The authors of one study cite a 1424 letter to a bishop in Iceland by a couple who had married in Greenland in 1408 and needed to provide proof of their marriage after relocating. However, we lack documentary evidence of other such cases of migration from Greenland to Iceland or any documents that explain the fate of any of the Greenlanders still on the island afterward.

By the last decades of the Greenland Norse, the kind of choices the Greenlanders made suggest desperation rather than simply adaptation. Was shifting to an almost all-seal diet really adaptation, or was it desperation? It is difficult to imagine why the Norse would have abandoned sheep and goats when they already had so few supplies of food, unless they had little or no choice in the matter. Perhaps some significant number of Greenlanders survived by moving back to Iceland, but even if that had occurred, it would not disprove the role of a deteriorating climate in contributing to such migration. The Greenlanders did not make up a transient population, nor were they the equivalent of fishermen or hunters who set up seasonal camps, hunted or fished, and then left. To the contrary, they had survived in the far western North Atlantic for more than 400 years, sinking enormous effort and investment into putting down their settlements. In even the best-case scenario, the cooling climate would have contributed to the end of the Norse presence in Greenland.

The same cooling trend also posed challenges to the larger settler population of Iceland. As in Greenland, the influx of Norse colonists during the MCA placed stress on the local environment. Settlers cut down most of the island's trees, and grazing prevented new tree growth and damaged soil, as the loss of vegetation caused erosion. Volcanic soils blew away easily, leaving much of the island as desert. Indeed, it is still possible to see the deserts created during this period.

Already vulnerable because of their depletion of resources, the settlers of Iceland faced a major threat from the LIA. They gave up marginal farms and reduced their already very limited farming. Sea ice lingered, blocking access to the ocean from northern ports. With fewer resources and the arrival of the plague, population fell, according to many estimates, though there is no consensus on exact numbers. Icelanders also faced volcanic eruptions. In 1362, Oraefajokull erupted, forcing the abandonment of farms in the region in southern Iceland.

Despite these challenges, Icelandic society, unlike the settlements of Greenland, survived the LIA. A larger population and closer ties to Europe put the Icelandic settlements in a better position. The Icelanders adapted to the loss of other food resources by increasing their consumption of seafood, and they found a source of revenue by selling dried fish to meet growing European demand.

The Little Ice Age in Europe

Few societies faced the same peril that confronted residents of North Atlantic islands during the LIA, but its onset also posed broad challenges for

European societies. Even before the fourteenth century, the series of volcanic eruptions that may have caused the onset of cooling led to abrupt spikes in mortality. The eruption of a large volcano in the tropics led to a bad harvest in 1257–8 in England. A monk described the suffering:

> The north wind prevailed for several months . . . scarcely a small rare flower or shooting germ appeared, whence the hope of harvest was uncertain. . . . Innumerable multitudes of poor people died, and their bodies were found lying all about swollen from want. . . . Nor did those who had homes dare to harbour the sick and dying, for fear of infection. . . . The pestilence was immense—insufferable; it attacked the poor particularly. In London alone 15,000 of the poor perished; in England and elsewhere thousands died.[17]

The climatic shift of the fourteenth century brought dramatic consequences for many Europeans when famine struck between 1315 and 1322. Heavy rains caused erosion and prevented planting. Numerous contemporary observers remarked on the extraordinary rains of 1315 and on the cold summer and fall. The rains persisted daily for five months. The rain and damp ruined crops and swept away soil, especially in some of the newly settled areas created by expansion during the prior Medieval Warm Period. Swollen streams and rivers carried away mills, bridges, and whole villages. Another year of heavy rains and failed harvests followed in 1316.[18] Rains, flooding, and the cold winter of 1317–18 caused both crop failures and hardship for farm animals who went short of food.

Previous population growth exacerbated famine for communities that already lived near the subsistence level. There were bread shortages in France and Flanders. People from the German countryside made their way to beg at towns along the Baltic.[19] Some communities disappeared as the residents left. Contemporaries complained about disorder and lawlessness. Destruction of seeds slowed recovery after 1317. Crop yields sank, and mortality from disease rose in a weakened population. In the 1320s, continuing cold weather and parasites killed off many sheep, damaging the important woolen industry in England.

The cooling phase caused particular hardship for societies in areas that shared some of the same qualities: either a northern location or a high elevation. In Norway, the LIA contributed to the abandonment of many farms. Some 40 percent of all farms were abandoned after the onset of the LIA. Farmers could no longer reliably grow grain above about 1,000 feet, and this made dependent communities so vulnerable that many residents left for opportunities elsewhere.[20] The pattern of deserting villages extended to other regions such as Britain and German lands. Thousands of villages in England were abandoned. A few today, such as the deserted village of Wharram Percy in North Yorkshire, have become tourist attractions. Germany, too, had its abandoned villages, called *Wüstungen*.

The relationship between climate change and outbreaks of plague remains the subject of debate. Climate fluctuations would in principle have affected the areas of Central Asia where plague existed among reservoirs of animals. In one scenario, periods of high precipitation first boosted plague activity among wild animals in grasslands. Warm springs and wet summers increased the presence of plague among great gerbils in Kazakhstan. Once these favorable conditions ended, gerbil population fell sharply, in turn forcing fleas to find new hosts among other animals, including humans.[21] However, another study does not show a correlation between climate and plague.[22]

With outbreaks of plague, cooling was certainly not the only factor leading to the abandonment of villages: European population fell sharply from the black death or the bubonic plague between 1347 and 1353. Carried west from Asia, the plague followed trade routes to the West. It reached Constantinople and Alexandria by 1347. In Alexandria in Egypt, the plague killed up to 1,000 people per day. In the fall of 1347 the plague arrived at Sicily, and by 1348 had reached northern Italy, where the mortality was devastating. "And I, Agnolo di Tura . . . ," a resident of Sienna wrote, "buried my five children with my own hands . . . so many died that all believed it was the end of the world."[23] At Avignon in southern France, Pope Clement VI sat between two fires to seek to avert the contagion. In Paris, 800 people died per day in late 1348. In all at least one-third of Europe's population died, though some estimates point to a much higher death toll of as high as 60 percent.

Beyond regions where the LIA severely hampered farming, it is difficult to separate the effects of cooling and of population loss on abandoned villages. The pattern of abandonment likely began in some regions in the early fourteenth century before the black death, though there is debate about the exact timing. Desertions in some regions, such as Denmark, also took place after the plague, peaking in the early fifteenth century.[24] Even if the LIA did not always cause abandonment of villages, it gave rural residents reason not to return to a landscape that had previously been farmed. In regions afflicted by plague, cold summers prevented or slowed recovery. It made sense not to move back to villages like Wharram Percy, especially if other land was available.

Near mountains, cooling posed a more direct threat to nearby farms and villages. Expanding glaciers moved downslope toward high Alpine villages. Long before tourism and the winter and summer sports industries, Alpine communities were poor and isolated. Lack of access to sources of iodine led to high rates of disability. The advance of glaciers during the LIA in some cases choked off valley floors, creating ice dams, and when ice dams broke, flooding ensued. In the Saas Valley in the Valais in southern Switzerland, the Allalin Glacier created just such an ice dam in 1589. In 1633, a flood that resulted when the lake burst through the dam, caused disaster for the valley's inhabitants: "half the fields were buried in debris and half the inhabitants were forced to emigrate and find their miserable bread in some other place."[25]

The Little Ice Age in East Asia

In contrast to the cold temperatures that dominated northern latitudes, shifts between droughts and periods of abundant precipitation, or pluvials, drove climate change and social responses in Asia. A switch from persistent La Niña conditions to a climate pattern resembling El Niño accompanied the transition from the MCA to the LIA. During this time the ITCZ shifted several hundred miles to the south, resulting in a weaker summer monsoon in Southeast Asia. Other internal climate oscillations, such as the NAO and the Pacific Decadal Oscillation, likely interacted with the ENSO cycles in the early LIA. There was severe and prolonged drought in the middle of the thirteenth century. The monsoon weakened in the fourteenth century, and the region suffered from severe drought in the early fifteenth century.

The weaker monsoon was especially damaging for the Khmer State in Cambodia, which depended on an elaborate system of supplying water and food to the sprawling city of Angkor. A vast irrigation system included hundreds of ponds and carried water through canals to reservoirs. There was no compact urban core, and the pattern of settlement resembled that found in the city-states of the classic Mayan era. Rice fields cleared from forest helped feed the population. The very scale of exploitation led to deforestation and erosion.[26]

The climate shifts posed a major threat to the entire water infrastructure at Angkor. Sharp fluctuations in precipitation both reduced the supply of water for farming and damaged the vast and critical irrigation and water-supply system. The combination of drought and short intense monsoon rains produced heavy deposits of sediment, thereby blocking canals before heavy rains taxed the limits of the system.[27] Evidence that the Khmer tried to rebuild canals shows an effort to respond that may not have ultimately worked.

Another possible scenario sees gradual decline at the Khmer core and the move of elites toward the Tonle Sap and southeast to the Mekong Delta as the result of factors including the growth of nearby states and the pursuit of trade opportunities. Indeed declining population at Angkor Wat never meant complete abandonment of the temple complex.[28] In a multicausal explanation for Khmer decline, elite movement could have contributed to disrepair of infrastructure making the city more vulnerable to climate fluctuations.[29]

In sum, these climate disruptions weakened the economic underpinnings of Angkor Wat just as the Khmer Empire faced external challenges, including conflict with the Kingdom of Siam of the Thai. Mongol conquests of Yunan in southern China propelled Thai migrations south. Thai from the Kingdom of Ayutthaya in what is now southern Thailand began to push into the Khmer Empire. Thai forces occupied the Khmer capital of Angkor several times, before sacking the city in 1444.

In East Asia, the weakening of the monsoon also affected China and the Yuan dynasty ruled by the Mongols. Toghon Temur who ruled the Yuan dynasty 1333–68 saw China face repeated crises: alternately severe droughts and major floods. Floods in the 1340s caused many deaths. Epidemics and famine struck China. These crises thwarted Yuan efforts at relief even as they created ideal conditions for banditry and for rebellion. Indeed, Zhu Yuanzhang, the founder of the Ming dynasty, which followed the Yuan dynasty, waged his early campaigns during years of severe drought in the late Yuan dynasty.

Other societies in Indo-China suffered major setbacks during the same period. Along with Mongol invasions, the state at Pagan in what is now Myanmar, or Burma, faced repeated rebellions and invasions in the thirteenth and fourteenth centuries.[30] Dai Viet, centered in the north of what is now Vietnam, experienced uprisings by peasants in the 1340s along with invasions. As in the previous period of state expansion, retreat and decline stemmed from several factors. Success itself created new problems when expansion placed stress on resources and the local environment. In Dai Viet, land shortages helped prompt uprisings. At the same time, these societies faced external military threats, in particular from Thai migration. The climate shift combined with these other factors to accentuate and intensify the crisis facing Southeast Asian states. Rainfall did not shrink all at once, but droughts became more frequent with markedly low rainfall between 1340 and 1380.[31] Much of the fifteenth century was also comparatively dry, and shorter-duration, intense monsoons also reduced the amount of water that could be easily stored and used.

Little Ice Age in the tropics

As the examples from Southeast Asia suggest, variations in precipitation characterized the LIA. A general cooling is evident in most of the tropics in the seventeenth century, but precipitation patterns are more variable. Evidence from lake cores from East Africa suggests complex effects. Sediment from Lake Edward indicates a pattern of droughts between 1450 and 1750, but records from Lake Naivasha point to a wet period.[32] Sediments from Lake Kitagata and from Lake Kibengo in Uganda show droughts lasting a century at 1100, 1550, and 1750. This pattern indicates a combination of drought in western East Africa but moisture further east.[33] On Mount Kilimanjaro, the small Furtwängler Glacier may have formed during the LIA.[34]

These fluctuations in climate may have influenced societies in Africa. In western Uganda, for example, population increased after 1000 CE. Settlements surrounded by large earthworks grew in the fifteenth and sixteenth centuries. A period of higher rainfall aided such concentration. The inhabitants appear to have practiced agriculture as well as some pastoralism. Settlement became more dispersed around 1700 with the abandonment of earthworks. During this phase of the LIA, higher elevations became drier. This shift could have contributed to increasing pastoralism, and to greater social differences.[35]

Climate fluctuations had the potential to move the boundary of the Sahel again. The zone in which millet was cultivated shifted north with more precipitation, but fell back south in drier times. Between 300 and 1000 CE, the zones moved north before falling back. In West Africa in the southern Lake Chad Basin, drought may have contributed in the fifteenth century to the relocation of the political center of the Kanem Empire toward Bornu.[36] Changing patterns of rainfall would also have likely affected relations between peoples of the savanna, or Sahel, and the Sahara.

In one model, increased precipitation would have enabled farmers to move further north. Such conditions would also have led to a more northern boundary for the tsetse fly, thus limiting the activities of mounted warriors. During drier periods, in contrast, states of the Sahel such as Mali would have been able to push further south.[37] In principle, this model is logical, but it requires firm data about historical precipitation patterns. Another interpretation points to a general drying trend in which closer contacts between the savanna and the Sahara eased communication and commerce, including the slave trade.

Climate fluctuations during the LIA and the MCA may also have benefited and challenged states in southern Africa. The state of Mapungubwe (900–1300 CE) provides a key example. Located near the meeting point of the Shashi and Limpopo Rivers where modern-day South Africa Botswana and Zimbabwe meet, the Kingdom of Mapungubwe emerged in the tenth century CE. Explanations for state expansion focus on economic factors, cattle raising and trade, and on climate. Generally increased precipitation of the MCA contributed to the growth of the Mapungubwe state. The population of the capital of Mapungubwe rose to some 9,000.[38]

In the very early fourteenth century CE, the population of Mapungubwe abandoned their city, but complex society in the region endured with the rise of the Great Zimbabwe state to the north. Severe drought associated with the onset of the LIA appears to have contributed to the demise of Mapungubwe. Changes in trade along with climate may have weakened the Mapungubwe state. There is dispute over the onset of the LIA in southern Africa, but climate proxy data obtained from baobab trees show an early fourteenth-century drought. Such drought may have undermined the legitimacy of rulers who held rain-making ceremonies.[39]

Seventeenth-century crisis

The overall colder conditions during the LIA (1400–1850) were most pronounced during the seventeenth century. The cooling and climate variability of the LIA in the late sixteenth and seventeenth centuries came at a time of widespread upheavals and crises in many areas. Thomas Hobbes, the English political philosopher who decried rebellion and the collapse of authority, gave expression to pessimism in his book *Leviathan*, published in 1651:

> there is no place for industry, because the fruit thereof is uncertain: and consequently no culture of the earth; no navigation, nor use of the commodities that may be imported by sea; no commodious building; no instruments of moving and removing such things as require much force; no knowledge of the face of the earth; no account of time; no arts; no letters; no society; and which is worst of all, continual fear, and danger of violent death; and the life of man, solitary, poor, nasty, brutish, and short.[40]

Hobbes was most interested in upholding a powerful sovereign authority, but his language suggested a far-reaching, broad crisis that extended beyond politics.

Later historians applied the notion of a general seventeenth-century crisis to European history. Numerous wars and rebellions took place during the seventeenth century. Between 1618 and 1648, the Thirty Years' War ravaged German lands and adjacent regions. England experienced protracted tension between the king and some of his subjects that escalated to civil war and revolution in 1642, and in 1649, the Parliament had King Charles I executed. Another revolution, the Glorious Revolution, unseated King James II in 1689. In France, an aristocratic rebellion called the Fronde broke out between 1648 and 1653 during the very early years of the reign of Louis XIV. In the low countries, the Dutch rebelled against the Habsburg dynasty, which also faced rebellions elsewhere in Iberia and Italy.

Standard historical narratives described and explained the crisis without focusing in particular on climate, and indeed there were obvious political and religious causes for conflict. Centralizing monarchs had already begun their attempts to curtail the power of feudal lords well before the seventeenth century. In the case of Europe, the Protestant Reformation and the ensuing Catholic or Counter Reformation introduced a new cause of division. Many of the wars that afflicted seventeenth-century Europe were at least in large part religious wars such as the Thirty Years' War, which began after Protestants in Prague in Bohemia threw the pope's emissaries out of a window in the Defenestration of Prague. Catholic armies gathered by the Holy Roman Empire suppressed Protestant revolts, but then confronted intervention in support of Protestants by Denmark in 1625 and, more successfully, by Sweden in 1630. Dynastic rivalries also contributed to war as the intervention of France and its Catholic monarchy against the Catholic Habsburg family in the long final phase of the Thirty Years' War.

At the same time, the renewed shift toward cold weather placed societies under multiple forms of stress. Climate change during this phase of the LIA contributed significantly to the risk of famine. The LIA was not a period of uniformly cooling temperatures, and the shift to a relatively warm period in Europe between 1500 and 1550 helped boost population. A renewed cooling phase from the late sixteenth into the seventeenth century then raised the risk of famine. It was not just cooling but also greater variability that damaged crops. Less reliable weather associated with the LIA was especially dangerous after prior population growth.

An upsurge in famines and epidemics took place in Europe between 1550 and 1700. In France, the cold humid period that stretched from the summer of 1692 through early 1694 gave rise directly to famine.[41] Some 10 percent of the population of northern France died in the famine of 1693–4, and the rate of death was even higher in the Auvergne region in the interior of the south. The late 1690s in Scotland were the period of "the Ills" of late harvests, crop failure, and food shortages.[42] Failed harvests brought famine in the 1690s as well in Scotland. The population fell by some 15 percent from a combination of death and emigration. The tacksmen with the excise (tacksmen managed estates and collected rents) wrote, "Many poor people were dying for want, and that the ground was manured in several places, and no Seed in the Countrey for sowing the same, which are two pregnant marks of Famine." The Privy Council in 1698 described dire conditions: "not only a Scarcity, but a perfeit famine, which is more sensible than ever was known in this nation."[43]

Epidemic disease struck frequently during the LIA. There were outbreaks of the plague in Italy and France in 1629–30, in southern Italy in the Kingdom of Naples from 1656 to 1658, and in England in 1665. Samuel Pepys memorably described the outbreak of plague in London in his diary. At the end of August 1665, he wrote:

> Thus this month ends, with great sadness upon the public through the greatness of the plague, everywhere through the Kingdom almost. Every day sadder and sadder news of its increase. In the City died this week 7496; and all of them, 6102 of the plague. But it is feared that the true number of the dead this week is near 10000—partly from the poor that cannot be taken notice of through the greatness of the number, and partly from the Quakers and others that will not have any bell ring for them.[44]

The frequency of outbreaks diminished in most of Europe in the eighteenth century, though the plague did not disappear altogether until the early nineteenth century.

As the plague grew more rare, other epidemic diseases caused greater fear, in particular smallpox. In the eighteenth century, the annual death toll in Europe from smallpox reached into the hundreds of thousands. It killed one of ten children in Sweden and even more in Russia.[45] In the Americas and in distant regions of northern Eurasia, outbreaks of smallpox brought even higher mortality for native populations who came into direct or indirect contact with Europeans. Thus a smallpox epidemic in the Americas between 1775 and 1782 afflicted settlers, troops in the Revolutionary War, and Indian populations, some of whom lived far from the war zone.

The interaction between the epidemics, including the plague, and the LIA was indirect. Cooling in general and bad weather in particular did not directly cause plague epidemics. A cold spell could potentially have reduced the breeding of fleas that carried the plague. However, famine contributed to epidemics in other ways. First, malnourished people were in general more likely to die from disease. Second, poor harvests forced peasants off the land in search of food in towns, and crowding contributed to the more rapid spread of disease. Famine in Cumberland and Westmorland in northwestern England in 1597 was followed by plague in 1598 and 1599.[46]

The overall effects of ill-health and poor nutrition can be seen in a decline in average height. The study of skeletons from northern Europe indicates a striking loss in average stature of some 2.5 inches between the High Middle Ages and 1700. The height of any individual depends on many factors, including genetics and nutrition, but shifts in average height across population serve as a proxy or measure of general health. Along with disease and climate change, other factors, such as the growth of towns, could also have affected average height.[47]

The crisis was also demographic. Harvest failure, disease, and war heightened insecurity and disrupted harvests. In Europe, population fell in most regions except England and the Dutch Republic. Population dropped in Central Europe, and the population of Castile in Spain fell sharply during the middle of the seventeenth century.

Human responses to conflict and climate change influenced the likelihood of famine. In the face of a broad common challenge, outcomes varied depending in part on comparative wealth, administrative effectiveness, and transportation. France fared worse than England from similar climate conditions, partly because of difficulties transporting supplies to the interior of France and partly because poor relief in England may have been more effective at directing food to the needy. War also aggravated the effects of dearth. In France, troops requisitioned grain during the Nine Years' War of 1688–97 at a time of famine. Military purchases of grain exacerbated the crisis by raising the price. Government responses also influenced the level of catastrophe within Britain. In Scotland, the Privy Council, the council that advised the king, struggled to provide adequate relief.[48]

Famine in Scotland helped shape the future history of Ireland because the harvest failures and food shortages boosted Scottish emigration to Northern Ireland. As many as 50,000 traveled west across the Irish Sea to Ulster and the numbers peaked in the 1690s: one pamphlet from 1697 indicated that 20,000 had left for Ireland since 1695.[49] In particularly hard-hit areas of Scotland, as much as 15 percent or more of the population left for Ireland. Some may have re-immigrated, but the surge of immigration more firmly anchored the Scottish, or as they would become known, the Scotch Irish, in Northern Ireland. The famine, therefore, encouraged one of the major pulses of migration that would for centuries thereafter make Northern Ireland a Protestant majority region within an island with an overall Catholic majority.

The damage inflicted by famine in Scotland and the country's struggles to recover immediately preceded and gave additional incentive for the creation of the United Kingdom. England and Scotland were separate countries that shared a common monarch when King James VI of Scotland became King James I of England in 1603 after the death of Queen Elizabeth I in 1603. In 1707, the countries united. There were multiple causes for the union of England and Scotland. Along with an English desire for greater control over Scotland and for more manpower at a time of frequent war, the aftereffects of famine also gave the Scots an additional reason to agree to union.

Amid the seventeenth-century crisis, some regions and countries actually prospered. The Netherlands reached its height of power and influence during this period.[50] In the 1560s Dutch rebels against Spanish power were able to take advantage of colder weather to defend their nascent republic. The Little Ice Age by itself did not determine the eventual success of Dutch efforts to separate from Spain, but a cold phase of the Little Ice Age both contributed to the breakout of Dutch rebellion against Spain and, on balance, hampered Spanish military campaigns to suppress rebellion. The Dutch Republic showed resiliency to the same climate fluctuations that damaged many other states. The Dutch faced challenges from the cold phase of the Little Ice Age, but in many instances through chance or by their own responses found advantages. Prevailing easterlies aided Dutch fleets in their naval battles against the English. Dutch commerce thrived during the Little Ice Age. Storms and thick ice threatened vessels, but Dutch merchants earned high profits by selling grain stockpiled in warehouses.[51] The history of the Netherlands stands out for this period because of achievements in commerce and culture, but the Netherlands was not unique in showing resiliency to cooling: in Finland, farmers shifted farming methods and planted more rye, thereby protecting yields.[52]

The Netherlands grew as an economic intellectual and cultural center during the Little Ice Age. A more sweeping interpretation of the period in Europe sees roots of modern Western divergence from other regions in responses to cooling. In this scenario, food insecurity and disruption of grain harvests increased the likelihood of conflict, but also sped economic shifts toward more open commerce, trade and colonization.[53] The timing of the Little Ice Age and European expansion is certainly intriguing, but a change as profound and startling as the comparative rise of Western power had multiple possible causes.

Applied originally to early modern Europe, the idea of a seventeenth-century crisis can be extended to cover a broader array of societies across much of the world in the Northern Hemisphere. Revolts and rebellions struck not just European states but also China and Japan as well as India. Persistent warfare extended beyond Europe to China. Heightened mortality was also widespread, with population losses in China.

Any individual rebellion, civil war, uprising, famine, or other political or social upheaval stemmed from many causes, but the idea of a general crisis that affected numerous far-flung regions points to a deeper problem, and climate change helps explain that pattern. The fluctuations and cooling of the seventeenth century created an added burden for many societies that intersected with other conflicts and tensions. In southeastern Europe and western Asia, the Ottoman Empire faced major disruption after a long period of expansion. The end of the MCA by itself did not undermine the Ottoman realm. On the whole, the Ottoman Empire thrived during the early modern era. The Ottoman Turks won notable victories in the Balkan Peninsula in the fourteenth century, taking the city of Adrianople/Edirne and Bulgaria. Ottoman forces fought Serbs at the Battle of Kosovo in 1389. The Ottomans controlled far more territory than the embattled remnant of the Byzantine Empire, and in 1453, Sultan Mehmed II besieged and captured Constantinople. After this victory, the Ottomans continued to build their empire, though they never managed to capture Vienna, the stronghold of the Habsburg dynasty in Austria. The expanding empire experienced strong population growth. This was not simply a product of conquest. Between the late fifteenth and the late sixteenth centuries, the Empire's overall population grew. The Ottoman authorities developed an effective system for provisioning the Ottoman lands.

The Ottoman realm had proven to be resilient, but fluctuations in climate and cold spells during the late sixteenth and the seventeenth centuries posed a major challenge to the Ottoman state. After experiencing strong population growth, the empire suffered from a deteriorating climate. The LIA in the Mediterranean region gave rise to more frequent and more severe droughts.[54] The period from the late sixteenth through the first half of the seventeenth centuries marked the driest era in the last 500 years.[55] Drought destroyed crops, and the cold winters killed off much of the livestock. Much as in Central Europe, wine production fell in the late sixteenth century in Ottoman Bosnia, where farmers shifted to growing plums for brandy.[56]

Multiple problems, including the dry and cold climate, interacted with each other to create crisis in the Ottoman Empire during the late sixteenth century. Droughts and harsh winters in the 1590s led to famine. Meanwhile, the empire's efforts to collect revenue to fight wars with the Habsburgs made the shortages even worse. Soldiers in the Ottoman army in Europe endured misery, and conditions also deteriorated in the Ottoman lands in western Asia. Ambassador Vernier described the dire conditions in Constantinople in February 1595. Food was short—a "scarcity born of the evils of the weather."[57] Cold weather took a heavy toll on livestock, and depleted animals fell in large numbers to disease. People began to leave the countryside, from areas including western Anatolia, in search of food.[58] Crowding in towns, in turn, helped spread epidemics.

The crisis contributed to revolts, including the Celali Rebellions. Named for the leader of an earlier uprising, rebellions broke out in the late sixteenth century. In 1596, the requisition of sheep helped prompt rebellion. Banditry increased, and by 1598 the entire district of Larende in south-central Anatolia had fallen into the hands of bandits or rebels, including madrasa students, or *softas*. The Celali rebels defeated Ottoman forces and plundered much of Anatolia until their final defeat in 1609.

The empire survived only after suffering population loss and uprisings. Population fell in many areas of Anatolia by the 1640s, and a renewed period marked by cold weather and drought in the late seventeenth and the early eighteenth centuries slowed a rebound. It would not be until 1850 that the Ottoman Empire reached the population levels from around 1590.[59]

Once again, the disruption in climate intensified hardships caused by military campaigns. Ottoman forces advancing on Vienna in 1683 experienced cold and rainy conditions before suffering defeat. Swelling streams and rivers swept away bridges, mud made roads almost impassable, and the carts in the Ottoman army's wagon train frequently broke down. The Ottoman cavalry also had to wait until grass was available for foraging, so the entire campaign slowed. The Ottoman force finally began its siege of Vienna in July, but a relief force broke the siege in September. This would be the last time that the Ottoman Empire tried to take Vienna.

In India, the Mughal dynasty gained and consolidated power in the sixteenth century and ruled over a large state until collapsing in the early eighteenth century. During the seventeenth century, Mughal India suffered from several severe droughts. The Mughal sultan Shah Jahan responded to the drought of 1630–2 by providing relief with both food and money.[60] Shah Jahan's successor, Aurangzeb, engaged in long, exhausting campaigns in southern India, and the Mughal state fragmented soon after his death in 1707. Indeed, the cost of seeking to expand and the difficulty in holding together an extremely diverse empire likely contributed more than climate to the erosion of Mughal power.

Seventeenth-century crisis in China and Japan

The crisis of the seventeenth century affected empires in both West and East Asia. In China, the fall of the Ming dynasty was one of the pivotal events of the seventeenth-century crisis. After overthrowing and replacing the Yuan dynasty of the Mongols, the Ming dynasty experienced a long period of economic and demographic growth. Between 1393, the date of a census, and 1600, China's population increased from some 60 to 85 million people to somewhere between 150 and 200 million. Trade within the Empire grew, spurred by the renovation and restoration in the early fifteenth century of the Grand Canal that linked the Yangtze and Yellow Rivers. With this project the Yongle emperor aimed to improve the supply of grain to Beijing. Agricultural output also increased during the Ming dynasty, and large numbers of settlers moved into southern China. Externally, the Ming dynasty saw the greatest period of overseas activity in Chinese history, culminating in the voyages of Admiral Zheng He, who traveled throughout the Indian Ocean and Southeast Asia between 1405 and 1433.

Despite expansion and population growth, the Ming dynasty collapsed during the crisis of the seventeenth century. Standard historical narratives ascribe Ming decline to a combination of external attacks and internal conflicts. In the late sixteenth century the Ming dynasty campaigned against the Hmong in the south and dispatched soldiers to Korea to assist in fighting the Japanese invasion by Hideyoshi just before the Ming dynasty faced a new powerful threat from the Manchu. The Manchus, or Jurchens, lived to the northeast of China. Much like the Mongols, they prized horse riding and archery, though many Manchus practiced agriculture. In the early seventeenth century, the Jurchens, or Manchus, gained greater military power under a Jurchen chieftain named Nurhaci. He unified the Manchu tribes by creating a banner system of military units and campaigned south of the Great Wall. As the Manchu pushed down from the north, rebellions hollowed out Ming power within China. One of the rebels, a soldier named Li Zicheng, campaigned in the early 1640s in areas of China suffering from drought and famine. His forces moved into regions where a large part of the population had died. He struck toward Beijing, where the last Ming emperor committed suicide in 1644. As varied rebels fought for power, Manchus gained the support of a Ming commander, moved south, and claimed the Mandate of Heaven as the rulers of the new Qing dynasty, which survived until 1912 as the last dynasty in Chinese history. Qing forces fought rebellions by Ming holdouts, winning final victory in 1681.

Beset by external and internal foes, the Ming dynasty also suffered from climate fluctuations. By the late sixteenth century, aridity and the expansion of deserts undermined a system of military farms vital for supplying military forces along the northern frontiers. Cooling and aridity reduced the overall food production in the late Ming era, and droughts in the seventeenth century contributed to rebellions.[61] Drought between 1614 and 1619 was so severe that the *History of the Ming*, an official history from the eighteenth century, described the land as burned. During the drought of 1640, desperate peasants from Shandong province were reduced to eating bark. People even ate corpses. In 1641, the Grand Canal dried up in Shandong. Cold also struck much of China, including provinces in the south. Describing the misery in Henan, an official wrote, “The people all have yellow jays and swollen cheeks; their eyes are the color of pigs’ gall.”[62] Desperate people flocked to cities in search of food, and there were reports of cannibalism. An account from Shanghai described the disaster:

> Massive Drought
>
> Locusts.
>
> The price of Millet soared.
>
> The corpses of the starved lay in the streets.
>
> Grain reached three-tenths to four-tenths of an ounce of silver per peck.[63]

This drought in the final years of the Ming dynasty may, according to tree ring data, have been the most severe in five centuries in eastern China and possibly since 500 CE. The frequency of drought in northern China was up to 76 percent higher during the late Ming compared to its earlier years.[64] Yields of key staple crops fell and prices rose.

Depleting China's resources, drought, and cooling also increased the pressure on the Ming dynasty along its frontiers. The late Ming era saw severe drought and cold conditions in both the Mongolian and Manchurian steppes. Thus, the Manchus attacked China even as drought damaged their pastoralist economy.[65] The Manchus also developed agriculture resources in the comparatively moderate climate of the Liaodung Peninsula in northeastern China, just northwest of Korea. Finally, the Manchus, who were more experienced campaigning in cold, gained a further military edge over the Ming during this cold and dry phase of the LIA.[66]

If drought and cooling by themselves did not determine Ming collapse, they imperiled the ruling dynasty in several ways. The climate fluctuations weakened military garrisons. The cool weather and droughts undermined the system whereby military garrisons along the northern frontier grew their own provisions. Ming authorities in the late sixteenth and early seventeenth centuries responded by increasing spending on the northern garrisons, which in turn depleted the empire's overall finances. Aridity undermined agricultural production, leading to famine. The outflow of desperate peasants from farmland in turn provided a pool of recruits for the numerous rebel movements.[67]

As in Europe, the Little Ice Age in East Asia contributed to crisis, but in some instances also showed the capacity of societies to respond effectively. Japan during the era of the Tokugawa shoguns displayed resiliency to the Little Ice Age. The shoguns or hereditary military commanders held political power in Japan despite the presence of emperors. Resiliency for Japan took different form than in the Netherlands. Just as the Netherlands engaged in extensive European and global trade, Japan under the shoguns drastically reduced contacts with the outside world. Dutch traders in Japan had to stay on Deshima/Dejima, a small artificial island, in Nagasaki Bay. Japan was not immune to the shocks of the Little Ice Age: a harsh winter in 1641–2 caused famine and contributed to uprisings. The Shogun Iemitsu responded with a raft of measures, including providing food, opening up government stores of rice, and reducing taxes, and the daimyo, or great lords, followed the Shogun's instructions and example. Tokugawa Japan was also underpopulated going into the seventeenth century because of the effects of civil wars in the sixteenth century, and the shogunate's policy of isolation from foreign affairs largely avoided the costs of war, by far the most expensive undertaking for any early modern state.[68]

North American colonization in the Little Ice Age

In North America, European settlers faced the challenges of an unfamiliar and often severe climate. After the conquest of Mexico, Spanish adventurers and in some cases missionaries ventured north but found the conditions extreme, as did competitors from other European countries.[69] They complained not only of cold and snow but also of drought. The cold of the LIA also affected native societies. Indeed, one interpretation of Iroquois migration west suggests that cold weather pushed the Iroquois to find new homes further south.

In eastern North America, English settlers experienced unaccustomed extremes.[70] Indeed, English writers voiced fears of the possible negative effects of an unfamiliar climate on the identity and character of English settlers. They soon discovered that even southern North America was colder than they expected. Further north, the length and severity of winter cold surprised French explorers such as Samuel Champlain, and English colonists had similar experiences in Newfoundland. However, colonization continued, and promoters and advocates of European settlement shifted to stress the capacity to adapt to North America.

In some cases colonization failed altogether. Before Plymouth and Jamestown, an English expedition established a colony at Roanoke in what is now North Carolina in 1585, and a new group of settlers arrived in 1587. War prevented the resupply of the colony, and when an English ship finally returned in 1590 all the colonists had vanished. The exact cause for the demise of Roanoke was never determined: the colonists may have fallen ill or been taken in war with local Indians, though no direct sign of violence was visible. The last group of colonists arrived during a severe drought in 1587–88, which may have contributed to the failure of the colony.[71] In 1607, an attempt to establish a colony at Popham Maine, the Sagadahoc colony, failed.

Even the settlements that survived could easily have failed. The English effort to establish a colony at Jamestown almost foundered from the start. The settlers suffered from disease and famine. They arrived right at the start of one of the longest droughts and coldest periods in centuries. Drought destroyed crops and severe cold burdened the settlers. After a severe winter in which the James River almost froze across, the colonists faced food shortages, and drought persisted into 1609. The next winter, the settlers dug for roots, and, according to one account, ate corpses or even engaged in cannibalism.[72]

Multiple factors contributed to the failings of any single English, French, or Spanish settlement or to the setbacks experienced by any expedition, but cumulatively, the tenuous early efforts to place a European footprint on North America reflected the difficulty of these undertakings during a peak phase of the Little Ice Age. The same extremes also posed challenges for native societies encountered by European and heightened competition for scarce resources.[73]

In New England, the Pilgrims encountered severe cold. They were unprepared for the first winter. As the journal of a plantation recounted from December 1620, "considering the weakness of our people, many of them growing ill with colds, for our former discoveries in frost and storms, and the wading at Cape Cod had brought much weakness amongst us, which increased so every day more and more, and after was the cause of many of their deaths."[74] The harsh winter was not the Pilgrims' only source of woe: they had arrived too late to plant crops. Only about half of those who had landed at Provincetown survived the first winter. Under any circumstances, the Pilgrims would have faced major challenges, but the climate of the LIA intensified the harsh and long winters.

Long after they had firmly established their presence in the New World, cold spells during the LIA created hardships for English settlers. After decades of life in New England, settlers in the late seventeenth century sensed that their climate had grown colder. In a 1699 almanac it was remarked, "The seasons not as they used to be; the Summers turned into Winters; and the Winters embittered with Hardships, which in the memory of many have not been known."[75]

The settlers had already gained firm military control over southern New England, during King Philip's War of 1675–6, but heavy snows increased Indian military power in more northern areas. In the 1690s, Abenaki bands took advantage of heavy snows to raid English settlements. On winter hunts, the Abenaki traveled across the deep snowpack in the interior by snowshoe. They killed moose, ate the meat, and used virtually the entire animal. During exceptionally cold winters of the 1690s and early 1700s, they extended these winter hunts.[76]

Long-distance raids caused consternation and fear among English settlers. A raiding party of French and Abenaki, for example, traveled more than 300 miles to attack Salmon Falls along the modern border between Maine and New Hampshire. Unable to move through deep snow, the English could not even attempt pursuit. As Cotton Mather, a prominent Puritan minister and author, recounted, "Through the disadvantages of their feet by the snow, they could not make no hand on it."[77] In January of 1692, Abenaki raiders attacked York, Maine, killing some 50 colonists and taking another 100 captive. They struck even farther south, attacking Haverhill on the Merrimack River in Massachusetts in March 1697 and Andover in early March 1698.

In one of the most ambitious and, from the English perspective, devastating raids, a party of French and Indians raided Deerfield in northern Massachusetts in the Connecticut River Valley in February 1704. The high snows made it possible for the attackers to climb over Deerfield's defensive wall. Inside Deerfield, they killed some 50 settlers and took some 112 captive, including the Puritan minister John Williams, his wife Eunice, and five of their children. Two other children were killed during the attack on Deerfield, and Eunice was killed during the march to Canada. John Williams, on reaching Montreal, was ransomed or redeemed, and four of his children were eventually returned, but his youngest daughter, also named Eunice, stunned the Williams family by staying with an Indian family and marrying an Indian. When John Williams later visited her, she showed no interest whatsoever in returning to her family or to English settler society.[78]

Attacks on farms and barns after poor harvests threatened the English settlers' food supply during the coldest time of year. Increase Mather, a leading minister, wrote of the lack of food and fear of famine. "The calamity of war still continues. In N. E. great fears . . . Tis such a time of scarcity as the like never was these 50 years."[79]

Confounded by the Abenaki raids, the English settlers made military gains by adopting the snowshoe. New laws required that soldiers keep snowshoes. English soldiers could now campaign deep into Indian hunting grounds. The Abenaki could no longer winter in any close proximity to English settlements.[80] Winter no longer provided the Abenaki with the same advantage.

The same climate conditions that sometimes stunned Europeans also presented challenges to native peoples, but as in Eurasia, the LIA had mixed effects in the Americas as some peoples adapted to a colder climate. In the far north in the Foxe Basin in the Canadian Arctic the Thule developed a strategy for caching or storing walrus meat, thereby preserving their food supply.[81] Severe winters, early frosts, and droughts posed threats to agriculture, but in the St. John River Valley of New Brunswick, the Maliseet took advantage of their knowledge of microclimates, flooding patterns, and soil conditions to grow maize during the Little Ice Age. They also planted maize that could be harvested before ripening, suitable for planting in a region with a short growing season.[82]

In Central America, Spanish conquest and the Columbian exchange created extraordinary shocks for native societies, but in Mexico indigenous agricultural practices helped to build resilience during much of the Little Ice Age. Only with a shift to a monoculture to cultivate agave did human practices combine with a climate extreme marked by cold and flooding interspersed with drought to accelerate degradation of the landscape.

Cultural and social effects

The experience of living through a period of fluctuating climate, cooling, and famine influenced culture and society. The cultural and social responses to the LIA have been best documented for the cases of Europe and European settler societies. Powerful cultural images and preferences make sense if we recall the different climate of the LIA when Europeans lived with more snow and ice. Speed skating, for example, is one of the top national sports in the Netherlands. At the Sochi Winter Olympics in 2014, the Dutch collected twenty-three medals in long-track speed skating. The country has numerous professional speed skating clubs and long-track speed skating rinks. This prowess at speed skating is obviously modern but draws on an older tradition, a fascination with skating dating back to the LIA.

Why would skating take root in a country that today does not have much naturally occurring ice? The myriad of works by Dutch master- and lesser-known painters give us a strong visual glimpse into life in the low countries in the early modern era. The subject matter is varied. We see rustic windmills and peasants, proud burghers (the members of the affluent urban middle class), scenes of domestic life of both men and women, still lives attesting to abundance, and tales from the Bible, among many other subjects. The artistic output also reveals scenes of ice, of people walking over and skating on the ice. Such images do not in themselves provide the key evidence for tracking the LIA in early modern Europe, because artistic choices can reflect market preferences, but they still help us imagine life in a colder climate.[83] We also know that periods of icing over have grown rare. Every year Dutch skating fans hope for a run of the Elfstedentocht, or eleven-stage race, a 120-mile speed skating race along the province of Friesland. The race was first held in 1909, but it can only occur when the ice is thick enough. In all, it has been held fifteen times, but the last race was held in 1997, and the race has taken place only three times since 1963. (It took place in both 1985 and 1986.) A cold winter in 2012 raised hopes for a running, but conditions did not hold for long enough.

Ice also provided the setting for urban fairs. In London a frost fair was held on occasion on the frozen River Thames. The Thames froze at least twenty-three times between 1309 and 1814, and at least five frost fairs were held—the last in 1814. The frost fair provided an opportunity to carry on trade at times when ice blocked navigation on the river. Along with the sale of food and drink, the frost fair gave rise to stunts, such as the marching of an elephant across the river. There were other causes for the freezing of the Thames, including the structure of Old London Bridge, which slowed the flow of water, but the frost fairs evoke the colder climate of the LIA (Figure 5.2).

Climate fluctuations did not simply create different living conditions but also had more complex cultural effects. Some saw failed harvests as

FIGURE 5.2 *Winter Landscape with Ice Skaters, Hendrick Avercamp,* c. *1608. Source: Purchased by the Rijksmuseum Amsterdam with support of the Vereniging Rembrandt, https://www.rijksmuseum.nl/nl/collectie/SK-A-1718.*

punishment for sin. All manner of unusual occurrences—the northern lights, heavy snows, and natural disasters—could be interpreted as signs of such punishment.[84] The LIA amplified the magnitude of some of these phenomena. If confounding or disastrous events could be attributed to sin, it made sense to look for sinners. Sin could be general and widespread, the cause for a heavy penalty on all, but it could also be assigned to particular groups, especially to those already held in mistrust, scorn, or fear.

It can, however, be difficult to identify a direct connection between climate shock and scapegoating. In much of Europe, Jews were the chief religious outsiders, but European anti-Semitism or hostility toward Jews predated the LIA. Enthusiasm for the First Crusade helped stir up popular attacks on Jews as the local resident non-Christians in towns along the Rhine River in 1096. The onset of the LIA came with a surge in anti-Jewish policies and acts, including the expulsion of Jews from England in 1290 and from France in 1306. French monarchs periodically expelled and recalled Jews until Charles VI expelled Jews in 1394.

In the early modern era, writers in German lands held Jews responsible for a variety of ills. Thus, Martin Luther described Jews as "a heavy burden like a plague, pestilence, and pure misfortune in our land."[85] But Jews by this time were not usually accused of having caused bad weather. Animosity built against Jews, however, as exploiters who took advantage of others' hardships. An illustration from 1629, entitled *Der Wein Jud*, or *The Wine Jew*, depicted a Jew who carries a flag marked "monopoly" and rides behind a devil. Here there is a hint of blaming Jews for a poor climate. Captions and images in the scene refer to bad weather as in the biblical passage, "I will command the clouds not to rain."[86]

The search for scapegoats for bad weather led more directly to a striking surge in accusations of witchcraft. For many Europeans, bad weather signaled sorcery and witchcraft. Jordanes de Bergamo, master of theology at Cortona in Italy, asserted, "by the power of words and signs the Strigae (witches) can produce hail and rain and things of this kind." A Papal Bull by Pope Innocent VIII on the evils of witches served as the introduction to a 1486 handbook on witches entitled *Malleus Maleficarum* (*For the Hunt of Witches*). Pope Innocent VIII stated, "It has indeed come to our ears . . . many persons of both sexes . . . have blasted the produce of the earth, the grapes of the vine, the fruits of the trees . . . vineyards, orchards, meadows, pasture-land, corn, wheat, and all other cereals." Others shared the pope's belief. King James VI of Scotland wrote in 1597 that witches raised "storms and tempests."[87]

Charges of witchcraft and punishments increased during the LIA and in particular during the cold phase of the late sixteenth and seventeenth centuries, with a major surge in the hunt for witches and trials of witches. The lower temperatures match up statistically with the punishment of witches. The pursuit of witches escalated in Central Europe in the 1560s. No less than sixty-three women were burned as witches in Wiesensteig in Baden Württemberg in Germany in 1563. Witches were also persecuted in Scotland and England in the 1560s.[88]

The attacks on witches reached a new peak between 1580 and 1620. Cold weather and crop failures produced famine. The cold wet springs and stormy weather caused particular hardship in elevated or remote locations. An epidemic of witch burning ensued. The numbers of witches burned at the stake were extraordinary: more than 1,000 in the Vaudois region of the Republic of Bern in Switzerland and more than 800 between 1580 and 1595 in the Duchy of Lorraine, with 2,700 executed by 1620. Trier burned witches too: more than 350 between 1581 and 1595. [89]

The connection between bad weather associated with the LIA and witch hunting was especially strong in the late 1620s in small German principalities. A severe late frost struck in May 1626, and contemporaries blamed witches, as an account from Franconia reported: "Everything froze, [some-thing] which had not happened as long as one could remember, causing a big rise in prices . . . As a result, pleading and begging began among the rabble, [who] questioned why the authorities continued to tolerate the witches and sorcerers destruction of the crops. Thus the prince-bishop punished these crimes."[90] There were killing sprees in other German states: in the territories of the Prince-Bishop of Bamberg 600 were burned. As these examples indicate, the wave of witch killing was strongest in small territories rather than in large states or larger towns and cities.[91]

Bad weather during the LIA also provided the setting for the most famous outbreak of witch trials in New England in Salem, Massachusetts. Fourteen women and five men were convicted and hung in 1692. Giles Corey, aged seventy-one, was pressed with stones to try to force a plea and died of the injuries. Others died in custody. Several others were tried in 1693. Analysis of these trials reveals many possible economic, psychological, and other causes. The LIA did not all of a sudden make a hunt for witches in Salem inevitable. However, the cold climate and hardship combined with war to cause hardship on the eve of the outbreak of persecution.[92]

Adaptation

Human societies affected by the LIA confronted real challenges and hardships. For some living at the end of long trade routes or on the margins of cultivation, the fluctuating climate could force retreat or lead to even worse consequences. The crisis was severe enough to topple elites in Ming China and speed migration from affected areas such as Scotland. Climate change combined with war to reduce population, at least temporarily, as well as overall health. Where population had grown, such as in China, greater climate extremes and variability placed large numbers of people at risk. At the same time, human societies also made adaptations to shelter, clothing, and their use of energy during this period. The LIA, in historical perspective, demonstrated both human vulnerability to climate fluctuations and also resilience and adaptation.

As temperatures cooled, Europeans in at least some regions made warmer clothes. In Iceland, women changed the way they produced woolen cloth. Icelandic women had woven wool throughout the Middle Ages. The textiles served not only as a main export to Europe but also as a form of currency within Iceland. As temperatures cooled during the period between the sixteenth and eighteenth centuries, Icelandic women spun plied homespun, better suited to the colder climate because adding threads produced a denser and warmer fabric. Greenlanders had employed a similar strategy as early as the fourteenth century, though historians have asked why they did not shift to wearing furs like the Inuit, whose communities survived the LIA on Greenland. In the case of Iceland, the greater use of plied warp threads indicated much the same response to climate fluctuations. An end to the use of textiles as currency combined with cooling to provide a reason to make this change.[93]

Europeans also made numerous changes in building and construction to create warmer homes. Some of these innovations were part of a broader trend toward improvement, but they also provided greater comfort in the colder climate of the era. Glass windows protected against drafts. Home furnishings, such as featherbeds, became more common in the sixteenth century.[94]

The cold weather encouraged Europeans to dress warmly. In general, even among elites, fashion shifted toward heavier fabrics. They sought out furs for coats and hats. Taste and fashion obviously influenced the choice of clothing, but so too did the search for protection from the cold (Figure 5.3). The historian Wolfgang Behringer gives insight into the importance of warm clothing for Hermann Weinsberg of the German city of Cologne. Weinsberg had a special nightdress made for himself with a filling of fox fur.[95]

The hunt for furs, in turn, threatened the species most valued for making warm clothing. Europeans had their own supplies of fur, but by the sixteenth century had depleted supplies of fur. They extinguished the beaver population in many regions, and other sources of fur, even rabbit, grew rare and expensive.[96] Beavers were only reintroduced in Scotland in a single forest in 2009–10, and beavers were spotted in England in the wild for the first time in centuries in 2014. They had become extinct in England in the sixteenth century.

The demand for fur in Europe encouraged imperial expansion and trade for fur from more distant sources. Russian expansion east across Siberia in the sixteenth and the seventeenth centuries opened up a vast trade in sable. Siberia would yield other resources, gold and silver, but fur provided the most immediate lucrative source of income. The conquest and colonization and the fur trade went hand in hand. Petr Beketov, a Cossack officer, reported

FIGURE 5.3 *A Lady Writing*, c. 1665 *(oil on canvas). Bridgeman Images.*

in 1633 on his expedition up the vast Lena River in Eastern Siberia. After more than two years, Beketov and his men "brought under your Sovereign Tsarist mighty hand on the Lena River many Tungus and Iakut lands."[97] They collected tribute in furs and pelts and built a fort at Yakutsk, which developed into the chief port on the Lena River. The sable trade grew rapidly. The number of pelts sent west to European Russia reached 256,837 by 1698 and 489,900 by 1699.[98] The trade remained lucrative for generations. The naturalist Peter Simon Pallas described in 1779 how "sables" "grow more common the more eastward you go; & at the same time their furr is more valuable, the more to the north & east or in the highest mountains they are bred."[99] Fashion tastes reinforced by the LIA drove the market supplied in part by Russian expansion to the east.

Demand for fashionable protective clothing similarly provided an incentive for commercial expansion to the west in North America. Traders sought varied furs and pelts: fox, marten, mink, bear, even raccoon, and muskrat, but above all beaver pelts.[100] Fashion and warmth combined to create the demand for beaver. Beaver hats were ubiquitous, and a beaver lining made a warm coat for the burghers and elites of Europe.

French, Dutch, and English traders traveled vast distances in pursuit of beaver pelts. English traders traveled along the Connecticut and Delaware Rivers, but soon exploited all readily available beaver in the Connecticut River Valley. From Canada, English traders eventually also went much farther north into Hudson Bay, where England and France competed for power. Soon after arriving at Quebec in the early seventeenth century, the French began to buy furs and pelts from the Huron, and the trade soon brought tens of thousands of pelts a year from the interior to French outposts. From Quebec City, the center of the French fur trade shifted west to Montreal. French traders traveled along the St. Lawrence and through the Great Lakes into the heart of North America.

The slaughter of beavers, in turn, affected the environment. The killing of hundreds of thousands of beavers led to many wet areas drying up. In all up to 50 million animals were killed. In principle, the near extinction of beaver in many areas might have affected the flux of methane and CO_2 from ponds enough to decrease CO_2 concentrations in the atmosphere.[101]

The fur trade also reshaped Indian societies. Competition for pelts encouraged conflict and war. Contact with Europeans increased over time, providing access to all manner of wares, including weapons and alcohol. These contacts also opened up paths for the exchange of disease and for the spread of devastating epidemics into the interior of North America.

The LIA did not destroy major centers of civilization, but across the Northern Hemisphere economic development and cultural change during the LIA hastened improvements to home heating. Europeans of the early modern era adopted the fireplace built into a sidewall. This provided many benefits over the previous practice from the Middle Ages of letting smoke from a fire escape through a hole in the roof. In a further improvement for efficient heating, Europeans began to use closed stoves. Possession of a heated and clean room also brought social and cultural changes, and not just for elites. In Germany this new room, called a *Stübe*, became the center of indoor peasant life during the cold times of the year.[102]

We rightly associate the Industrial Revolution with coal, but well before the Industrial Revolution, heating brought a shift away from burning wood in several regions. Fuel consumption and the desire to stay warm fed demand for firewood. Firewood prices rose in Europe during the sixteenth century, and theft of wood increased.[103] Fuel prices for both wood and coal continued to rise throughout Europe in the eighteenth century.[104] Shortages of timber and cold weather provided incentive to continue to improve heating. In northern Europe, tiled stoves installed in the sixteenth and seventeenth centuries used wood more efficiently and retained heat longer than fireplaces.[105] In the houses of the wealthy, designing wings of rooms made it possible to close off areas of a building to reduce heating costs and fuel consumption.

In England, intensifying demand for wood for heating, building, and for nascent industry depleted the supply of timber. In 1783 a count of five woods in England showed a drop from 232,011 "timber trees" in 1608 to 51,500.[106] Trying to compensate for the shortages, England imported wood from North America, but prices continued to rise.

In North America, colonists cut down trees in large numbers to build, to clear land, and to warm themselves. As a visitor to a Virginia estate at Christmas 1686 observed, "It was very cold, yet no one ever thinks of going near the fire, for they never put less than a cartload of wood in the fireplace & the whole house is kept warm."[107] In 1770, a farmer named Landon Carter in Virginia's North Neck wondered where wood would come from in the future: "I must wonder what succeeding years will do for firewood. We now have full ¾ of the year in which we are obliged to keep constant fires; we must fence our ground in with rails, build and repair our houses . . . and every cooking room must have its fire the year through."[108]

Demand for wood and rising prices spurred a shift to greater exploitation of other fuel sources, mainly coal. Coal had long been employed on a small scale in England and had been mined on a larger scale in China for centuries. Coal was already a major fuel source in parts of northern China during the Song dynasty. In the capital, Kaifeng, coal was the main source of fuel.[109]

However, the rate of expansion of coal extraction was unprecedented as Britain shifted to coal as its chief fuel source even before the start of the Industrial Revolution. Thus, total coal production in Britain rose from 0.2 million metric tons in 1550 to 9 million metric tons by 1800.[110] As of 1550, some 35,000 metric tons of coal were shipped from Newcastle to London, but the total rose to 560,000 metric tons by 1700.

A colder climate also supported enterprises such as the ice industry that today would not only be costly but also be impossible in many regions where they once flourished. In a world without refrigeration, the ice trade from the nineteenth into the early twentieth centuries made possible a first wave of long-distance transport of perishable goods and food. A Massachusetts businessman named Frederic Tudor cut ice from ponds and packed it in sawdust to slow melting. Tudor's business lowered costs when Nathaniel Wyeth, who became Tudor's foreman, created a horse-drawn plow to cut ice.[111] Remarkably, given the climate of the early twenty-first century, ponds from eastern Massachusetts supplied much of the ice for Tudor's enterprise. Harvested natural ice allowed both for shipping ice and for transporting ice and perishable goods inland.

Abrupt climate change in the Little Ice Age

Several periods of abrupt climate change occurred within the long time span of the LIA. Volcanic eruptions led to sharp episodes of cooling. Individual eruptions did not have the same consequences as the sequence of volcanic eruptions that may have helped initiate the LIA in the first place, but might explain fluctuations within the LIA.[112]

The individual eruptions during the latter stages of the LIA led to marked cooling for discrete periods. One such episode of abrupt change stemmed from a volcanic eruption in Iceland. In June 1783, lava started to erupt from the Laki fissure to the southwest of the Vatnajökull Glacier in southern Iceland. The eruption continued until February 1784, releasing 120 Tg of sulfur dioxide into the atmosphere. Jon Steingrimsson, a pastor in a nearby village, provided a vivid account of the Laki eruption. "It began," he wrote, "with the Earth heaving upwards, with a great screaming of noise and winds from its depths, then splitting asunder, ripping and tearing as if a crazed animal were tearing something apart." The stench was overpowering: "The foul smell of the air, bitter as seaweed and reeking of rot for days on end, was such that many people, especially those with chest ailments, could no more than half-fill their lungs of this air, particularly if the sun was no longer in the sky; indeed, it was most astonishing that anyone should live another week."[113] As the lava approached the village, Steingrimsson remained resolute in his church: the lava stopped flowing just before reaching the church. He became renowned for what was called the "Fire Sermon."

In Iceland, the Laki eruption killed up to 20 percent of the population. Many people and animals died from fluorine poisoning. Three-quarters of the sheep and half of the horses on the island died. The losses were so devastating that the Kingdom of Denmark, which had gained rule over Iceland, considered relocating the island's entire population. The effects of Laki were visible thousands of miles away. Benjamin Franklin wrote of "a constant fog over all Europe, and a great part of North America." A series of cold winters followed with effects as far away as Brazil.[114]

In the early nineteenth century, the eruption of Tambora produced another period of cooling; other volcanic eruptions in the early nineteenth century may have contributed to cooling effects.[115] The Tambora volcano is located on the island of Sumbawa in Indonesia to the east of Bali. It exploded on April 10, 1815. The eruption continued until the mountain collapsed, losing some 1,400 meters of elevation. Thousands died immediately. Those who survived the immediate blast then faced tsunamis, walls of water 12 feet high, pushed by the tremors of the eruption. The eruption and the tsunamis destroyed crops, and tens of thousands of people died afterward from famine (Figure 5.4).

The early nineteenth-century eruptions, including Tambora, affected climate far from the eruption site. In South Asia, sulfate gases slowed arrival of the monsoon, causing drought and famine. In North America, the aftermath of Tambora led to summer frosts in New England, where people described a year without summer. The year without summer stood out for the low temperatures of cold snaps. After a deep frost in May, frost struck again in

FIGURE 5.4 *Contemporary illustration of the eruption of Tambora volcano in the Moluccas islands, 1815 (engraving). © SZ Photo / Bridgeman Images.*

June and as much as a foot of snow fell. The summer also brought drought. There was a cold spell in early July and a frost in the middle of August. The miserable summer ended with yet another frost in late September. Vermont farmers turned to eating nettles. The failed harvests accelerated migration to the Midwest. More than 40,000 settlers left the east for Indiana.[116]

In Europe, crops failed as well. Stormy weather afflicted Ireland in the summer of 1816.[117] Repeated rain ruined waterlogged crops. "There never was such distress and want of money known in any former times," wrote Daniel O'Connell, a prominent Irish nationalist political leader.[118] Heavy rains fell as well in England, France, and German lands. A German farmer in Lower Franconia reported rain for five solid weeks. As daily rains continued in Switzerland, rivers broke their banks. Lake Geneva expanded until residents of some parts of the city of Geneva had to use boats as their only means of transport.[119] Contemporaries noted frequent rain in Czechoslovakia, but the effects may have been less severe than in other regions of Europe.[120]

High prices for grain led people to try to gather any possible foodstuff, and many turned to ersatz food or to roots. Economic distress added to the number of beggars. Cases of theft increased. The hunger led to poaching. With famine conditions, some in Austria turned to a sect, the Poschlianer, named after a Catholic priest.[121]

The crop failures, rising prices, and hard times increased political discontent. Britain had only just triumphed over Napoleon. Britain along with its allies defeated the French emperor at Waterloo in 1815, but the years after this victory brought political unrest. The cause was not simply the climate shock, but the effects of Tambora intersected with ideological reasons for resentments. Radicals in England demanded reform of Parliament. In 1819, British cavalry charged on a crowd of 60,000 who had gathered at St. Peter's Field near Manchester to hear the Radical speaker Henry Hunt. The troops killed eleven and injured many more in what was soon dubbed "Peterloo" in angry reference to the great victory a few years before at Waterloo. Elsewhere, riots and protests proliferated in France in 1816 and 1817. Crowds blocked the transport of food on roads and forced immediate sale. In the Kingdom of Bavaria, authorities resorted to the military to quell unrest.[122]

The poor harvest caused by the climate shock of Tambora boosted emigration to the Americas. In Ireland, crop failures led to a surge of emigration to the United States.[123] Displaced peasants begged for food. The winter of 1816–17 was so severe that the weather impeded grain imports to regions such as the Rhineland and Switzerland, already suffering from the aftereffects of the previous cold rainy summer. People sought to leave: for the Americas or even for Russia.[124] Five million people left German lands, many from Southwestern Germany, for North America in the nineteenth century. Climate change counted as one of several causes for their departure, but the effect was most noticeable in the years immediately following Tambora. In 1846, a hot and dry summer contributed to a wave of migration.[125]

The stormy weather made an impression on travelers, including some of England's leading romantics. Lord Byron, Percy Bysshe Shelley, and Mary Wollstonecraft Shelley visited Switzerland in the summer of 1816, and the gloomy weather may have influenced Wollstonecraft's classic *Frankenstein*. The influence was even more direct in Byron's poem "Darkness":

> The bright sun was extinguish'd, and the stars
> Did wander darkling in the eternal space,
> Rayless, and pathless, and the icy earth
> Swung blind and blackening in the moonless air;
> Morn came and went—and came, and brought no day.[126]

In China, Tambora may have weakened the Qing dynasty. This dynasty founded by the Manchu had generally prospered during the eighteenth century. The empire experienced strong population growth. In its external relations, the empire stabilized its frontier with seminomadic peoples to the west and north. The Qing dynasty gained a decisive military edge over Mongols. The Qianlong emperor in the mid-eighteenth century responded to the persistent threat posed by the Zunghar, or western, Mongols, by ordering his commanders to destroy the western Mongols. The emperor ordered massacre: "Show no mercy at all to these rebels. Only the old and weak shall be saved."[127]

China in the late eighteenth century was a powerful and confident state, as the British discovered when they tried to expand trade. A diplomatic mission of 1793 led by Lord George Macartney sought to persuade China to engage in more trade. The mission brought a selection of wares intended to impress and entice the Chinese: clocks, watches, telescopes, Wedgewood pots, and paintings. However, the Qing dynasty declined to make any changes. Indeed, the Qianlong emperor in a letter to George III stated that Britain had nothing China needed: "we possess all things. I set no value on objects strange or ingenious, and have no use for your country's manufactures."[128]

By the late eighteenth and early nineteenth centuries, however, the Qing dynasty began to encounter major troubles. China suffered from an economic slump and ethnic tensions, and abrupt climate change added to the Qing dynasty's difficulties. In particular, Tambora caused famine in Yunan in southwestern China.[129] The crop failure may have encouraged desperate farmers to turn away from staples, such as rice, to poppy cultivation for the opium trade. They thereby found a reliable and lucrative cash crop.[130]

On the whole, the effects of climate change were less severe in the late Qing dynasty than in the earlier late Ming era. Northern China's climate cooled between the 1780s and the 1830s, but the drop was not as pronounced as during the late Ming period. Qing administration also proved more capable than the Ming government in organizing relief, and migration within China was greater in the Qing than in the Ming dynasty.[131]

In southern Africa the massive Tambora eruption accentuated a period of social and political instability. Historians used the name Mfecane, or Difaqane, to describe a period of war and migration in the early nineteenth century, in which bands on the move displaced many peoples, and the Zulu state consolidated and expanded power. Precipitation patterns help explain this sequence of events. Thus higher rainfall in the late eighteenth century combined with the cultivation of maize to first boost population just before southern African populations confronted droughts at the end of the eighteenth century and in the early nineteenth century.[132] Historians debate the social and political interaction with climate. The term "Mfecane" (crushing) attributed migration and war to the rise of Zulu power under Shaka, but these trends did not stem solely from the Zulus.

The eruption of Tambora combined with the effects of previous volcanic eruption to exacerbate drought in southern Africa. Models of the effects of Tambora show pronounced cooling and drought in southern Africa, damaging both the cultivation of maize and the raising of livestock. Analysis of tree ring records from Zimbabwe confirms a dry period in the early nineteenth century.[133]

Summary

The LIA challenged many states and civilizations while in the end revealing the growing resilience of the most advanced complex societies. Despite debates over the exact timing, there is strong evidence of a cooling, though the precise timing of the most pronounced cool phases varied in different regions. In the North Atlantic and Europe, the earliest period of the LIA was most severe for those living at high latitudes or at a higher elevation. In East and Southeast Asia changes in precipitation associated with the shift of the ITCZ contributed to crisis in regions such as the Khmer Empire.

The strong cooling phase beginning in the late sixteenth century that extended into the seventeenth century is one of the best documented phases of the LIA. Cooling contributed to what historians have described as a general seventeenth-century crisis. Many other factors, including dynastic and religious conflict, contributed to such a crisis, but cooling exacerbated famine, and in China, aridity added to the burdens facing the late Ming dynasty.

The long period of the LIA, at the same time, also saw increasing decoupling between climate and some of the most advanced societies. The Netherlands, for example, prospered during the seventeenth-century crisis, and early modern European states varied greatly in their capacity to respond to the threat of famine.

CHAPTER SIX

Humans take over

- An energy revolution
- Industrial Revolution
- Carbon and climate
- Nineteenth-century droughts
- World Wars and Fordism
- Globalization
- Breaking constraints
- Toward dependency

The Industrial Revolution remade human societies and transformed the relationship between humans and Earth's climate. Previous human societies extracted and exploited sources of energy, modified landscapes, and changed the local environment. In some cases, human societies possibly altered the composition of the Earth's atmosphere sufficiently to begin to influence climate,[1] though the question of whether preindustrial societies significantly affected Earth's climate is a subject of debate. The Industrial Revolution built on these trends but intensified the extraction of resources and accelerated the remaking of landscapes and local and regional environments to an unprecedented degree. Industrialization initiated a series of increasingly vast waves of economic change that started with the Industrial Revolution itself, expanded with the rise of a new form of production and consumerism in the twentieth century, and spread still further with the globalization of industry in the late twentieth and early twenty-first centuries. These successive waves of economic growth all drew primarily on energy produced from fossil fuels and led to marked shifts in Earth's atmosphere. Human activity became the main agent of climate forcing.

Industrialization paradoxically increased both human resilience and vulnerability to climate change. New sources of power and technology advanced a long-term trend toward growing human independence from climate fluctuations, but at the same time the Industrial Revolution and the global expansion of modern production and consumption heightened risks from climate change.

Over several thousand years, human societies took advantage of the comparatively stable and warm climate of the Holocene. People increasingly engaged in intensive agriculture on a scale that would have been unimaginable to hunter-gatherers of the past. The shift to farming carried costs, such as the decline in general health and the rise of epidemic disease, but massive cultivation also sustained dramatic increases in population. Elites and states that gained the ability to extract a surplus from large-scale farming created an imposing physical infrastructure. The Roman aqueducts, the great canal of China, the temple complexes of the Maya, the shrines at Angkor—all these and more depended on the creation and collection of a surplus on a new scale. Holocene societies also developed extensive trading networks, but the bulk of the population typically focused on cultivation.

Within the generally stable climate of the Holocene, human societies still experienced climate fluctuations. Small shifts to generally more favorable conditions for cultivation and communication aided in the rise of population in the Roman Empire, the Han dynasty, and the Mayan classic era, among other cases. In Europe, the Medieval Climate Anomaly contributed to the expansion of borders of cultivation within Europe and made it easier to colonize the North Atlantic. Favorable climate did not determine a particular outcome for any single civilization, whether Rome, the Han dynasty, the Mayan city-states, or many other complex societies, but comparative warmth and reliable precipitation created conditions that assisted growth and expansion. In contrast, fluctuations that produced colder and/or drier conditions posed challenges to Holocene societies. The kind of shifts experienced during this era did not inevitably lead to decline and collapse—thus many societies adapted to the fluctuations of the Little Ice Age. However, climate shifts of this order could place sufficient strain on a given society to contribute to full or partial collapse as in the case of Chaco Canyon, abandoned entirely by its population, or the Maya, who endured but only after leaving some of the largest cities and sites of the classic era.

During the Holocene, complex societies generally became more resilient in the face of ordinary climate fluctuations. The beginnings of this decoupling can be traced back to the Bronze Age, and the further development during the Iron Age of transportation networks, storage facilities, and states with the capacity to monitor and respond to grain shortages all provided greater resiliency. The Qing dynasty before its decline could stockpile grain and reduce taxes in areas affected by drought and famine. Climate fluctuations caused hardship in affected regions, particularly for those already living close to the margins, but did not normally imperil complex societies and civilizations.

Resiliency to climate fluctuations typical of the Holocene continued to increase during the early modern era as agricultural improvements raised crop yields, and rising trade enabled commercial centers to diversify their food supply. Trade networks that brought together crops produced during different times of the year and in different places produced greater food security. Parts of Europe escaped a pattern of subsistence crisis even before the end of the LIA. By the middle of the seventeenth century, most of England no longer suffered massive famines. In Southeast Asia complex societies also gained greater resilience in the face of droughts.[2]

Even as Holocene societies developed greater ability to adapt to and to respond to fluctuations in climate, population growth provided greater potential for catastrophe. The contrast with the prehistory of *Homo sapiens* is striking. Early humans faced all manner of deadly perils and challenges, but even though cold snaps led to the extinction of human lineages, a small population of hunter-gatherers could disperse in a way that was far more difficult for much larger populations of farmers and peasants. In the best-case scenario, such farmers responded to famine by migration, if a new region for settlement was available as in the case of the Scotch Irish. In a worst-case scenario, people died by the millions, especially when human responses accentuated the effects of climate fluctuations.

An energy revolution

The early modern era saw the start of a revolution in energy use. Late Holocene societies in general required more resources. They increased crop yields, broadened trade networks, and generated higher demands for water and energy sources. Several engaged in intensive phases of economic and technological development without moving to full-scale industrialization.

There were several preindustrial economic powerhouses, including China, the Netherlands, and England. In China, the Song dynasty vastly expanded its production of iron and coal. The residents of the capital Kaifeng turned to coal as their main source for fuel.[3] However, this burst of economic and technological development slowed. The Qing dynasty also achieved high rates of demographic and economic growth, but this combination did not continue into the late eighteenth century.[4]

The Netherlands in the late sixteenth into the seventeenth centuries also experienced a surge of demographic, economic, and technological growth. Large parts of the population began to work outside of agriculture. For energy, the population increasingly turned to peat. The excavation of peat from regions, including the future site of what is now Amsterdam's Schipol Airport, created new lakes.[5]

England was yet another society that turned to new energy sources as it used up available resources. As early as the late Middle Ages, shortages of wood for making charcoal caused a downturn in making iron and glass in southeastern England. Coal became the chief fuel source for heat, and coal mining increased in the seventeenth century. Urban growth, especially in London, spurred demand. London's population increased from some 200,000 in 1600 to between 575,000 and 600,000 a century later, and the growing number of city dwellers obtained heat mainly by burning coal. The writer and novelist Daniel Defoe, author of *Robinson Crusoe*, remarked on the "prodigious fleets of ships which come constantly in with coals for this increasing city."[6]

Coal shipments from Newcastle in the north of England increased steeply from the sixteenth century through the seventeenth century. Early mines were often part-time enterprises operated by farmers, but the scale of mining increased in the northeast of England in Northumberland and Durham and in the midlands in Staffordshire. Mining also began to increase in southern Wales.

Miners at first dug coal deposits from close to the Earth's surface. Extracting coal to meet rising demands, miners soon exhausted the veins closest to the Earth's surface. They had to dig down, more than 100 feet and as deep as 300 to 400 feet by the early eighteenth century. This made mining more costly and more dangerous. It was necessary to line the roofs of tunnels and mines with lengths of lumber to support those roofs, and the deeper mines also required shafts for ventilation.

The challenges of extracting increasing quantities of coal as an energy source led to a pivotal revolution in generating energy. The problem at one level was very simple: English and Welsh miners produced more coal by digging larger and deeper holes in the ground, but water flows downhill. Even when miners could breathe and escaped roof collapses or explosions, flooding was a constant problem. The solution to this basic problem, however, was anything but simple. To pump water out of shafts, mines employed horses, but this was an inefficient, slow, and expensive method, especially as mine shafts sank to greater depth. In short, Britain risked running short of coal to meet the rising demand of consumers and of nascent industry without a better method to remove water from deeper mines. Attempts to create new machines to solve this problem dated back to the seventeenth century, but failed. In the 1660s the Marquis of Worcester may have invented some kind of mechanism, possibly an early version of a steam engine, to raise water.

In 1712, Thomas Newcomen, an ironmonger from Cornwall, working with a plumber named John Calley, invented a steam engine to pump out mines. They unveiled their device at a coal mine in Staffordshire in the West Midlands of England. It was costly and inefficient, wasting most of the energy used for power as heat, but it was still vastly better at pumping out mines than any previous method. A single Newcomen engine could replace pumps powered by as many as fifty horses.

The Newcomen engine proved critical to the future of industry and the climate in two ways. First, it sustained the growth of coal mining. Over the next twenty years more than 100 Newcomen engines were installed, contributing to the continued growth of mining output in the eighteenth century. Second, the engine derived its power from coal shoveled into a firebox that heated up the boiler. The engine employed coal as the power source to aid in mining more coal to produce more power. The effects on the Earth's atmosphere were small, but relying on a fossil-fuel-powered machine to extract more fossil fuel set a path for the future.

In the 1760s, James Watt, a Scottish inventor and engineer, set to work to improve on the steam engine design after he was asked to repair the University of Glasgow's Newcomen engine. Pondering on the problem, he took a walk in 1765 on the green in Glasgow. Suddenly, he recognized a solution: "the idea came into my mind, that as steam was an elastic body it would rush into a vacuum, and if a communication was made between the cylinder and an exhausted vessel, it would rush into it, and might be there condensed without cooling the cylinder."[7] Today, a statue of Watt stands on the Glasgow green to memorialize the moment. Acting on his insight, Watt created a separate condensing chamber for steam. The new steam engine was vastly more efficient than the Newcomen engine, and Watt's improvement ushered in an era of rapid further innovation (Figure 6.1).

Matthew Boulton, a Birmingham manufacturer, bought out Watt's bankrupt business partner, and in the 1770s Watt and Boulton began to manufacture and sell the improved steam engine. By 1800, they had produced some 450.[8] Boulton also encouraged the use of steam engines beyond mining. In 1785, Richard Arkwright made the first use of steam in manufacturing textiles, initially to pump water. Over the next twenty years, a series of inventors developed and made improvements to the power loom. Boulton himself minted coins using a steam-powered manufacturing process.

Industrial Revolution

Beginning in the late eighteenth century, Britain combined the use of the steam engine powered by coal with industrial growth and agricultural improvement to break with all previous economic and demographic development in world history. The Industrial Revolution displaced some workers and subjected many others to miserable working conditions, as protesters and reformers soon noticed, but it also raised productivity to unprecedented levels. Where other technologically advanced societies had previously hit plateaus in growth, Britain experienced decades of growth in output, productivity, population, and, eventually, in per capita income, pioneering the emergence of an entirely new kind of society, never before seen in world history (Figure 6.2).

FIGURE 6.1 *Newcomen steam engine invented by Thomas Newcomen in 1712. It consisted of a pump designed to reduce water steam in the galleries of mines. Engraving. Tarker/Bridgeman Images.*

A series of industries grew rapidly. In the first stages of industrialization, dating back to the 1780s, textile manufacture expanded at an exponential pace. This would set a pattern for later-industrializing countries that often began in manufacturing clothing and textiles before branching out into other areas. Think, for example, of the countries of origin on the labels of clothing such as caps, sweatshirts, and knit shirts, and it is possible to see how this pattern persists to the present.

FIGURE 6.2 *Commencing industrialism. A Lancashire town in the early nineteenth century. Source: Wellcome Library, London.*

Industrialization in the early nineteenth century also saw rapid growth of iron production. The application of steam power to a railway engine in the 1820s brought a further burst of innovation and industrial expansion. Britain, soon followed by other Western countries, quickly acquired an extensive railway network, which both vastly speeded up transportation and communications and provided a further source of demand for increasing iron output. Between 1830 and 1850, 6,000 miles of railways were built in Britain.

English cities of the nineteenth century gained fame for their rapid growth and dynamic markets and cultural life, but also became notorious for filth and squalor. In England, the cities of the midlands and north, places like Birmingham and Manchester, became major cities in decades. Birmingham, today England's second-largest city, is located in the West Midlands some 125 miles north and west of London. In 1800, Birmingham was a large bustling town with a population of more than 73,000. It continued to grow as a center for metal work, among other industries, and by the census of 1851 Birmingham had become a city of more than 233,000 residents. Manchester is located still further north, around 210 miles from London. A market town that had also been a center of weaving in the early modern era, Manchester began to industrialize in the late eighteenth century. Its close proximity to the port of Liverpool, a little more than 30 miles to the west, gave Manchester easy access to imported cotton. There were already sixty-six cotton mills by 1821, and hundreds of mills dotted the city landscape within a few decades.[9] The economic transformation remade Manchester into a city. The population increased from a little more than 70,000 in 1800 to more than 300,000 people by 1851.

The new industrial cities, renowned for their production, also became known for the costs of industrialization. The standard of housing for workers who migrated from the countryside was low. Burning coal created a stench and haze. In 1854 in his novel *Hard Times*, Charles Dickens described the conditions of a prototypical northern English city, which he dubbed Coketown, rather than using the name of an actual city. As Dickens put it, Coketown

> was a town of red brick, or of brick that would have been red if the smoke and ashes had allowed it; but as matters stood, it was a town of unnatural red and black like the painted face of a savage. It was a town of machinery and tall chimneys, out of which interminable serpents of smoke trailed themselves for ever and ever, and never got uncoiled. It had a black canal in it, and a river that ran purple with ill-smelling dye, and vast piles of building full of windows where there was a rattling and a trembling all day long, and where the piston of the steam engine worked monotonously up and down, like the head of an elephant in a state of melancholy madness.[10]

Coketown may have stood in for Manchester, but it could have represented many growing cities. The technology would change, but the basic pattern of pollution of air and water would be repeated in many industrializing cities of the twentieth and twenty-first centuries.

London was by far England's largest city, and it became the world's largest city. As Britain's capital and economic center, it already had a population of some 1 million in 1801 and grew to a size of more than 6.2 million by 1901. London was a major manufacturing center, but it was not an industrial city dominated by factories in the same way as a city such as Manchester. The poor crowded its slums, but it was also a center for consumerism. The rising population and mix of economic activity all required more energy. The newly industrializing society boosted the further extraction of fossil fuels. Manufacturing, heating, and transportation depended on the supply of copious amounts of coal, especially with the construction of railway lines and the London underground.

Despite critiques, the new industrial society acquired a reputation as an engine of progress. It was clear to the British themselves of the mid-nineteenth century that they already lived in a very different era from the past. A Great Exhibition held in 1851 dramatized the progress of the day. This event, officially titled the Great Exhibition of the Works of All Nations, became a kind of prototype for what would later become world's fairs. The main exhibition space, a towering structure built of glass called the Crystal Palace, reminded every visitor and passerby of the singular achievements of industrialization. Within the hall, visitors saw numerous displays of machines and items of trade. Crowds flocked to the Great Exhibition, so many that a travel agent named Thomas Cook began to arrange for special trains, and more than 6 million people attended.

From England, the Industrial Revolution spread to other countries in Europe, North America, and East Asia. In the early nineteenth century, industry sprang up in New England beginning with technology brought over from England. Output rose sharply in the mid- and late nineteenth century in mill towns such as Lowell, Massachusetts. On the European continent, Germany industrialized. Coal production in the Ruhr region of northwestern Germany increased from 1.7 million tons in 1850 to 11.6 million tons in 1870.[11] In the late nineteenth century, the newly united German empire became a leader in steel production and in new economic sectors, such as the emerging chemical and electrical industries.

Industrialization also began to take root in Japan after the United States forced an opening to trade. Since the seventeenth century, Japan's shoguns, or military commanders, had closed the country off to almost all trade with the West, but in a show of force, Commodore Matthew C. Perry of the US Navy arrived with warships at the city of Edo, what we today know as Tokyo, in 1853. This initiated a period of rapid change and political struggle in Japan that led to the revival of rule by the emperor, the Meiji restoration, by 1868. Imperial restoration inaugurated a period of rapid modernization, during which Japan sought foreign expertise and began to industrialize. The Imperial Charter Oath of April 1868 stated, "Knowledge shall be sought throughout the world so as to strengthen the foundations of imperial rule."[12]

Industrialization brought a further acceleration of extraction and combustion of fossil fuels. This was not because the first industries relied exclusively or even primarily on steam power. As the landscape created by the early Industrial Revolution still reveals today, water mills played a prominent role: surviving mills in old mill towns cluster around rivers, but mills began to make increasing use of steam power during the nineteenth century. In Britain, a sharp decline in the cost of steam power in the 1830s and 1840s eased a shift in power sources. The use of coal-powered steam engines increased in British textile production. In the United States, early mills typically used water as the power source, but steam engines gained increasing use after the Civil War.

The birth and development of the Industrial Revolution brought multiple fateful effects for the future history of climate. Industrialization expanded and intensified use of fossil fuels at an exponential rate. By 1815, Britain consumed fifty times more coal per person than did France and more than thirty times more coal per person than did Germany.[13] Coal remained by far the dominant fossil fuel throughout the nineteenth century, but industrialization also accelerated extraction and use of oil and gas. Oil drilling began in Pennsylvania in 1859, and the development of the gas-powered internal combustion engine created a new market for oil. In 1861, Nikolaus Otto, a German engineer, built a gasoline-powered engine, and in

1876 he designed a four-stroke internal combustion engine. Gottlieb Daimler in 1885 invented a high-speed engine, and by 1913 Henry Ford began the mass production of the Model T. The new engines increased demand for oil and created a pathway for a future of soaring energy use for transportation.

Electrification further raised demands for fossil fuels. Well into the industrial era, many households remained dimly lit. Gas lighting spread rapidly in the nineteenth century. A series of inventors in the late nineteenth and early twentieth centuries figured out how to generate electric power and how to use electricity to produce light. Thomas Edison, for example, invented the first commercially viable incandescent light bulb. The spread of artificial lighting transformed the way people lived and worked and also further boosted demand for fossil fuel, and in particular for coal. For most of the industrial era, coal-fired power plants provided the chief means to produce electric power.

Along with oil and coal, natural gas also emerged as a vital fossil fuel. Natural gas is found in swamps and also in oil fields. It was often burned off, but wide-scale use of natural gas took off after the Second World War and continued to increase into the twenty-first century as a major fuel for power plants and for home heating.

The exponential increase in the use of fossil fuel made more energy available to people of industrial societies than humans at any other point in world history to that point could have imagined. The extraction and consumption of fossil fuels brought unprecedented speed of travel. Trips by rail were so fast that early reports raised concerns about the possible ill effects of speed on health. Railway travel did have the effect of compressing time and space, and so too did travel by steamship. The ordinary citizens of an industrial society could travel great distances with comparative ease. Britons, Germans, Americans, and others whose countries industrialized also came to enjoy undreamed of access to heat and light. Industrialization by the late nineteenth century and far more so by the twentieth century supported mass production that allowed mass consumption.

Power and speed also had their deadly side, and not just for the victims of workplace injuries or transport accidents. The industrialization of war made the battlefield ever more deadly for large numbers of combatants and civilians. More people than ever before could travel great distances, more people than ever before could purchase clothing or household items, and more people than ever before could die in quick succession on the battlefields of the First World War and the Second World War.

Carbon and climate

British industrialization in the late eighteenth and early nineteenth century shaped the future of climate forcing, not so much because the early industrialization of one country dramatically remade the chemistry of the atmosphere, but because Britain created a path that many other countries would follow, collectively raising the level of CO_2 in the air. Our understanding of what is now dubbed the "greenhouse effect" emerged around this time. Joseph Fourier, a French mathematician and physicist, was the first to describe the role of the atmosphere in heating the planet in 1824. He determined that Earth's temperature would be much colder if it lacked an atmosphere, and drew an analogy to earlier experiments that had demonstrated a warming effect created by a glass cover placed on a box.[14] Thirty-five years later, an Irish physicist named John Tyndall began experimentation on the radiative properties of various gases, revealing that gases such as water vapor and CO_2 were strong absorbers of heat. He proposed that changes in the concentration of these gases could be linked to climate change. The idea that climate could shift was not widely accepted at the time, though it had been promoted by naturalists such as Louis Agassiz. Building on the work of earlier naturalists and based on his own observations in the Alps, Aggasiz proposed in 1837 that the Earth had experienced a great "Ice Age." The scientific community at first rejected his ice age theory, but it gained broad acceptance by the 1870s.

Pivotal research on greenhouse gases came from Svante Arrhenius, a Swedish Nobel Prize–winning chemist and physicist, who in 1895 presented a paper on the effect of CO_2 and water vapor on Earth's temperature. He later helped clarify the role of the water vapor feedback. Arrhenius calculated that CO_2 accounted for 21°C of the warming above the level that would exist if the Earth lacked an atmosphere, and found that water vapor generated by such warming contributed an additional 10°C. Arrhenius also noted in 1904 that the increase in CO_2 by industry could in turn increase the temperature of the atmosphere. Scientists at the time largely dismissed his idea: the thinking that prevailed before the mid-twentieth century was that the effect of human activity was either too small to override that of natural forces or that any such increase in CO_2 and associated warming could actually be beneficial. By the 1930s, temperature measurements across the United States, which began in the late nineteenth century, indicated an overall warming. In 1938, Guy Stewart Callendar, a steam engineer, attributed the warming to increasing levels of CO_2 in the atmosphere. He wrote that few "would be prepared to admit that the activities of man could have any influence upon phenomena of so vast a scale," but that in his study, he hoped "to show that such influence is not only possible, but is actually occurring at the present time." He concluded that "the return of the deadly glaciers should be delayed indefinitely" due to the warming.[15]

Views about the effect of increasing CO_2 from industrial activity began to shift in the 1950s. Roger Revelle, a scientist at Scripps Institute of Oceanography, and Hans Suess from the US Geological Survey, proclaimed that "human beings are now carrying out a large scale geophysical experiment of a kind that could not have happened in the past nor be reproduced in the future. Within a few centuries we are returning to the atmosphere and oceans the concentrated organic carbon stored in sedimentary rocks over hundreds of millions of years."[16] They noted the lack of information available at that time to predict changes in climate as a result of fossil fuel combustion, particularly if it continued to increase at an exponential rate, and they called on geoscientists to gather that information.

Revelle teamed up with Charles Keeling, also of Scripps, and Harry Wexler of the US Weather Service, to begin direct measurements of atmospheric CO_2. The first such measurements began at Mauna Loa, Hawaii, in 1958, and measurement has since expanded to many locations. Atmospheric CO_2 concentration prior to 1958 has been determined based on air bubbles trapped in ice cores: values in 1850 were 285 ppm,[17] on the upper end of the natural range observed during glacial/interglacial swings. The spread of industrialization to European countries and to North America raised CO_2 concentrations to above 300 ppm by the eve of the First World War. The first measurements at Mauna Loa in 1958 showed that concentrations had risen to 315 ppm.

Nineteenth-century droughts

As humans increasingly became an agent of climate change, natural fluctuations continued to affect the growing population. Drought in particular posed tremendous challenges. North America suffered repeated droughts during the late nineteenth century. A severe drought took place from the mid-1850s through the mid-1860s, inflicting severe damage on bison, which were already being killed in large numbers. Another severe drought struck in the 1870s, and there was renewed drought in the 1890s.[18]

The El Niño–driven drought of the late 1870s also affected other regions, including China, Korea, India, Brazil, Egypt, and southern Africa. Between 1877 and 1879, Brazil's Great Drought caused economic devastation. *Retirantes*, or internal refugees, from the northeast, faced with the very real prospect of starvation, fled the *sertao*, or interior: the population fell by some 90 percent over two years. These refugees from drought faced further hardship and danger as they crowded into cities on the coast.[19] In India, the death toll reached some 6 to 10 million. Those who could make their way out of the famine districts left for regions such as Ceylon (now Sri Lanka).[20] Drought also spread across northern China in the late 1870s. A famine that lasted from 1877 to 1879 led to between 9 and 13 million deaths in northern provinces.[21] One of the provinces most affected was the Shanxhi province on a high loess plateau. Poor roads made it especially hard to send relief to Shanxhi. Prints in the Shanghai press publicized the extent of the disaster and described eating of the dead and even cannibalism.[22]

Declining administrative effectiveness contributed to the disaster. In the past, the Qing Empire had established a host of measures for responding to famine. Officials purchased grain at low prices and sold it at low prices in times of dearth. They also canceled taxes in afflicted regions. Administrators investigated local conditions and worked with local elites to provide relief.[23] But by the late 1870s, years of severe external challenges and internal disruption had weakened the Chinese state. Less than fifty years after rejecting overtures from Britain for more trade, China found itself unable to stop British attacks during the Opium War of 1839–42 and suffered further losses in a second war of 1856–60. China also experienced massive internal rebellions and uprisings, including the Taiping rebellion of 1850–64 that made vast areas of China into a war zone and left millions dead.

The Qing Empire that confronted the El Niño droughts of the nineteenth century was much less effective than the Chinese state of prior generations at averting famine. The years of war and unrest damaged the system of grain stores. The stocks of grain were less than half those during the pinnacle of the Qing dynasty.[24] Defeats also shifted priority from disaster relief for civilians toward provisioning the military.[25] Qing consensus broke down in the face of competing threats. Proponents of a movement to strengthen China sought to discourage the Qing leadership from diverting money designated for improving coastal fortifications toward famine relief.[26] Facing these competing needs, the Qing court sought both to continue self-strengthening and to provide famine relief.[27]

The death toll from the El Niño–induced famines was both horrifying and striking because it appeared to reverse the trend toward greater human resiliency in the face of climate fluctuations. Both drought and the human response affected the chances of survival for the population in famine districts. In India, British officials took pride in their campaigns to gain donations for Indians, but some of these same officials found British policy unsettling. Some Indians themselves questioned the British commitment to averting famine. Surendranath Banerjee, an early Indian nationalist, argued that the outcome of drought showed the shortcomings of the government. "But we are told that famines are due to drought; to the operation of natural causes, and governments and human institutions are powerless to avert them. We ask—is drought confined to India? . . . Other countries suffer from drought; but they do not suffer from famine."[28]

The nineteenth century ended with another severe El Niño event. In 1898, drought struck northern China. Many peasants found themselves without food. Desperate for food, people ate bark and almost anything else they could find. Some even sold children. The scale of the disaster struck an observer, "There is an awful famine pending in the interior. All along our route . . . is a desert and the mortality by famine this year must be enormous."[29] The drought and famine in China contributed to the uprising known as the Boxer Rebellion. The Boxers, who engaged in martial arts and followed a blend of spiritual and religious practices, objected to rising foreign influence in China and to the activity of Christian missionaries.

Indeed, the Boxers attributed the bad weather to foreign influence. Boxer posters pledged, "when the foreigners are wiped out, rain will fall."[30] The rebels attacked missions and missionaries as well as railroads and telegraph lines. They gained the support of the dowager empress Cixi and marched to Beijing, where they besieged foreign diplomats before an international military force with troops from countries, including Japan, Russia, France, Britain, the United States, and Germany, took the capital in August 1900.

The El Niño event produced another severe drought in much of India. A meager monsoon in the spring of 1896 caused grain prices to rise, and by autumn grain riots broke out. "Already grain riots are common," R. Hume, an American missionary, wrote in a letter to the *New York Times*. Julian Hawthorne, son of the novelist Nathaniel Hawthorne, described seeing corpses from a train window: "There they squatted, all dead now, their flimsy garments fluttering around them."[31] Nearly three-quarters of a million people died in Bombay. The viceroy, Lord Elgin, estimated that some 4.5 million people died, but many more died even after rains returned in 1898. Renewed monsoon failures struck India. Famine resumed in 1899, and millions died across India.

The famines of the 1890s and early 1900s that caused misery in India, China, and northeastern Brazil as well as the Dutch East Indies and Africa stemmed from the interplay of climate shifts and human actions. The El Niño events created widespread drought. The lack of rain was the first cause of crop failures and food shortages, but the death toll and misery were especially striking at a time when Western countries had achieved the ability to avoid subsistence crises. Climate fluctuations still affected water supply and crop yields, but the industrializing world had already gained far greater ability to withstand the typical climate fluctuations of the Holocene without suffering immense death and misery. Why, then, did so many die in China, India, northeastern Brazil, the Dutch East Indies, and east and southern Africa?

Along with the El Niño events themselves, the erosion of local governance and the power of Western ideologies exacerbated the effects of drought. The timing and location of these famines at the very height of imperialism have also raised scrutiny of the role of the Western world, and in particular of Britain. Britain in India did not neglect relief entirely but adherence to free trade and a desire to keep down administrative costs aggravated the effects of famine. Indeed, India continued to export grain during famine in 1878, 1896, and in 1899–1900.[32]

In the Philippines, also suffering from drought, the American military, which had just taken the islands from Spain in the Spanish-American War, reduced the food supply to crush a native insurgency. US forces destroyed rice stores in 1900—the resulting deaths were predictable. Colonel Dickman, who prepared notes on the Philippines, stated, "many people will starve to death before the end of six months." Famine, in turn, raised mortality from epidemics.[33]

Political responses also intensified famine in Brazil. Port cities along the Brazilian coast sought to block off displaced people from the interior. Many of the displaced sought refuge with a religious movement led by Conselheiro, a preacher who built what he termed a holy city. Fearful that this movement pursued radical aims, forces dispatched by the Brazilian government attacked the movement at its base at the town of Canudos in the Bahia state. There was little evidence that the people at Canudos intended any kind of rebellion against Brazil, though migration to the city created the potential to reduce the labor supply. A Catholic guard defended Canudos, repelling federal troops repeatedly until a renewed federal assault took the town.[34]

In southern and East Africa, imperial rule magnified suffering by accelerating the movement of disease. Rinderpest, a virus extremely deadly for cattle, decimated herds, damaging local economies. Rebellions in Africa in the late nineteenth and early twentieth centuries against the newly implanted European imperial regimes blamed the newcomers both for the disease that killed off Africans' cattle and for the increasingly unreliable climate. "These white men are your enemies. They killed your fathers, sent the locusts, this disease among the cattle, and bewitched the clouds so that we have no rain," a religious leader informed warriors who rebelled against the British in what is now Zimbabwe in 1896.[35] Drought reinforced resentment against colonial rule to generate uprisings in Mozambique and in Tanganyika (now Tanzania). German forces in East Africa responded to the Maji Maji rebellion of 1905–7 with a campaign of extermination.

In Namibia, then–German Southwest Africa, rinderpest arrived in 1896 and soon killed off most cattle of the Herero people. An inoculation program introduced by Germany had very mixed results, initially killing off many of the vaccinated cattle. The epidemic along with heavy loss of human life to malaria and typhus severely damaged Herero society and contributed to a Herero uprising in 1904. The German commander in Namibia, General Lothar von Trotha, responded with a campaign of destruction that historians have widely identified as an early twentieth-century case of genocide.

ENSO-related droughts continued into the twentieth century, such as the ones that occurred during the 1930s and 1950s in North America. Just as prevailing La Niña conditions caused drought in North America during the MCA, cold tropical Pacific Ocean waters set the stage for these twentieth-century droughts. In the case of the Dust Bowl, which led to large agricultural losses in the Great Plains of the United States and forced the migration of millions of people, poor land management practices amplified the effects of La Niña drought. Thus the plowing of fields to plant wheat crops, which are vulnerable to drought, along with poor cultivation methods, led to soil degradation. When drought struck in the early 1930s, soil loss through wind erosion generated large dust storms. The veil of atmospheric dust blocked sunlight and amplified the effects of drought.[36]

World Wars and Fordism

Well into the industrial era, El Niño events created more dramatic fluctuations in climate than did climate forcing caused by human activity, but the scale of industrialization continued to increase in the twentieth century. Energy consumption and production increased in several stages. Economic growth was not linear. The First World War inflicted significant damage on the world economy, and recovery was uneven during the 1920s. The US economy boomed, but European states saw more modest growth. The stock market crash of 1929 and the Great Depression of the 1930s brought the greatest modern crisis of the world economy and capitalist system. Even with this blow and sharp contraction of industrial output in countries from the United States to Britain to Germany, the imprint of industrialization was sufficient that CO_2 levels in the atmosphere continued to rise, though at a rate much slower than in the second half of the twentieth century.

The Second World War brought both rising economic output and devastation. The Great Depression ended for good in the United States. Gross domestic product (GDP) rose some 70 percent between 1940 and 1945, and the country emerged from the war en route to becoming a global superpower. At the same time, the countries in the war zone suffered immense damage to infrastructure and, by the war's end, to production.

Within a few years of the war's end, the world economy entered a period of rapid growth in which mass production and mass consumption of consumer durables rose at a steep pace through an economic model called Fordism. Henry Ford had first introduced such production with the manufacture of the Model T in the early twentieth century at a price low enough to make it possible for large numbers of people to purchase the vehicle. The combination of mass production of consumer durables, items of significant expense designed to last for more than a short time, mass consumption or purchase of such durables, and wages sufficient to make those purchases possible became known more broadly as Fordism, but Fordism was still in its nascent phases before the Second World War.

After the Second World War, Fordism as an economic model became much more firmly established across the industrial world. Household purchase and ownership of motor vehicles were still a rarity in Europe in the late 1940s, but became commonplace by the 1960s and continued to rise thereafter. Similarly, purchase and ownership of other consumer durables such as washing machines rose at an exponential pace. The 1950s and 1960s on the whole saw unmatched economic growth rates. In postwar West Germany, people spoke of an "economic miracle." The prime minister of Britain, Harold Macmillan, declared in 1957, "Indeed let us be frank about it—most of our people have never had it so good. Go around the country, go to the industrial towns, go to the farms and you will see a state of prosperity such as we have never had in my lifetime—nor indeed in the history of this country."[37] Britain's economy was actually growing at a slower rate than that of many of its European peers, but the broader point was still true. Postwar Fordism raised output and standards of living, and this new standard of living increased demand for fossil fuels. Overall coal output rose, though the precise contribution of different countries and regions varied. In some regions, coal production actually declined because of shifts in production to new competitors and the replacement of coal by oil and gas for many uses, in particular for transportation. In the United Kingdom, coal output peaked before the First World War. In the Federal Republic of Germany coal output began to decline with a shift to other energy sources, including oil and gas. At the same time, coal production and consumption soon began to increase in China. In the United States, coal production grew through the early twentieth century, fluctuated in the era of the world wars and depression, and rose sharply again in the second half of the twentieth century.

Oil output and consumption increased at a faster pace than coal production throughout much of the twentieth century. In the United States, demand for oil boosted drilling and exploration and the emergence of oil as a major extractive industry in regions such as Louisiana and Texas. Global demand for oil increased at an especially fast pace after the Second World War, bringing wealth to major oil producers, in particular Saudi Arabia, where exploration yielded major finds in the 1930s.

The Industrial Revolution, the exploitation of fossil fuels on a new scale, and the expansion of industrialization based on fossil fuels together raised levels of CO_2 in the atmosphere. The collection of measurements at Mauna Loa began during the postwar economic boom. The first annual average measurement was 315.98 ppm, marking a significant rise from the 285 ppm determined from ice cores for the very early stages of the Industrial Revolution.

As noted by Callendar in the late 1930s, temperatures had been increasing, but then leveled off in the 1940s into the 1970s. Explanations for the pause in warming included a lower solar output, volcanic eruptions, and global cooling associated with orbital variations. As scientists debated the cause of this plateau, measurements emerged from the Southern Hemisphere that revealed that this phase may have been a Northern Hemisphere phenomenon rather than a global signal. This led some scientists to suggest that particulates emitted by factories, which effectively block sunlight, were responsible for the plateau given that most of the industrial activity was concentrated in the Northern Hemisphere. This hypothesis gained support in later decades, but by this time, the CO_2 accumulating in the atmosphere overrode the cooling effect of pollution.

Globalization

A further burst in industrialization and in fossil fuel extraction and use came with global industrialization in the late twentieth century. This was not the first era of globalization, but it was the first period that saw industry take root across much of the world. The mature industrial powers confronted new economic challenges in the 1970s when the era of rapid and robust postwar growth ended, but overall global industrial output reached new heights in the late twentieth century into the early twenty-first century. The globalization of industry transformed economies around the world. Much of East Asia experienced extraordinary rates of industrialization. Japan, the first Asian industrial powerhouse, entered a period of sustained high economic growth from the postwar years through the 1990s. Japanese companies such as Toyota, Sony, and Honda became household names around the world. In 1990, Japan was the world's third leading exporter. Japan's economic prowess was such that the country's success led to anxiety in some quarters. Others saw the country as a model for lessons, with the theme of a book from 1979 entitled *Japan as Number One: Lessons for America.*

Export-led growth provided a similar path for stunning economic expansion elsewhere in East Asia, including in South Korea and in Taiwan. At the end of the Korean War in 1953 South Korea was an impoverished country. Indeed, though it may have seemed impossible to imagine in the early twenty-first century when North Korea had become synonymous with deprivation and malnutrition for the mass of its citizens, South Korean per capita GDP was no higher than that in North Korea in the 1950s. South Korea's economy remained comparatively small for years after war, but export-led growth led to sharp increases in GDP in the south from the 1970s into the 1990s and early 2000s. South Korean manufacturing firms, such as Hyundai, became major global exporters. This economic model proved attractive to political leaders and economists seeking to boost output and growth elsewhere. A host of other Asian countries, including Malaysia, Indonesia, and more recently Vietnam, to note only some of the examples, similarly pursued export-led growth.

By far the most stunning case of export-led growth occurred in the People's Republic of China. Years of struggle between Communists and Nationalists as well as varied warlords, interspersed with harsh Japanese occupation before and during the Second World War, ended with a Communist victory in the Chinese Civil War of 1945–49. The Nationalist forces of the Kuomintang fled to Taiwan. On the Chinese mainland, the Communist Party led by Mao Zedong created the People's Republic of China.

Once in power, Mao and the Communist Party collectivized agriculture and sought to boost heavy industry through centralized planning. In the late 1950s, Mao sought to accelerate the process through the Great Leap Forward. The Communist Party exhorted peasants to produce steel in backyard furnaces. Much of the steel produced was useless, and the diversion of labor exacerbated famine. Tens of millions of people died. After the Great Leap Forward, Mao regained the political initiative in the Chinese leadership through a program of Cultural Revolution that lasted from 1966 until his death in 1976. At this time, China still had a mainly agricultural economy that accounted for a small share of world exports, far below the level of countries such as Japan, West Germany, and the United States.

China's economy and the future of the world's climate took a decisive turn with the leadership of Deng Xiaoping. A longtime Communist who took part in the Long March when the Communists escaped destruction after attacks by their Nationalist rivals, Deng later faced persecution during the Cultural Revolution as a so-called capitalist roader. He accomplished a remarkable political comeback in the 1970s. Deng was determined to maintain Communist power—he crushed the democracy protests at Tiananmen Square in Beijing in 1989, but he also aimed to remake China's economy. To that end, he sought foreign investment and created free-trade zones. This initiated decades of strong export-led growth that made China into the world's leading exporter by 2009 and the world's largest trading nation by 2013.

Domestically, the transformation of China brought rapid urbanization and new patterns of consumption. Though large numbers of Chinese remained in rural areas of the interior, dozens of cities experienced sharp population growth. The proportion of Chinese people living in urban areas increased from 13 percent in 1950 to 57 percent in 2016. Along with world-famous cities such as Beijing and Shanghai, Chinese citizens flocked to scores of other cities, barely known in the wider world: Changsha, Hefei, Quanzhou, Xiamen, Hangzhou, Shenyang, Zhengzhou, Suzhou, Xi'an, and Chongqing, along with many others.

Rapid urbanization became the norm in the late twentieth and early twenty-first centuries. India is a complex example because it never enjoyed the same rate of export-led growth as East Asia countries such as Japan and China. At the same time, India, though it retained a vast rural population, did see urban growth, and Indian per capita GDP rose sharply in the early years of the twenty-first century. Urbanization, in turn, encouraged shifts in consumer tastes and preferences and mass consumption.

The expansion of industry and mass consumption has been especially striking in Asia in recent decades, but this has been a global trend despite the fact that more than a billion people globally lived on less than $1.25 a day as of 2011. Urban population soared in Latin America. Africa also saw a surge in urban population, as did the Middle East and North Africa.

The contrast with the start of the twentieth century was striking. Exact comparison of urban population can be difficult to make, given different standards for measuring the population of a city or of a larger metropolitan area, but the overall trend was clear. In 1900 when London was the world's largest city, ten cities in all had a population of at least 1.4 million, the population of Philadelphia—then the world's tenth largest city. In the early twenty-first century, there were more than 200 cities with populations greater than 1.4 million and close to thirty cities with populations larger than that of London in 1900. Industrialization and globalization had given rise to a world of many Londons, not necessarily as measured by cultural influence or economic might but as determined by population. The higher average standard of living in London meant that the average person there, as elsewhere in the Western world, had a higher energy footprint, but the rest of the world was intent on catching up.

In principle, urban living can reduce the carbon footprint per person in an industrialized society through more use of mass transportation and small living quarters, but the shift from the countryside to cities led to greater power consumption. Residents of growing cities who could afford to do so purchased the same kind of appliances that their Western counterparts used. China, for example, had only 4 million refrigerators in service in 1985, but by the end of the twentieth century, ownership of a refrigerator was fast becoming the norm in Chinese urban households. The story was much the same for ownership of other appliances such as televisions or washing machines. Refrigerator ownership also began to rise swiftly in India, though the rate of ownership has remained far below that of China.

In hot climates, residents and citizens of industrializing and urbanizing societies also frequently purchased air-conditioning units. Air-conditioning sales rose at a clip of 20 percent per year in India. In urban areas of India, walls of air conditioners popped out of windows. The percentage of homes in urban China with air-conditioning rose from 8 to 70 percent between 1995 and 2004.[38] Explosive growth in the ownership and use of air-conditioning seemed set to continue for the foreseeable future. Air-conditioning also boosted the development of new suburbs. New housing estates and developments on the outskirts of cities such as Shanghai or Guangzhou promised all the amenities of an affluent lifestyle. Residents would drive cars and use all major appliances. Air-conditioning would keep them cool. With artificial cooling for residents of cities and suburbs, rapidly modernizing countries followed much the same path previously forged by cities of the United States. How large would the population of the Houston or Atlanta metro areas be without air-conditioning?

The growth of air-conditioning not only raised energy consumption but also threatened to release large quantities of a family of potent greenhouse gases into the atmosphere. Production of chlorofluorocarbons (CFCs), commonly referred to as Freon, began in the 1930s for use in refrigeration, air-conditioning, and as propellants for aerosol cans. CFCs absorb heat thousands of times more efficiently than CO_2, but their major environmental impact is found in the upper layers of the atmosphere, where they react with our protective ozone layer. The Montreal Protocol, an international agreement to phase out the use of ozone-depleting CFCs, went into effect in 1989. Hydrochlorofluorocarbons (HCFCs) and hydrofluorocarbons (HFCs) replaced CFCs, helping preserve the ozone layer, but their global warming potential remains. While they currently contribute to less than 1 percent of human-caused warming due to their low abundance, it is estimated that their continued use could amount to nearly 10 percent or more of anthropogenic warming by midcentury if not phased out.

In China and in large swaths of the industrializing and urbanizing world, cities and industries typically depended on coal for their power supply. China's cities became famous, not only for dynamic growth but also for exceptionally bad air pollution. To make the air breathable for the Beijing Olympics in 2008, authorities temporarily closed down nearby factories. At the US Embassy in Beijing, air quality measurements were regularly so bad that they exceeded 500 on the Air Quality Index scale, a metric that originally placed 500 as the maximum value. January 2013 produced a reading of 755 at the embassy. The city of Harbin in northeastern China was virtually shut down in October 2013—many schools and some roads were closed—when smog became so thick that the level of particulates in the air reached forty times the level regarded as safe by the World Health Organization. Such readings have attracted attention, but Indian cities have actually produced even worse air quality readings with a daily average peak in January 2014 of over 400 with spikes above 500.

Heavily dependent on ever-higher use of fossil fuels, the global expansion of industry and urbanization generated steeply rising emissions of CO_2. When measurements began in 1958, atmospheric CO_2 was already higher than previously observed in the ice core record at 315 ppm. Global levels surpassed 410 ppm in 2020 (Figure 6.3).

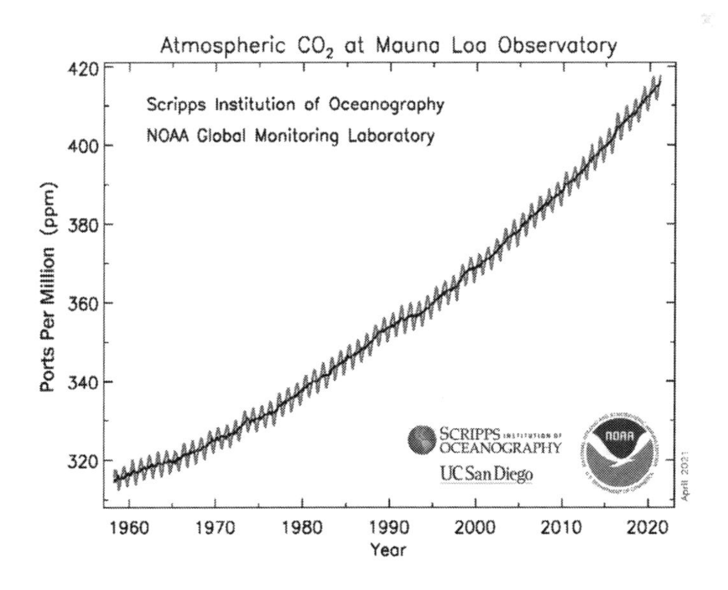

FIGURE 6.3 *Atmospheric CO_2 at the Mauna Loa Observatory. Source: NOAA, https://www.esrl.noaa.gov/gmd/webdata/ccgg/trends/co2_data_mlo.pdf.*

Breaking constraints

Industrialization, urbanization, and economic development powered by fossil fuels both built on and remade the human relationship with climate. Complex societies of the early modern and modern era had already developed more resilience in the face of climate shifts. Agricultural improvement, greater administrative capacity, and better transportation brought the era of major subsistence crises to an end in some societies such as England and Qing China at its height.

The new era of global industrialization took these trends to an extreme. Population centers became far more removed from areas with a climate suitable for providing ample food. Preindustrial population growth in cities like London depended on provisioning the city. Animals were driven through the streets to slaughterhouses. Indeed, the practice did not come to an end until well into the industrial era. Many were brought to the city's Smithfield Market. In his novel *Oliver Twist*, Charles Dickens gave an idea of the dimensions of the market:

> All the pens in the centre of the large area, and as many temporary pens as could be crowded into the vacant space, were filled with sheep; tied up to posts by the gutter side were long lines of beasts and oxen, three or four deep. Countrymen, butchers, drovers, hawkers, boys, thieves, idlers, and vagabonds of every low grade, were mingled together in a mass; the whistling of drovers, the barking dogs, the bellowing and plunging of the oxen, the bleating of sheep, the grunting and squeaking of pigs.[39]

Whereas animals could be driven on foot to London, vast population centers in the twentieth and the twenty-first centuries increasingly emerged at great distances from areas with a suitable climate for raising crops or animals. The process began relatively early in the United States. As the center of agriculture shifted West in the nineteenth century, the growing cities of the Atlantic seaboard depended on a vast network of food distribution. Crops could be cultivated near Eastern cities, though not at as low a cost as in new farming areas in the Midwest. The twentieth century saw the rise of urban and suburban conglomerations in areas with climates utterly unsuited for supporting a large population. This was the case in rising Sun Belt cities in areas of the West and Southwest.

Las Vegas provides just one example of how development became separated from climate. Las Vegas has an average rainfall of just over 4 inches a year, well below the level classified as desert, and it was thinly settled in its early years: the city's population was all of twenty-five in 1900. By 1960, the population increased to more than 64,000, and by 2010 Las Vegas had more than 584,000 residents. The city derived most of its water supply from Lake Mead, the vast artificial lake created behind the Hoover Dam, which was completed in 1936.

Las Vegas was not an anomaly: we now live in a world with many population centers that have stretched the connection between human communities and the presence of water. China has its own desert cities and plans to build more: in 2012, a development company announced plans to flatten mountains to build a new metropolis outside Lanzhou in northwest China. Cities in China's dry western Xinjiang region experienced rapid population growth.

Intensive extraction and use of fossil fuels increased the carrying capacity of many regions far above preindustrial levels. The Industrial Revolution helped enable massive population growth. The 1801 census of Britain calculated the population of England and Wales at 8.9 million and that of Scotland at more than 1.6 million. By 1901, the figures stood at 32 million for England and Wales and 4.47 million for Scotland. Indeed, the share of world population made up of people predominantly of European ancestry peaked around this time.

Overall world population increased during the twentieth century from 1.6 billion in 1900 to 2.55 billion in 1950. The rate of expansion actually increased during the second half of the twentieth century as world population reached more than 6 billion in 2000. This growth in population was possible because of a burst in agricultural productivity and output, dubbed the Green Revolution, that made intensive use of fossil fuels. This Green Revolution brought increased distribution and use of new seeds, pesticides, fertilizers, and machines. Key fertilizers combine nitrogen with hydrogen obtained from natural gas, and improved irrigation often required fossil fuels to produce power for pumps. The addition of fertilizer leads to emission of nitrous oxide, a powerful greenhouse gas. In all the Green Revolution added more greenhouse gases to the atmosphere and increased energy inputs into agriculture by up to 50 to 100 times.[40]

Intensive use of fossil fuels also sustained the service economy, which has typically become by far the largest economic sector in maturing economies. Whether in England, the United States, Germany, or any of the early industrial powerhouses, far more people now work in services, in offices, hospitals, or schools, than in factories or mines, but these service sector employees still depend on power generated to a large degree from fossil fuels. Transit for office workers has relied almost entirely on fossil fuels, as has the power for heat and light in office buildings, though that is changing in the countries such as Germany that have advanced the furthest in installing solar power. Air-conditioning has increased hand in hand with the development of the service economy around the world. In Singapore, for example, the service sector contributed some 73 percent of GDP in 2011. Air-conditioning helped make possible the rise of Singapore as a major service sector hub. When asked to identify the twentieth century's most important invention,

Lee Kuan Yew, Singapore's first prime minister, who led the country for decades, selected the air conditioner.

Toward dependency

The Industrial Revolution and global economic development based on fossil fuels transformed human societies, making it possible to build vast cities in the desert, air-condition office workers by the tens and hundreds of millions, and allow farm workers equipped with fertilizer, pesticides, pumps, and tractors to grow enough food to feed many billions, but these same trends have also reversed the historical trend of several centuries that brought greater independence from climate. For thousands of years, human societies derived benefits from the comparatively stable climate of the Holocene. But after decades of increasing carbon emissions, these same societies faced the growing challenge of living with an increasingly less stable climate.

The level of carbon emissions increased dramatically. The United Kingdom, the world's first industrialized country, led emissions of carbon dioxide in 1850 with 123 metric tons. In 1900, the United States was already the world's largest CO_2 emitter, producing 663 tons at a time when the United Kingdom emitted 420 tons of CO_2. The postwar Fordist era brought great growth in carbon emissions to 2,858 tons in the United States,

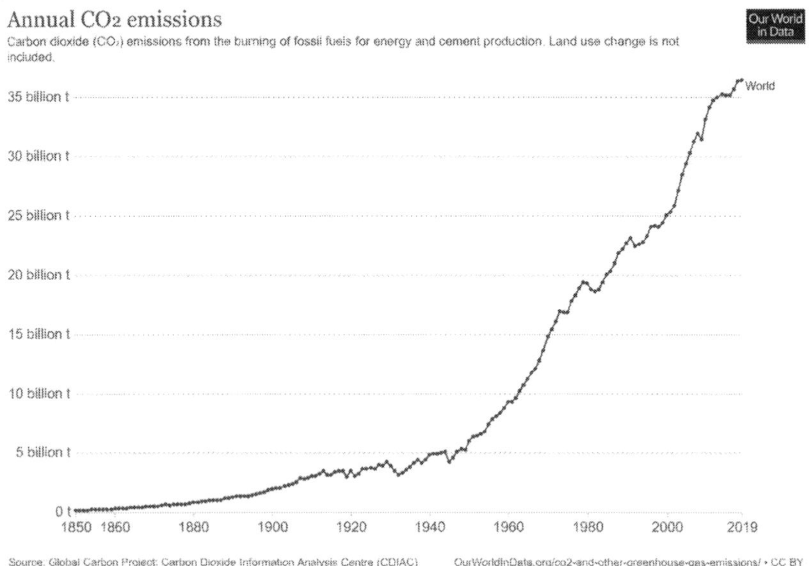

FIGURE 6.4 *Global carbon dioxide emissions, 1850–2020. Source: Global Carbon Project: CDIAC, https://ourworldindata.org/grapher/annual-co2-emissions-per-country?tab=chart&time=1850..latest&country=~OWID_WRL®ion=World.*

with large emissions in other industrial powers such as West Germany with 814 tons.

Meanwhile, a focus on heavy industry in the East Bloc led to emission of 891 tons in the USSR by 1960. Export-led growth raised Japanese carbon emissions to 914 tons by 1980. The era of global industrialization further raised emissions rates: China became the world's largest producer of CO_2 in 2011 with 9,511 tons. India, though much less central to world manufacturing, was also on path to becoming a major emitter with 1,800 tons of CO_2.[41] (Figure 6.4)

Cumulatively, the emissions produced by multiple waves of industrialization, globalization, and population growth made human societies more vulnerable to climate shocks. For thousands of years, human societies had increasingly managed to decouple prosperity from the level of climate change in the mostly stable Holocene. But with industrialization, humans had created a higher level of climate instability with greater extremes. As the next chapter shows, the effects varied greatly by region, and some societies were far more vulnerable than others.

CHAPTER SEVEN

The future is now

- Arctic warming
- Glacial retreat
- Precipitation extremes
- Wildfires
- Shifting biomes
- Rising seas and coastlines
- Oceans
- Climate conflict
- Climate inequity
- Climate migration

In the early twenty-first century, a mountain of climate data and increasingly sophisticated climate models show rapid climate change and a foreseeable future of pronounced and accelerating change. Average global surface temperatures make the trend very clear. As of 2021, there had not been a month with average global surface temperatures below the twentieth-century norm since February 1985. In other words, no child, teenager, or young adult had ever experienced such a month. We will soon live in a world in which even middle-aged people have no personal experience of twentieth-century climate norms. Indeed many images of popular culture and advertising, of skating outdoors or of snowy fields are built on an illusion that equates norms of the past with exceptional conditions of the present. Even in a rapidly warming world, individual regions can experience cold and even record cold at times, but the ratio of record-high temperatures to record-low temperatures has rapidly shifted toward record highs.

The strong El Niño of 2015 and 2016 contributed to a period of many months in a row in which each month's global average surface temperature was the highest on record. The El Niño contributed to the spike in warming, but temperatures during the 2016 El Niño reached levels unprecedented from any other recorded El Niño event. Through August 2016, the National Oceanic and Atmospheric Administration (NOAA) reported record high global average temperatures for sixteen months in a row in a data set of 137 years; that trend lasted until September 2016, which registered as the second warmest ever.[1] The years 2016 and 2020 rank as the two warmest years on record, and overall, 2011–20 ranks as the warmest decade on record, containing nine of the ten warmest years since 1880 (Figure 7.1).[2]

Human-driven climate change has measurably increased average global temperatures. This does not mean a steady linear curve in which warming takes place at the same rate everywhere. Instead, the overall warming trend is leading to faster temperature shifts in particular regions, while causing a general warming trend globally. Regions in high latitudes, in particular, are warming at an especially fast rate.

The world's ocean is seeing the same trend as regions on land. The sea surface temperature of the ocean has been steadily rising since the 1970s, with more pronounced warming since the 1980s. The high heat-absorbing capacity of water, combined with the immense volume of seawater on the globe, has allowed the world's ocean to take up an estimated 90 percent of the increase in heat caused by global warming. The buildup of heat in the ocean has, in turn, brought far-reaching effects, including changes in ocean circulation, rising seas, and climate feedbacks.

From the Arctic to the tropics, these climate trends are transforming the natural world, including the biomes or communities of plants and animals

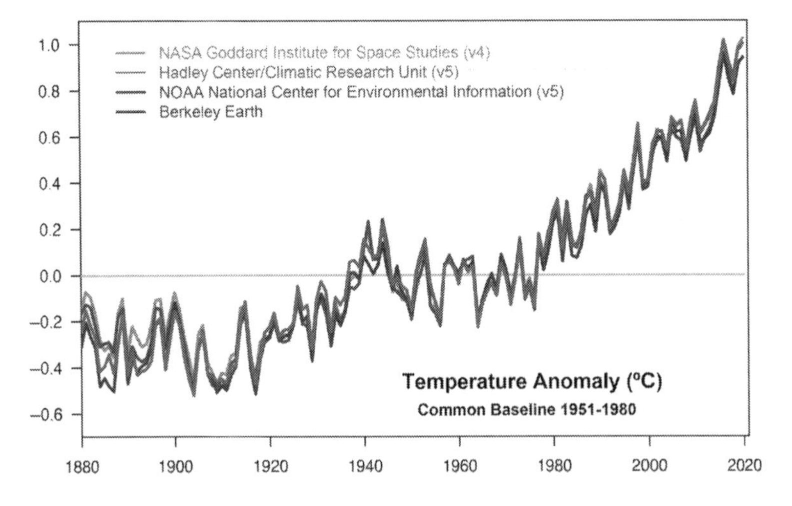

FIGURE 7.1 *Temperature anomaly relative to 1951–80 baseline. Source: NASA: https://www.nasa.gov/sites/default/files/thumbnails/image/temp-2020_comparison-plot.jpg.*

in a shared climate zone. In turn, climate change is amplifying many threats, from wildfire to flooding and drought, and magnifying multiple sources of stress for human societies, including disease, conflict, and inequality.

Arctic warming

The effects of climate change to date can be subtle, but in some regions, the impact of climate change is already so dramatic that it is hard for anyone to overlook it. The warming trend and its effects have been especially pronounced at higher latitudes, including the Arctic. Residents of these regions have already seen remarkable and highly visible effects. In Alaska, coastal erosion has placed many communities at risk, and as of the early twenty-first century, some Alaskan villages were actually trying to come up with plans for relocation. Dwindling sea ice reduced protection from waves and winds of storms. Where much of Alaska faces rising sea levels with less protection from storms, the land here is actually rising in areas where nearby glaciers have been losing mass. The decrease in weight causes isostatic rebound, or the rise of land after the weight of ice is reduced or removed altogether. Near Juneau, Alaska, one property owner built a golf course on land that had once been underwater.[3] Melting glaciers are also creating conditions for landslides as glaciers recede in mountainous regions. The melting can lead to slow slumping of the land, but sudden massive landslides have also occurred in the region of Glacier Bay National Park (Figure 7.2).

Each year, NOAA issues an Arctic Report Card—the 2019 *Report* provided overwhelming evidence of rapid climate change. Arctic sea ice was thinning. Greenland's ice sheet was losing 267 billion metric tons of ice a year. The tundra was greening in areas from Canada, to the North Slope of Alaska to the Russian Far East. The thawing of permafrost was on track to release 300–600 million tons of net carbon a year into the atmosphere.[4]

Communities along the Bering Sea provide an example of the pervasive and far-ranging effects of an Artic with diminished ice. Indigenous peoples of the Bering Sea saw the months with sea ice drop from some eight to only three to four months a year. The warming waters threatened rich fisheries that have benefited from cold nutrient-rich water, and thinning ice made it dangerous or impossible to engage in crabbing or in hunting, and the loss of sea ice exposed coastal communities to greater storm surges. Routes across the ice became impassable, cutting communication and routes for travel between isolated communities. In a statement, entitled "Voices from the Front Lines of a Changing Bering Sea," a group of elders wrote

> We are losing so much of our culture and connections to the resources from our ocean and lands. Store-bought groceries and meat, which are flown in by airplane at great cost, are continually replacing our local,

> fresh and nutritious subsistence foods. Our food security is even more at risk as our young people increasingly face high rates of unemployment with few job opportunities, and are less likely to own a boat for hunting and fishing.[5]

Within the Arctic, climate change is altering the tundra biome. This climate region is characterized by very cold temperatures, short growing seasons, and vegetation composed of grasses and shrubs. Because of the cold temperatures here, permanently frozen soil, known as permafrost, is extensive. Much of the permafrost is now melting because of rising temperatures.

Permafrost in the tundra and other cold climate regions currently stores vast amounts of carbon, equivalent to almost twice as much as is currently in the atmosphere. The melting of permafrost therefore creates a scenario to further amplify warming: melting permafrost releases CO_2 as well as methane, which is about twenty-five times more efficient as a greenhouse gas compared to CO_2. In the tundra of northwest Siberia, the melting of permafrost has apparently given rise to a series of mysterious craters with a diameter of up to 1 kilometer. The precise mechanism responsible for creating these craters is a matter of debate. In one hypothesis, the warming has released methane, which then exploded under pressure. A second explanation attributes the formation of the craters in part to the rapid melting of ice. In northern Canada, melting permafrost causes landslides as earth slumps, pushing mud, and silt into waterways. Climate change is also creating drier conditions in the tundra. Warmer temperatures increase evaporation, and lower snow falls decrease the supply of water. Northern Alaska, for example, is becoming drier. Many lakes in the tundra are vanishing.

In the belt of vast boreal forests south of the tundra, the melting of permafrost has produced a phenomenon called drunken forests or drunken trees, in which trees list or tilt. Such drunken forests are visible in places

FIGURE 7.2 *(a) Muir Glacier in Glacier Bay National Monument, 1941. (b) Muir Glacier in Glacier Bay National Park, 2004. Source: http://nsidc.org/data/docs/noaa/g00472_glacier_photos/images/pair_example_ highres.jpg.*

FIGURE 7.3 *Drunken forest, Canada. Photo Credit Dr. John Cloud, NOAA Central Library.*

such as Denali National Park in Alaska as well as in Canada and in Siberia. Melting permafrost has also created new wetlands. A landscape dubbed thermokarst has appeared in some regions as ponds form due to melting permafrost. In areas of human settlement, collapsing ground undermines and damages roads, power lines, and buildings. Houses sink and tilt as the permafrost beneath them melts (Figure 7.3).

Glacial retreat

Along with higher latitudes, higher elevations are also especially sensitive to the effects of climate change to date. High-elevation mountain ranges often contain biomes different from nearby lower-lying areas. Even at lower elevations, mountains often provide habitats for particular species and have a major effect on regional river systems.

Rising average temperature in mountain regions on average reduces snowfall and shrinks glaciers. Temperature and precipitation interact to affect the rate at which glaciers retreat or grow. Thus, in principle, a glacier could expand in a modest warming period as long as sufficient precipitation fell as snow. Conversely, glaciers may not expand during colder periods if there is not enough precipitation. Warming has become the dominant factor affecting glaciers in recent decades, to the point where most glaciers are receding worldwide.

Glaciologists have documented the retreat of glaciers in many regions of the world, but this is one of many effects of climate change that anyone can easily perceive. Suppose you visited or saw an imposing glacier in the 1970s, 1980s, or 1990s, and had the chance to pay a return visit twenty, thirty, or forty years later. In case after case, the shrinking of the glacier would immediately be evident to the eye as a dramatic change in the local landscape. Returning visitors to Glacier National Park in Montana can see the retreat of glaciers for themselves. The same phenomenon is easily visible in the Alps: glaciers remain, but most have visibly receded at a rapid pace.

Glaciers in mountains in the tropics, from New Guinea, to East Africa, to the Andes, are also in retreat.[6] As recently as the late 1980s New Guinea's highest peak, Puncak Jaya, which stands 4,884 meters, or just over 16,000 feet in height, had five ice fields. However, two had vanished by 2009, and the remaining three had receded sharply. At the Quelccaya ice sheet in the Peruvian Andes, ice that took 1,600 years or more to form has melted in just 25 years.[7]

The melting of ice and permafrost at high elevations affects people in many ways. Glacial retreat in the Andes led to devastating floods in the middle of the twentieth century.[8] Ongoing melting leaves lakes trapped behind debris that exposes communities in countries like Bolivia to the risk of flooding.[9] Warming permafrost also poses dangers. Frozen ground beneath the surface acts as a kind of glue, holding together slopes that rise at rates that seem impossibly steep to the naked eye. Melting permafrost can lead to the sudden collapse of soil, and in the mountains that creates a higher danger of landslides. In 2006 a section of the east face of the Eiger, a famed Alpine peak near the mountain town of Grindelwald, Switzerland, fell down. Rockfalls in the high mountains are obviously nothing new, but the increased risk posed by melting permafrost creates greater dangers for mountaineers. Some climbing routes have become too dangerous to take.

Landslides that plummet into mountain lakes and reservoirs pose yet another risk. They can create the equivalent of a small tsunami, which can lead to floods and damage hydroelectric facilities as well as threaten nearby houses and communities in narrow mountain valleys. Landslides block roads and railways—even if a road or track is empty at the moment of rockfall, the debris shuts down roads and rail lines temporarily.

Melting glaciers and shrinking snowpack alter the mountain landscape and call into question the very identity of some locations long identified with ice, but the widest effects spread far beyond high mountain ranges. Shrinking glaciers and reductions in snowpack threaten to diminish water supplies in many regions of the world. In the tropical Andes glaciers contribute to water supply as well as to hydroelectric power, but glaciers are rapidly shrinking. A farmer in Peru described the problem: "The snow keeps getting farther away," says Melgarejo, a farmer worried about his livelihood. "It's moving up, little by little. When the snow disappears, there will be no water."[10] The water supply will not disappear entirely, but it will drop. The shift in water supply can be sudden, because glacial melting can actually lead to an increase in water downstream from glaciers, before the supply suddenly falls off.

Warming has combined with illegal wells and diversion of water to shrink lakes in high plateaus in Bolivia and in Iran. Lake Urmia, a saltwater lake at an elevation of more than 4,000 feet in Iran, has shrunk by 80 percent in thirty years. In Bolivia, drought, along with mining and extraction of water for agriculture, has choked off a reliable supply of water for Lake Poopo, a salt lake at 12,000 feet.[11]

In the Himalayas, large quantities of water are stored in the glaciers and snowpack. Indeed, the region has sometimes been described as the Earth's third pole. In fact, the amount of glacial ice still does not match that found in Alaska and Canada, but the concept points to the importance of the Himalayas as a source of water for much of South Asia, Southeast Asia, and East Asia. This is an extremely heavily populated region with India, Pakistan, China, and the countries of Southeast Asia, so the glaciers that stretch from the Hindu Kush through the Karakoram to the Himalayas make up one of the important water sources for a large fraction of the Earth's population. In all, some 1.3 billion people live in the basins of rivers fed by Himalayan snows and ice.

Glaciers are already shrinking and thinning in many areas of the Himalayas. In Jammu and Kashmir where the streams that feed into the Indus River begin, glaciers are diminishing, and the same trend is taking place at the starting points for the Ganges and Brahmaputra River. As elsewhere, melting can temporarily increase water supplies, making flooding more likely. But in the long term, the overall reduction threatens hydroelectric power and human populations as well as plants and animals that require water.

Reductions in snowpack already affect agriculture in major farming regions of the world. In the United States, California produces a large proportion of total food and is also a major exporter for agricultural products. It not only leads the United States in total agricultural output but is also the chief producer of a host of crops, including almonds, avocados, broccoli, grapes, lemons, lettuce, peaches, plums, strawberries, tomatoes, and pistachio nuts. It also leads the United States in dairy production. Remarkably, however, much of California does not have abundant rainfall. Rainfall in the Central Valley, which runs down the center of California, varies from some 20 inches a year in the north to desert conditions in the south. California's agricultural industry depends heavily on water from the snowpack in the Sierra Mountains in the east. In both the winters of 2013–14 and 2014–15 that snowpack fell well below previous norms. Measurements in the spring of 2015 determined that snowpack contained only 5 percent of the average amount of water, and reconstructions of tree rings indicated that this was likely the lowest amount in 500 years.[12] Heavy snows returned to northern California in the winter of 2016–17, but recent water shortages may be a harbinger of things to come in a future where continued warming deprives the Sierra Mountains of much of their snowpack.

Even when snow falls, warm springs accelerate melt, intensifying drought. In the four corners region of Utah, Colorado, New Mexico, and Arizona snowpack built up in the winter of 2019–20, but a warm spring rapidly melted the snow, ending hopes for a pause to long-term drought in the region. Moderate to extreme drought, as identified by the United States Drought Monitor, covers most of the area, including the lands of the Navajo nation.[13] Since the turn of the century, warming from climate change pushed what might have been into a moderate drought into a megadrought comparable to droughts that had severely challenged complex societies in the historical past.

Precipitation extremes

In multiple biomes, climate change is bringing precipitation extremes of heavy rains or snows and drought. Much of the world's population lives in or near temperate biomes. These biomes can vary greatly in their level of precipitation, from humid to arid regions. More prosperous residents of temperate zones have in some respects been better able so far to insulate themselves from the effects of climate change, but across the globe, human-driven climate change reinforces a pattern of extremes. Climate change has increased the frequency of extreme precipitation events. A warmer atmosphere and warmer waters provide storms with more potential energy, whatever the season. Much of the world's population lives in or near temperate biomes. These biomes can vary greatly in their level of precipitation, from humid to arid regions. More prosperous residents of temperate zones have in some respects been better able so far to insulate themselves from the effects of climate change, but across the globe, human-driven climate change reinforces a pattern of extremes. Climate change has increased the frequency of extreme precipitation events. A warmer atmosphere and warmer waters provide storms with more potential energy, whatever the season.

Extreme weather events provide signals of climate change for residents of temperate regions. No single severe weather event can be attributed to climate change, but climate science has made rapid strides in the ability to attribute severe weather to climate change, usually in terms of probability. Even in cases when analysis does not connect heat to climate change, the general trend toward warming exacerbates the effects of warm weather, boosting high temperatures and increasing rates of evaporation.

Recent decades have seen a sharp upturn in the frequency of 100-year events, weather events that statistically should occur an average of once a century, or have a 1 percent chance of occurring any given year. Residents of Washington State, for example, have experienced multiple 100-year floods on rivers within several years. Flooding elsewhere has occurred with greater frequency than predicted. Rains in Britain in the summer of 2007 were some 20 percent heavier than any recorded since 1879, and Britain again experienced severe flooding in the winter of 2013–14—in Oxford rainfall was unprecedented in nearly 250 years. In this case, climate change appears to have been a secondary cause of heavy precipitation.[14] It is not yet possible to prove that a single flood is the result of global warming, but the increased frequency of events shows a trend toward greater risk. A flood in New York in 2007 came after heavy rains that would occur on average every twenty-five years, but that was only five years before the highly destructive Hurricane Sandy in 2012. New York governor Andrew Cuomo quipped in 2012 that "we have a 100-year flood every two years now."[15]

With greater extremes of precipitation, climate change has also increased risks of flooding in inland areas. The First Street Foundation found in 2020 that many communities in varied regions of the United States faced hidden flood risks. The proportion of properties at risk of flooding in many instances greatly exceeded the proportion affected by previous FEMA (Federal Emergency Management Agency) flood maps. In Chattanooga, Tennessee, for example, First Street put the proportion of properties at risk of flooding at 42 percent instead of the 4.2 percent listed by FEMA and in Buffalo estimated the percentage of at-risk properties at 27 percent instead of the only 0.4 percent listed by FEMA. Overall, First Street found the number of properties at substantial risk of flooding at 1.7 times that estimated by FEMA.[16]

The city of Chicago shows the increased risks of flooding. The vast city is built on a Prairie wetland. The French explorer and missionary Marquette remarked on the extent of seasonal flooding when he first visited the region. As the city grew, planners embarked on a series of vast and grandiose projects to try to control water, culminating in the construction of the Tunnel and Reservoir Project or Deep Tunnel.[17] Despite or perhaps because of these efforts and with increasing precipitation extremes, much of Chicago is today vulnerable to flooding. Thus officials from the Metropolitan Water Reclamation District of Greater Chicago confirmed that flooding affects areas outside of FEMA's current flood maps. Interviewed by the *New York Times*, Kevin Fitzpatrick of the Water District said that the finding that some 13 percent of Chicago properties face the risk of floods from rains "would not surprise me one bit."[18]

The frequency of heavy snowfalls has also been increasing in temperate areas even as overall temperatures warm. This at first may seem counterintuitive, but a warmer atmosphere results in more evaporation, which leads to more moisture in the air. This in turn can lead to abundant snowfall when temperatures cool in the winter. Five of the ten highest snowfalls recorded in Boston, Massachusetts, for instance, have taken place since 1997, and all five of the snowiest seven-day periods have occurred since 1996 over a time span with records dating back to 1891.[19] So long as it is still cold enough to snow, warming can lead to more extreme snowfalls, though continued warming will eventually reduce the likelihood of snow.

Climate change has increased the likelihood of extremes of both precipitation and drought. Thus, the risk of severe, prolonged drought has risen in areas already prone to drought. As is the case for extreme precipitation, no single period of hot dry weather can be attributed to climate change. On one level that is true, but it is already proving possible to attribute some extreme events to climate change. Thus numerous studies have linked Australia's exceptional heat of 2013 to human-caused climate change.[20] An initial analysis of the massive Russian drought of 2010 did not find a connection between a heat wave and global warming, but a study found a high probability that the event would not have occurred in a world without global warming.[21] Moreover, even when a warming trend cannot be proven to have caused an individual drought, warming exacerbates drought by accelerating evaporation. In California, heat contributed to drought in 2014 and 2015. Lack of precipitation initiated drying, but research suggests that human-caused warming sustained and magnified the drought.[22] In the upper Missouri, higher temperatures led to a drought in the early twenty-first century unprecedented in the last 1200 years.[23] Droughts are posing a major challenge to human societies in multiple regions of the world. A massive drought struck northeastern Brazil in 2013 and continued into 2015. This drought did not cause famine: in that respect Brazil proved to be resilient, but farmers from the region lost their crops and livestock. Some were reduced to grinding up cactus to feed cows.[24] The drought expanded in the southeast of Brazil as well, reducing hydroelectric power and threatening water supply in major cities. Falling production of Arabica coffee beans led to global increases in prices. Losses of water from leaks in water systems and theft contributed to the shortages, but high temperatures and low precipitation exacerbated the crisis. In desperation, the state government of São Paulo turned to pipelines, but reservoir levels fell through the fall of 2015. The El Niño of 2015–16 boosted water levels, but the long-term challenge persists.

In 2020 and 2021, Turkey saw water supplies running short after extreme drought. Urbanization and dam-building exacerbated a water shortfall caused by lack of precipitation. By early 2021, Istanbul, Turkey's largest city, faced the prospect of depleting all water within a few months, and other large cities confronted similar conditions.

In temperate biomes, climate extremes are having especially dramatic effects on areas that are already arid. China's western regions, for example, have suffered from extensive droughts. In China as well as in adjacent countries of Central Asia, pastoralists have struggled to find sufficient water and food for their herds. Drought has also damaged agriculture in the region. The government of China has gone so far as to resettle some people as "ecological migrants."[25] Severe dust storms have reached major Chinese cities, including Beijing.

The effects of climate change in tropical biomes remain somewhat uncertain. It is certainly possible to find evidence of recent precipitation extremes. In early 2013 regions of northern Bolivia suffered the worst floods in twenty years before again enduring severe floods, the worst in sixty years, in February 2014. “Some people were saying it was the end of the world,” an indigenous leader said. “We were flooded as never before and left under a metre and a half of water. The waters killed our crops—bananas, cassava, pineapples, avocados, everything—as well as our pigs, ducks and chickens.”[26] In January 2015 Malawi in southern Africa suffered from severe flooding that killed 176 people; injured many more; destroyed crops, animals, and homes; and displaced a quarter of a million people.[27] The flooding also raised concerns about water contamination and the spread of epidemic diseases. Here, as in other tropical regions, deforestation and high population densities exacerbate the damage caused by heavy rains.

Regions of the tropics have also experienced major droughts. In East Africa and the Horn of Africa, precipitation extremes have created threats for agriculture and residents. Climate change increased the likelihood of a 2017 drought associated with El Niño.[28] The absence of rain deprived pastoralists of water for their animals. Oxfam reported on families of pastoralists on the move in search of water. “This drought is slowly killing everything,” one said. “First it ‘swept away’ the land and the pastures; then it ‘swept away’ the animals, which first became weaker and weaker and eventually died. Soon, it is going to ‘sweep away’ people. People are sick with flu, diarrhoea, and measles. If they don’t get food, clean water, and medicines, they will die like their animals.”[29]

Heavy rains created other problems. In the spring of 2020, farmers in East Africa faced a threat to their crops from huge swarms of desert locusts. Environmental degradation and desertification contributed to the expansion of deserts where the locusts can breed, and rains, more likely with precipitation extremes, created favorable conditions for a surge in locust population.

In South Asia, extreme precipitation events have increased in Central India. It’s not just that more extreme rain events are occurring, but that the rains are becoming more intense. At the same time, dry periods between heavy rains may also be more likely.[30]

Wildfires

In biomes across the world, in the tundra, boreal forests, temperate zones, and tropics, climate change increases the risk of fires. With hotter and drier conditions, lightning strikes are more likely to set organic material such as peat on fire. A lightning strike in 2007, for instance, started a fire along the Anaktuvuk River in Alaska, creating the largest known tundra fire. Some vegetation has recovered, but a higher frequency of fires in this biome will release large amounts of carbon previously stored in soil.[31] The consequences of fire in the boreal forests farther south are also severe. In this region, climate change creates greater potential for massive fires, whether started by lightning or by humans. In 2016, the residents of Fort McMurray in Alberta, Canada, had to evacuate their city because of a large wildfire. This large fire in Alberta cannot be attributed solely to climate change, but the very hot dry conditions that aid the rapid spread of the fire will occur more frequently as global warming continues.

Fire on an unprecedented scale struck the Arctic in 2019 and again in 2020. In 2019, smoke from wildfires spread across much of Siberia, Alaska, Canada, and Greenland. Extreme heat covered much of Siberia and temperatures reached 100°F or 39°C in the Siberian town of Verkhoyansk. Trees in the Taiga (the vast belt of boreal forest), grass, shrubs, and in some places even the soil burned as peat-rich earth caught fire. The Worldwide Weather Network Attribution network calculated that climate change made the Siberian heat of January to June of 2020 at least 600 times more likely. Climate scientist Dr. Sarah Kew explained that this "would have been almost impossible without human-induced climate change." The Arctic fires released more carbon into the atmosphere in June 2020 than in any month in an eighteen-year record of measurement.[32] Heat and fire intensified feedback effects, by increasing overall carbon emissions and by releasing more methane previously stored in permafrost. Burning peat also released carbon into the atmosphere, and soot that was produced by Arctic fires absorbed more heat.

Threats from wildfires are increasing as well in temperate regions. In California, large wildfires in 2019 and again in 2020 tore through populated areas close to major cities and lowered air quality. The wildfires in 2020 in California and in the Pacific Northwest burned through regions accustomed to fire. In 2020, fires reached the Willamette valley of Oregon and suburbs of the city of Portland. Multiple factors cause such fires, but dry conditions and high heat provided more fuel to burn.

In Europe, severe wildfires have damaged regions, including the Mediterranean. In 2019, wildfires in the Canary Islands forced evacuations and led firefighters to declare that "we are fighting for our island!!!"[33] Wildfires caused evacuations in locales across the southern Mediterranean in France.

The wildfires that struck Australia in 2019–20 gained particular notoriety. High temperatures contributed to a massive fire season with wildfires trapping entire communities. In December, Australia experienced its hottest day on record with average high temperatures of 41.9°C or 107.4°F. Australia experienced its warmest year ever in national records going back to 1910.[34] Fires grew so intense that they generated their own weather patterns and even thunderstorms. The smoke and ash lowered air quality:

at one point, the Australian capital Canberra had the worst measured air quality in the world.

The fires led to numerous evacuations. In Mallacoota, a small town on the coast of East Gippsland, residents and tourists huddled along the seashore. On January 4, some 1,000 people and their pets got out of Mallacoota on a naval ship. In late December 30,000 people were advised to evacuate from East Gippsland, but then told not to do so because fires had made roads too dangerous. In early January 2020, Australian authorities told 240,000 people in the state of Victoria to evacuate.

The wildfires took a heavy toll on animals as well as people. It is difficult to confirm numbers, but one estimate put the number of animals and birds killed by the wildfires at some 1 billion. A 2020 report for the World Wide Fund for Nature put the number of animals killed or displaced at 3 billion. The Chief Executive for the World Wide Fund for Nature Australia said, "It's hard to think of another event anywhere in the world in living memory that has killed or displaced that many animals. This ranks as one of the worst wildlife disasters in modern history." In Australia, the wildfires caused deaths and injury to iconic animals, including the koala. Intense and widespread fires damaged plants and trees, many unique to Australia. Fires killed some of the remaining Nightcap Oak, a rare tree whose ancestors date back to the continent of Gondwandaland.[35] Trees such as the eucalyptus tree are adapted to survive and even flourish from periodic wildfires, but the new scale of Australia's wildfires threatens to upset that balance. Eucalyptus trees re-sprout after fire, but the intensity of recent fires raises questions about whether this will continue at previous rates.[36]

Climate change and human practices are also affecting wildfires in the tropics. In Brazil, droughts in the Amazon River Basin aided in the spread of wildfires in 2005, 2007, and 2010, and there were further wildfires in 2013 and 2014. Individual acts by people can start specific fires, but climate conditions have intensified fires that burn in the understory of the rain forest far below the top. Analysis of satellite data carried out by the National Aeronautics and Space Administration (NASA) found that low nighttime humidity made such fires more likely.[37] During an El Niño–related drought of 2015–16, large parts of Indonesian tropical rain forests in regions such as Sumatra and Kalimantan (the Indonesian portion of Borneo) burned, including areas that provided habitats for endangered species, such as orangutans. Here as elsewhere, land-use patterns exacerbated the problem. In this case deforestation is under way to clear land for palm oil. During the period of the fires, Indonesia became one of the world's largest emitters of carbon.

The year 2019 was a year of fire in the tropics from the Amazon to Southeast Asia. In the Amazon, the extension of ranching and of soybean cultivation led to a wave of deliberate fires to clear land. In Southeast Asia, demand for palm oil provided a similar incentive to burn forests. Brazil saw widespread fires in the Amazon in 2019. Government policies and human practices combined to increase environmental stress and raise greenhouse gas emissions. Climate change in itself did not cause fires, but South America overall experienced the second highest temperatures in the 110-year NOAA dataset. The practice of setting fires between June and December in the southern Amazon accelerated under a government eager to open up the rainforest for gold mining, logging, ranching, and soybean cultivation. Fires themselves released carbon, and the residue of dead trees made sizable parts of the area into net carbon emitters.

Shifting biomes

Climate change is shifting biomes and intensifying the effects of human encroachment on animal and plant habitats. Warmer temperatures affect forests in much of the American West. Most public discussion of warming focuses on high temperatures, but the warming trend has also led to higher daily low temperatures. The daily low temperatures affect the survival rates of larger animals as well as insects and ticks. In western North America, the trend toward higher minimum temperatures has increased the population of bark beetles. Beetles now feed on trees for much longer periods of time than in the past, and on trees at higher elevations as well as on younger and older trees. The age of trees and past efforts to stop forest fires have contributed to the scale of the infestation, but rising beetle populations are playing a part. Beetles have killed off large numbers of trees in locations including Alaska, British Columbia, Colorado, and Montana. Heat and drought also appear to be weakening aspen trees in Colorado, contributing to sudden die-offs. Further south, beetle infestations have killed trees in Mexico. Determining the precise interaction between beetles, climate, and deforestation poses a complex scientific problem, but in a best-case scenario, persistent drought places western forests under increasing stress and improves conditions for beetles to thrive.

Warming is shifting the range of many species to higher elevations. The range for species of birds in the Cerro de Panticola raidge of the Andes in southern Peru has moved higher up the mountains, but this has also had the effect of reducing the range in which many species can live. Thus, the range for the versicolored Barbet shrank by two-thirds.[38]

The rapid changes in biomes and in the earth's oceans are creating new threats for humans, animals, and plants. We have created and confront a very real danger of mass extinction. Along with climate change, the world also now confronts a related crisis of biodiversity. An international assessment by the Intergovernmental Science-Policy Platform on Biodiversity and Ecosystem Services in 2019 demonstrated extreme and rapid decline. The chair of the organization explained, "There is no question we are losing biodiversity at a truly unsustainable rate that will affect human wellbeing both for current and future generations."

Development, extraction, rapid depletion of natural resources across wide areas of the world, consumption and waste of water, and pollution have destroyed habitats for a myriad of species. Where they survive, remaining populations are concentrated into increasingly fragmented habitats. Climate change creates additional stress for these already vulnerable populations. Changes in temperature range and available foods displace both plants and animals, creating new patterns of contact.

Human encroachment on animal habitats not only endangers species that are already at risk of extinction but also increases the chances of pandemics through zoonotic diseases or diseases that cross from animals to humans. As humans move into animal habitats in order to build housing or to expand agriculture, they also come into closer contact with a range of animals, boosting transmission of zoonotic diseases. Such zoonotic diseases include HIV, Ebola, SARS, MERS, and notably (for humans around the world) quite possibly Covid-19. Climate change did not cause the spread of these diseases, including Covid-19, but climate change is combining with other threats to the environment to multiply possible pathways for diseases to cross from animals to humans.

Climate change also aids the spread of vector-borne diseases. These are diseases in which vectors, typically mosquitos, ticks, or fleas, spread pathogens or organisms that cause disease. Vector-borne diseases include malaria, dengue, yellow fever, Lyme disease, and many others. Such vectors can also spread pathogens that transmit certain zoonotic diseases. Vectors in general thrive in a warming world, though the relationship between a particular vector, temperature, and precipitation can be complex. Warming creates a longer season and a bigger range for the spread of vector-borne diseases.[39]

Climate change also creates at least the potential prospect of releasing pathogens that have long been locked in ice. In 2016, in Siberia, the hospitalization of more than 100 people and the death of a twelve-year-old boy was attributed by some to the release of anthrax caused by spores released from melting permafrost.[40] However, other factors, including the suspension of vaccinations and increase in reindeer population, may have caused the outbreak.[41]

Rising seas and coastlines

From the tropics, through temperate zones, to the Arctic, climate change has brought rising sea levels. Sea level has risen more than 200 millimeters, or around 8 inches, since 1880. The pace of sea-level rise has increased in recent decades, from an average of 1.7 millimeters per year during the twentieth century to nearly twice that rate, 3.2 millimeters per year, since 1993. There are two main contributors to global sea-level rise—one is the melting of ice caps and glaciers, which currently accounts for about two-thirds of the observed increase. The expansion of water as it warms, thermal expansion, accounts for the remainder of sea-level rise to date. The rates can vary locally due to land subsidence or rebound as well as gravitational effects. Along the Gulf of Mexico coast, sea level rises at a faster rate than the global average due to the sinking of land there.[42] Parts of Alaska, in contrast, see falling sea levels due to the continuing rebound of the land surface even after glacial-era ice sheets melted by 6,000 years ago.

The present-day melting of glacial ice adds a complex picture to regional sea-level rise. In regions with large ice sheets, such as Greenland and Antarctica, the gravitational attraction between ocean water and the massive ice sheets causes sea level to rise locally—the gravity from the ice sheets essentially pulls ocean water toward it. When this ice melts, it will add water to the ocean and therefore increase global sea level, but the loss of mass locally will remove this gravitational pull, and local sea level will drop in response. Locations close to the ice sheets will therefore experience less sea-level rise, or even a drop in sea level, while locations farther afield will see even greater sea-level rise. The locations most affected by this shifting of the Earth's mass will depend in large part on which massive ice sheets lose the most mass—those in Greenland or Antarctica. Adding to the shifting of gravity will be the crustal rebound associated with the mass loss.

The effects of sea-level rise to date have been most dramatic in low-lying areas, including small island states. In the Indian and Pacific Oceans, several island states already face a dire threat to their future. Kiribati, one of these threatened states, is made up of several atolls and reefs in the Pacific Ocean, approximately 1,000 miles south of Hawaii. The total population of approximately 102,000 people lives mainly in the chain of the Gilbert Islands, with the largest number on the island of Tarawa. Almost the entire area lies below an elevation of 5 meters above sea level, and some of the reefs and atolls barely rise above the sea. Rising seas in such a low-lying area are already damaging the water supply. The government of Kiribati has purchased land in Fiji to provide a possible eventual refuge when rising seas displace the population.

Tuvalu, a series of reefs and atolls in the Pacific Ocean between Australia and Hawaii, faces similar threats. Higher tides bring saltwater deeper into the islands. In 2014, Tuvalu's prime minister, Enele Sopoaga, described his country's predicament, "We are caught in the middle, and certainly in Tuvalu, we are very, very worried—we are already suffering." Sopoaga added, "It's already like a weapon of mass destruction, and the indications are all there."[43] There are also low-lying island nations in the Indian Ocean, including the Maldives. The islands that make up the Maldives have a population of some 400,000. The highest of these islands is not more than 2.4 meters above sea level.

Collectively these island nations have been termed small island developing states. They will not wash away entirely in the immediate future, but they face a major common challenge, as their organization stated: "As their population, agricultural land and infrastructure tend to be concentrated in the coastal zone, any rise in sea-level will have significant and profound effects on their economies and living condition."[44]

These and other low-lying islands contain only a small fraction of the world's total population. They therefore contribute very little to overall carbon emissions or to human-driven climate change. They demonstrate a larger problem that extends far beyond the comparatively small island states: many of the populations that are and will be most affected by climate change have played little role in creating the conditions that will shape their future. They have no chance, on their own, of making cuts in emissions on the scale that would be necessary to curb the worst-case scenarios for sea-level rise.

The threat is particularly great in low-lying island states, but sea-level rise is already affecting populations in other regions around the world. Indeed, sea-level rise is already having a disproportionate effect on human populations because such a large share of the world's human population, some 40–44 percent, lives in coastal areas. In both developed and developing countries, large populations dwell along the coasts. In the United States alone, many major cities are clustered along the seaboard on both the Atlantic and the Pacific coastline. As of 2010, some half of the total US population lived within 50 miles of the coast, and as of 2019 94.7 million lived in coastline regions. Much the same patterns of habitation can be found in Central and South America, Africa, Asia, and Europe. Some large population centers that do not lie directly on the coast are located in basins of tidal rivers. London is a prime example.

In many regions the most immediate effects have been felt with what is termed "nuisance flooding." Higher waters at high tides makes minor flooding more likely even without major storms. A 2020 report on high-tide flooding by NOAA observed, "Evidence of a rapid increase in sea level rise related flooding started to emerge about two decades ago, and it is now very clear."[45] Such floods may briefly close roads or force businesses and homeowners to purchase pumps to remove water from basements. This innocuous-sounding phrase is accurate insofar as it describes the nuisance that can be caused by minor flooding, but fails to convey the reality of a trend toward greater flooding that will create far more damage and risk than minor irritation and inconvenience. In places where a hurricane may have caused floods in the past, smaller events can cause similar effects, and a large hurricane would cause much greater flood damage than in the past.

The area around Newport News, Norfolk, and Hampton Roads in Virginia provides a striking example of the effects of sea-level rise. Local homeowners, businesses, and nonprofits must now regularly deal with flooding. Municipalities and homeowners are already raising buildings. Some, who can afford to do so, pay to have their houses raised up on jacks so that new, higher foundations can be poured. Local residents have come up with different ways to navigate their region as flooding closes roads. The combination of sinking land and rising sea level also creates problems for the US military, which operates the world's largest naval base at Norfolk, Virginia. The US Navy is raising piers. Defense contractors have also had to raise their power supplies to get them farther away from the water.

Such flooding has become increasingly common in northern Chesapeake Bay. In Annapolis, Maryland, nuisance floods that occurred about four times a year in the 1950s had increased to the point where they occurred about forty times a year by 2014.[46] In Washington, DC, nuisance flooding strikes with more frequency along the Potomac River, affecting neighborhoods such as Georgetown. A coastal community on an island in the Chesapeake is already experiencing dramatic threats from rising seas. A report from the US Army Corps of Engineers from 2015 found that just a third of the land area of Tangier Island in Virginia's area of Chesapeake Bay remained of the land that had existed in 1850.[47]

High-tide floods (HTF), sometimes called sunny-day floods, or, if they coincide with especially high tides, king-tide floods, occur in many regions. In 2019 alone, HTF records were set along the Texas coast and in Savannah, Charleston, and Miami. Large parts of the state of Florida are located at very low elevations, including the city of Miami, where sea-level rise exacerbates flooding. Miami is especially vulnerable because it is built on limestone, and water can easily seep up from below streets and foundations even as higher seas make flooding more likely during high tides. The problem is especially severe along Miami Beach, which is built on barrier islands. Along the Gulf of Mexico, coastal communities have been particularly hard hit, showing the fastest rates of sea-level rise in the United States. In southern Louisiana, where local sea-level rise tops 9 millimeters per year, floods frequently cut off the small community of Isle de Jean Charles. More than 1 million people in Louisiana alone live lower than 6 feet above the high-tide line.[48] NOAA noted that "the cumulative toll of impacts are becoming disruptive and damaging within many coastal communities." These included flooding or homes, erosion of roads, and damage to wastewater systems.[49]

For large populations in South Asia, rising seas create a far greater threat than mere nuisance. In Bangladesh, higher seas bring damage for both rural and urban populations. Rising waters at low-lying villages in the Ganges Delta have ruined freshwater supplies and increased the salinity of soil. Storms cause greater damage, and many villagers have found themselves forced to move. The encroaching waters have combined with other factors, such as economic incentives, to increase migration to Bangladesh's capital, Dhaka. The International Organization for Migration calculated that some 70 percent of migrants to Dhaka moved after some kind of environmental hardship.[50] Seasonal migration in Bangladesh between the countryside and the city has long provided rural residents with a source of income and food, but many onetime seasonal migrants are no longer returning to their former homes. This will not, however, provide a secure long-term refuge if sea-level rise continues unabated because Dhaka itself lies at an elevation of only a little above 50 feet above sea level, and parts of the metropolis, including slums filled with rural migrants, lie at even lower levels. Dhaka is not alone in facing a growing risk of floods. Many other major Asian cities, including Mumbai, Ho Chi Minh City, and Shanghai, lie along coasts.

The numbers of residents at risk from rising seas is far greater than even recent estimates because previous elevation measurements based on satellites often referred to the tops of trees and buildings rather than to the elevation at ground level, especially in built-up urban areas. Projections that incorporated data from a digital elevation model present a far more dire picture. Some 70 percent of those affected live in China, Bangladesh, India, Vietnam, Indonesia, Thailand, the Philippines, and Japan.[51]

In Southeast Asia, sea-level rise combined with damage to mangroves to displace residents of low-lying areas from agricultural land and inundate former fish ponds. Cutting down mangroves to create pools to raise prawns created further problems "Before, we had a good life with rice and fish," one village head observed. "Now we only have a memory of agricultural land." Local people have responded by creating rows of brushwood in shallow water to try to trap sediment.[52]

On the Mekong Delta, salt water intrusion is reducing soil quality and removing access to clean water in a key agricultural area for Southeast Asia. The coast is eroding along the southwest of the Delta. Climate change combines with the effects of dam-building up river and sand mining to propel many in the population of some 18 million to migrate out of the Delta.[53]

Africa is also home to many cities threatened by sea-level rise. Dakar, the capital and largest city of Senegal in West Africa, has suffered from numerous floods in recent years. Rising sea levels exacerbate the problems caused by rains. Mayors from other coastal towns in Senegal have reported repeated and persistent flooding. Lagos, the largest metropolitan area in Nigeria, also lies close to sea level. Much of the urban area stands at less than 2 meters' elevation. East African coastal areas face similar risks. Low-lying cities such as the ancient port of Mombasa in Kenya have experienced heavy flooding in recent years.

Oceans

Sea-level rise is one indicator of the climatic changes occurring in the ocean as a result of global warming. Much of the warming that occurs in the atmosphere is soaked up by the ocean, increasing ocean temperatures and ocean heat content globally. Surface waters have warmed by around 0.5°C since the 1970s, with an average warming rate of 0.11°C per decade.[54] This translates into an increase in heat content on the order of 100TW, nearly six times the amount of energy that humans use globally.

The increase in ocean temperature has implications not only for rising seas but also for marine life as well. Warming oceans have already led to movement in fish stocks. Anglers off the west coast of North America in 2014 noticed species of fish that typically would only be found well to the south. Waters off the coast of Alaska became so warm that fish such as skipjack tuna turned up, even though there had not been a documented case of such a fish appearing in Alaskan waters since the 1980s.[55]

Drought and warmth combine to threaten some fish. In California, low precipitation and high temperatures imperil the run of the Chinook salmon. Continued warming could eliminate the southern end of the range in which salmon have previously lived.

Warming oceans reinforce the effects of overfishing to further endanger already depleted fish stocks that are sensitive to water temperature shifts. Waters off of New England provide an example of one such ocean region. For years fishermen and federal regulators have disputed the condition of the cod fishery off the coast of New England in the Gulf of Maine. The once-ubiquitous species, the source for the name of Cape Cod, has become increasingly hard to find. In an effort to restore the cod population, NOAA has created increasingly sharp limits on cod fishing. This plan addresses the effects of overfishing, but warming waters may also be contributing to the movement of cod. Warming in the Gulf of Maine affects the balance of sea life. Along with cod, the population of northern shrimp has declined sharply, and federal regulators closed down the entire northern shrimp fishing season in the winter of 2014–15. The population of green crabs, an invasive species introduced from Europe, has increased sharply as the Gulf has warmed, and the green crabs have in turn reduced the number of soft-shell clams. Black sea bass, a fish species previously found to the south, have also been turning up in larger numbers. Some fishermen are concerned that the Black Sea bass will eat small lobsters, but Maine's lobsters face another threat from warming waters. In Long Island Sound, the lobster fishery has collapsed. This is the very southern edge of the range of the lobster. Pollution may have contributed to the disappearance of lobsters from these waters, but the lobsters are also suited for colder waters. If the waters of the Gulf of Maine continue to warm, lobsters could move away from the region altogether. This would certainly not be the most drastic outcome of human-driven climate change, but it would be a blow to an iconic animal and industry.

Warming waters have ravaged kelp forests in many regions. Eastern Tasmania has lost 95 percent of its once-abundant offshore kelp forests. Sea urchins moved in and ate much of the remaining vegetation. Warming combined with other factors, including a change in upwelling patterns and a disease that damaged the sunflower sea star, a species that eats sea urchins, to severely damage bull kelp forests off the coast of northern California. Populations of sea urchins increased exponentially, making a recovery of the kelp forests impossible.[56]

Warming of oceans has to date had different effects for individual species, but on the whole warming oceans have resulted in an overall reduction of yield, some 4.1 percent between 1930 and 2010 according to a 2019 study.[57] The effects of warming combine with overfishing to reduce populations of many species.

Warming waters have also given rise to more marine heatwaves, extended periods of anomalously warm sea surface temperatures. Such extreme warming events, which can post greater risk in some cases than steady warming, have severe impacts on marine species and ecosystems such as coral bleaching and mortality of organisms, including economically important fish. Associated with the observed increase in sea surface temperature, marine heatwaves have increased in frequency and duration since the 1980s, a trend that is attributable to human carbon emissions. Those trends are projected to accelerate into the future, with more frequent marine heatwaves increasing in intensity.[58]

Warming ocean waters have the potential to fuel more intense storms, especially tropical cyclones. The year 2020 was a record-setting year for Atlantic hurricane activity, with thirty named storms, including two back-to-back catastrophic cyclones that devastated Honduras and Nicaragua in November. Because such storms derive their energy from warm waters, the increase in sea surface temperatures could be expected to increase hurricane activity, but the available historical data on tropical cyclones poses a challenge in detecting trends. There is broad consensus that storm surge, or the increase in water level associated with a coastal storm, will reach higher levels on land due to global sea-level rise, resulting in more inundation. There is also growing confidence that heavy rainfall associated with tropical storms, such as the United States experienced with Hurricane Harvey in 2017, will become more common.[59] While frequency of hurricanes globally does not appear to be increasing—changes in wind shear, which also affects cyclone formation, may reduce overall hurricane numbers—there is evidence that the number and proportion of catastrophic storms (category 4 and 5 on the Saffir-Simpson scale) are on the rise.[60] This increasing intensity of storms, along with higher storm surge and heavier precipitation, combines to make hurricanes more destructive when they make landfall.

Another consequence of warming waters is the expansion of marine dead zones. Areas of ocean waters experiencing hypoxia, or low oxygen conditions, have been increasing rapidly since the 1950s. While some areas of hypoxia occur naturally, the size and number of these dead zones have increased in regions where excess nutrients are delivered to the coastal ocean. Nutrients are used on land for fertilizing crops, as well as lawns and golf courses—any nutrients that are not taken up by those plants are washed into streams, rivers, and, eventually, the ocean. These excess nutrients promote algal growth in surface waters. When the algae die and sink to deeper water, their remains are decomposed, and that uses up oxygen from deep waters. Compounding the effect of excess nutrients, coastal waters are typically stratified, or layered, especially in the summer. With stratified water, oxygen in the surface ocean cannot mix into deeper waters, and hypoxic conditions develop. Warm surface waters also contain less oxygen than colder waters, which magnifies low oxygen conditions further. Many animals such as fish can move to another location when faced with limited oxygen, but less mobile animals, such as crabs, cannot relocate readily and therefore perish under hypoxic conditions. On top of their impact on marine life, dead zones have negative economic consequences as well.

The effect of climate change on the occurrence of hypoxia is complex, but overall dead zones are expected to worsen as temperatures increase and precipitation patterns shift.[61] As ocean waters warm, they become more stratified—this inhibits the mixing of oxygenated surface waters into deeper layers. Warming can also lengthen the hypoxic season—currently a summertime phenomenon in many locations, the extended summer seasons will result in prolonged hypoxic conditions.

Precipitation also influences the development of dead zones. Areas that receive higher precipitation also see higher river runoff, which enhances the delivery of nutrients to coastal regions and also makes the waters more stratified. The most extensive dead zones that have occurred off the coast of Louisiana, the largest dead zone in the United States, typically develop during years of Mississippi River flooding. Efforts to model the effect of climate change on dead zones show that, with the projected increase in temperature and precipitation for this region, severe episodes of hypoxia will increase in duration and frequency.[62]

Off the east coast of Australia, the Great Barrier Reef provides one of the most striking examples of the threat of global warming to coral reefs. Warmer waters, human activity, and pollution combine to damage the reef itself and many of the organisms that depend on it. Climate change is vastly increasing the likelihood of ocean warming that contributes to bleaching events, which occur when the algae that live symbiotically within the coral reef structure are expelled due to rising temperature. The Southern Hemisphere summer of 2016–17 brought a major bleaching event to the Great Barrier Reef. Coral reefs in some regions have remained resilient, but episodes of bleaching show that climate change is already a threat to these remarkable formations.

Warming temperatures and associated bleaching events, along with pollution, are not the only threat to coral reefs. Ocean waters are becoming more acidic as a result of the rise in CO_2. As atmospheric CO_2 increases, so too does CO_2 dissolved in the surface ocean, though this uptake may decrease as ocean waters continue to warm. When CO_2 dissolves in water, it reacts with the H_2O to create carbonic acid (H_2CO_3), which then dissociates into ions. The overall result is an increase in the water's acidity, or ocean acidification, indicated by a decrease in seawater pH. The ocean has absorbed an estimated 30–50 percent of the CO_2 released by the burning of fossil fuels;[63] the acidity has, in turn, increased approximately 30 percent since the beginning of the Industrial Revolution.[64]

Acidification poses a threat to any organism with a shell made of calcium carbonate because the added acidity inhibits carbonate shell formation. The effects are already evident in areas such as Puget Sound, which has more corrosive waters due to the upwelling of CO_2-rich waters. In the Pacific Northwest broadly, acidic waters have started to dissolve shellfish. Larval oysters started to perish in large numbers as early as 2005. Area shellfish producers responded by adjusting the water acidity in hatcheries for oysters.

Climate conflict

Rapidly transforming biomes around the world, intensifying climate change has also placed human societies under increasing stress. Climate change acts to magnify both conflict and inequality. Strife, war, and gaps in wealth and opportunity stem from many causes, but as of the twenty-first century climate change amplified competition and conflict. Such rivalry took numerous forms, including the pursuit of mineral and fossil fuel reserves, disputes over water, and possibly even armed conflict.

In the Arctic, the potential for exploiting warming to extract natural resources has led to new competition. Canada and Russia sought to assert sovereignty in the Arctic. Both countries carried out military exercises. The claims for sovereignty inspired a new wave of interest in identifying the boundaries of continental crust, including the ownership of the Lomonosov Ridge, a submerged ridge between Russia and Greenland. Russia based its sovereignty claims on the Lomonosov Ridge, which, according to Russia, is an extension of that country. Canada made similar arguments about the ridge because its southwestern edge sits at the location of Ellesmere Island, a large island in the Canadian arctic. In 2007, a Russian submarine planted a flag deep in the ocean beneath the North Pole. In 2014, Denmark countered that the area around the North Pole was connected to the continental shelf of Greenland, which belongs to Denmark.

Climate change is also exacerbating competition for resources from oceans. Extraordinary heating of ocean waters exacerbates the effects of unsustainable fishing practices that have already reduced supplies of many species of fish and seafood.[65] As warming oceans shrink some populations of fish and seafood and lead others to move toward areas further north, conflicts emerge over access to fish and shellfish. Waters around Iceland, a country historically heavily dependent on fishing that even engaged in so-called Cod Wars or maritime altercations between vessels from Iceland's Coast Guard and the UK's Royal Navy, have warmed between 1.8 and 3.6°F over twenty years.[66] Iceland and the EU have disputed Iceland's quotas for mackerel. Iceland has increased its quota as mackerel populations shift west, angering Scotland and in particular the fishing industry in the Shetland islands.

Off the east coast of North America, in the Gulf of Maine, lobstermen from Maine and New Brunswick have disputed lines of demarcation as lobster have shifted north and east because of warmer waters. Both claim rights to lobster from waters near a small island called Machias Seal Island on the edge of both the Gulf of Maine and the Bay of Fundy.

In a world with wars that have produced extreme violence and suffering, it is easy to overlook these conflicts, but as oceans continue to warm, the movement of surviving fish species coupled with overfishing is likely to lead to more conflicts over fisheries, to so-called fish wars.

On land, drought exacerbated by climate change contributes to tension and discord in many settings around the world. Even in cases in which a particular drought cannot be attributed to climate change, higher heat increases evaporation. In developed countries drought has set off competition between water users. In the American West, shortfalls of water have pitted consumers, often urban and suburban residents, against farmers and agricultural producers. In Texas, farmers sued when regulators sought to restrict water supply in order to keep water flowing to residents and industries. Water rights have been divided up, with precedence often awarded to the oldest claims, but population growth and development have placed far more stress on water supplies, and drought has intensified emerging conflicts.

The lines of conflict are often complex. California, for example, is not the dominant almond producer just in the United States but also in the entire world. Production tripled since the 1990s with rising demand domestically and growing exports globally. Almond cultivation requires extensive water, but so too do other crops, notably alfalfa, and it is difficult to identify which crops are most deserving of water when supplies run short.

Drought has also intensified competition for water between major agricultural water consumers and Californians who wanted to protect the state's population of Chinook salmon. Warm shallow waters place the Chinook salmon at risk in the Klamath River system in northern California, even though much of the area's waters are now diverted to sustain cultivation of almonds, other nuts, and fruits in the San Joaquin Valley to the south. A winter of heavy snow may relieve the competition in the short run, but in a warming world, a return to drought is likely to lead to renewed tensions over water.

States also compete for water. In the western United States, seven states share water from the Colorado River Basin, but total demand exceeds supply. Arizona and California, in particular, dispute control of water from the Colorado River. Dams and irrigation divert so much water from the Colorado River that it no longer flows to the sea, but climate change has placed additional stress on the states, communities, and economic sectors dependent on the River. Lake Mead has not been full since 1983, but the

Colorado River's waters support agriculture, residents, and recreation, and shortages now pit these interests against each other. Areas of the southwestern United States and northern Mexico share many of the same water resources and competition for water is affecting both countries and crossing national boundaries. Mexican farmers in 2020 seized a dam in the state of Chihuahua to try to stop water from reaching the United States. Use of water taken from the Conchos River to irrigate corps exceeded the proportion permitted under a 1944 treaty. The Mexican government agreed to make up the water "debt."

Similar competition has intensified in many drought-affected regions and countries, including Brazil. Plans to divert water from the São Francisco River, a major river in eastern Brazil, to the northeast raised debate. Opponents charged that the project would benefit large agricultural interests over residents of the arid northeast. Brazil's largest cities have also disputed control over water. Drought and water shortages in late 2014 generated conflict over water resources between the cities of São Paulo and Rio de Janeiro. Rio objected to a São Paulo plan to tap a reservoir that feeds water to Rio.

Competition for water also pits countries against each other. In northeast Africa, Ethiopia and Egypt have disputed the use of the waters of the Nile River. The very first civilization to emerge in Egypt depended on the Nile and its annual flooding for irrigation. Modern Egypt, if anything, relies on the Nile even more now, not only for irrigation and water but also for electricity. The enormous Aswan Dam on the Nile has generated as much as half of Egypt's electricity. Plans for constructing a large dam in the headwaters of the Blue Nile, one of the two main tributaries of the Nile, in Ethiopia have caused alarm in Egypt as well as in Sudan. Named the Grand Ethiopian Renaissance Dam, it is projected to supply electricity, but Egypt has voiced concern both about the rate of fill and about loss of control over water supplies.

In 2011 construction started on the Grand Ethiopian Renaissance Dam that would provide power to as many as 65 million Ethiopians. Building a dam upriver sparked concerns in Sudan and Egypt about a diminished flow of water to regions that have been dependent for millennia on the Nile's waters. Egypt, in particular, has objected to the project. Efforts by the African Union, the UN, and by countries, including the United States, have failed to produce an agreement, in particular about what to do in drought and how to define drought conditions.[67]

In Southeast Asia, disputes over water have also emerged along the Mekong River. China has invested heavily in dam-building on the river and in Laos, Chinese-financed hydro projects have displaced villages. Cambodia, Laos, Thailand, and Vietnam faced droughts in 2019 even as the upper Mekong River experienced a "relatively average" wet season.[68] In 2020, Cambodia, Thailand, and Vietnam asked Laos to carry out a "transboundary impact assessment." Cambodia, in an official reply form urged the need to "protect and conserve the habitats, preventing the loss due to the impact from the Mekong River Cascade Dams, particularly for key fish species that require flowing water for migration."[69]

In the Western Asia and the Middle East, similar tensions have emerged between Turkey and Iraq as Turkey builds dams upstream from Iraq. Turkey is constructing numerous dams in its southeast as part of a Southeastern Anatolian Development project. The construction of the Ilisu Dam on the Tigris River in southern Turkey stands out for its size and potential impacts. The 440-foot-high mile-wide project cost some $2 billion and was projected to provide power for 1.3 million people. It displaced 70,000 people and flooded ancient fortress, ruins, and archaeological sites at Hasankeyf and cut off paths for fish to migrate upstream. The dam has raised concerns in Iraq that water levels in the Tigris downstream in Iraq could fall precipitously during dry years. Reduced flows of water would also threaten marshes in southern Iraq, especially as sea waters and salt levels rise.[70]

Water shortages have contributed to growing tension and even violence between residents of East Africa.[71] The precise connection between precipitation and conflict remains complex. Pastoralists in recent decades may have been more likely to enter conflict during years with plentiful rain when livestock raiding increases.[72] However, extreme deviation from the norm, either years of extreme drought or heavy rain, is associated with social conflict.[73] In Ethiopia and Kenya, pastoralists have increased their range in search of forage for their animals, but this search for food has also increased friction and conflict over water. Declining water levels in Lake Turkana of northern Kenya have led pastoralists to move farther in search of water, increasing the chances of conflict. Indeed, interviews conducted in 2014 by Human Rights Watch described such conflicts.[74]

In West Africa the use of the waters of Lake Chad has set off competition for a valuable dwindling resource. A once large, but shallow, lake in the Sahel grasslands on the south of the Sahara, Lake Chad has lost a large part of its surface area in recent decades—from approximately 25,000km^2 in the early 1960s to less than 1500km^2 today. Climate change has combined with human water use to lead to this overall reduction. Conflict between Cameroon and Nigeria over the lake ended with a 2002 decision by the International Court of Justice in favor of Cameroon, but pastoralists, farmers, and fishermen continue to compete for water in the region. Drought and the extension of desert are leading pastoralists to move farther south in search of grazing lands in central Africa.[75] The loss of water in the Lake Chad Basin made the region vulnerable even before the rise of the militant Boko Haram movement, which in 2009 initiated warfare to seek to create an Islamic state. Boko Haram carried out systematic attacks on farmers, creating famine.

As climate change accentuates extremes, drought has emerged as a cause not only of competition for water resources, or even of armed violence between local groups but also as a contributing cause of war. Analysis of the role of climate change in war and peace parallels the broader study of climate in human history. In place of merely seeing climate as background, ongoing research into causes of war now identifies climate change as a significant variable or possible cause. But as for climate history in general, a counterargument in security studies cautions against the idea that a particular change in climate necessarily leads to or determines a particular outcome.[76] Drought does not by itself make war inevitable or determine the outcome of any given war, but it increases the pressures for war in societies that are already suffering from other conflicts and facing other forms of instability. El Niño years, in particular, increase the chances for new conflict.[77]

Drought interacted with other factors to promote the rise of political conflict and of war in the Middle East. In December 2010, a wave of protests and uprisings in much of the Arab world started after a Tunisian fruit seller burned himself to death in protest against police corruption. This sequence of challenges to many regimes became known as the Arab Spring. Some governments engaged in harsh crackdowns, and in others, the struggle for power produced complex prolonged violent struggles between multiple forces.

The Arab Spring stemmed most immediately from political and social discontent with authoritarian and corrupt regimes, but growing discontent also rose from reaction to the effects of drought. A scholar of this problem described climate as a hidden stressor or as a "sudden change in circumstances or environment that interacts with a complicated psychological profile in a way that leads a previously quiescent person to become violent."[78] Countries, including Syria and Libya, had been suffering from severe drought for several years before the Arab Spring.

There were multiple paths from drought to political upheaval. Drought combined with population growth and a lack of effective government response to encourage large numbers of farmers and pastoralists to migrate to cities in Syria. At the same time drought in other regions of Eurasia contributed to instability in the Middle East. Grain supplies, already depleted by the Russian summer drought of 2010, fell when a drought in China caused the failure of the Chinese winter wheat harvest in 2011 and led China to expand its imports of wheat. The Chinese drought along with warm weather in other wheat-growing areas reduced supplies of wheat and drove up global prices. The price hikes hit the Middle East and North Africa especially hard, because countries in the region were already major wheat importers. Thus, Egyptians found themselves paying ever-higher shares of their income for wheat just as the protests of the Arab Spring gathered strength. Drought did not initiate the Arab Spring, but it multiplied the level of resentment and discontent.

In the Arabian Peninsula, climate change has interacted with other causes of conflict to contribute to destabilization, migration, and violence in Yemen. Climate change has amplified long-term water shortages. War in Yemen that pitted Houthi in the north against a Saudi-backed government in the south involved other forces and has led to fatalities, internal displacement, widespread food and water shortages, malnutrition, and outbreaks of diseases, including cholera.

Climate inequity

In societies around the world climate change is also accentuating inequality. After centuries in which complex societies built up greater resiliency to climate fluctuations, many residents of wealthy societies today may overlook some of the obvious signs of climate change. Gardeners who have cultivated plants for years have noticed changes in the growing season, and outdoor sports enthusiasts have detected changes in the seasons, but an affluent citizen of a Western society who spends hours each day in a climate-controlled house, vehicle, or office can easily miss these changes, especially if that person internalized changed conditions as the norm.

Climate change has long had varied effects for different groups of people. In early human prehistory, hunter-gatherers proved able to adapt to a wide range of local and regional environments. The rise of agriculture had mixed effects. Complex societies stored food, in principle becoming more resilient to climate fluctuations, but climate shocks could prove more damaging to societies with built-up infrastructure than to hunter-gatherers. In some circumstances, ruling elites that depended on the intense extraction of large quantities of resources could even find themselves most vulnerable to climate shifts.

More recently, the pendulum of greatest risk has swung away from elite groups. The continued development of advanced societies has for the time being placed the less affluent and less powerful at greatest risk. The most vulnerable, people with limited resources living in very low-lying areas, feel the effects first, and regimes that have already suffered from other major problems face the greatest shocks. No amount of wealth or power confers absolute and total immunity from the effects of floods, storms, or drought, but the wealthy can better protect themselves from suffering some of the worst effects. A resident of a small parcel of land on a deforested hillside is far more likely to suffer severe consequences from flooding than is the owner of a large, well-maintained house on a parcel of land with retaining walls. In similar fashion, a villager along the coast of Bangladesh is likely to have much less chance of recouping losses after flooding than is the case for a homeowner along the East Coast of the United States.

Around the world, climate extremes create disproportionate risks for communities and groups with the fewest resources to respond. India is one of many countries where climate change endangers the poor more immediately than the rich. Poor migrants from other parts of the country who craft dwellings in low-lying urban areas are especially vulnerable to rising seas or to flooding from rivers. In Mumbai, monsoon rains in July

2019 coincided with high tide to produce flooding of the Mithi River. In September, heavy rains again swamped much of the city. The poor, who live in informal settlements or slums, are also more vulnerable than wealthier residents to heat waves, and, with makeshift or minimal roofing, to damage from heavy rains. Stagnant waters left behind enhance mosquito breeding, leading to more cases of mosquito-borne disease.

Poor residents also suffer far more from water shortages. Poor maintenance of infrastructure and uncontrolled boring of wells combined with the effects of precipitation extremes to make water supply unreliable in the southern Indian city of Chennai. Large reservoirs almost emptied of water in 2019. Water supply to the residents of the city of 10 million was reduced to just a few hours a day. Those who could afford it paid to have water brought in on tanker trucks. Bangalore suffered from similar conditions. For many poor residents in these and other cities, the effort to simply obtain water took up every day.

In Bangladesh migrants displaced by climate change who move to slums in Dhaka often end up in low-lying areas susceptible to flooding. In July 2020, heavy rains submerged large parts of Bangladesh. Rains swelled the Brahmaputra River upstream pouring torrents of water into Bangladesh. Without attribution analysis, we cannot determine whether climate change made these events more probable, but they were consistent with precipitation extremes of climate change.

South Asia is not unique: across East Asia and the Pacific, climate change to date has created the greatest challenges and suffering for many of the poorest communities. A 2017 World Bank Report emphasized the burdens imposed by climate change, "the high risks associated with natural disasters and climate change" on the estimated 250 million residents of slums in East Asia and the Pacific.[79]

In Manila, capital of the Philippines, many who move to the city live in slums close to the water where they are exposed to flooding. In the city of Tacloban storm surge in 2013 from Hurricane Haiyan, known in the region as Yolana, drove water up the Leyte Gulf, causing destruction. The government planned to relocate some 200,000 residents, but with a large shortfall in actual new dwellings, many moved back.

Across the world, heat waves of long duration and greater intensity also bring the greatest threat for the poor. A 2015 heat wave in India killed at least 2,300 people, causing greatest harm to the elderly, the young, and to outside workers.[80] A study by the RAND Corporation found the greatest vulnerability to heat in regions of Central India with low rankings in health, education, and economic development.[81]

The uneven effects of climate change burden the poor in the most affluent countries as well. It's not that the wealthy are immune to rising seas, heat waves, or drought but at least in the short run, they can better cushion themselves against the effects of a rapidly changing climate. Rising heat poses immediate peril for the poor in cities experiencing increasing numbers of hot days. It's not just the lack of air-conditioning. Poor neighborhoods are often literally warmer than wealthier areas. A lack of trees in some poor neighborhoods accentuates heat waves. By worsening air quality, heat also intensifies health problems among poor residents.[82]

In the United States, the threat to health strikes people of color disproportionately. Blacks and Latinos are more likely to live in heat islands where lack of trees coupled with expanses of pavement and buildings trap and build up heat.[83] Greater heat exacerbates health problems from higher rates of asthma. Higher temperatures along with air pollution are also associated with more premature births and lower birth weights.[84] Extreme heat also causes higher mortality among Blacks as well as for the elderly. In Los Angeles, African Americans are twice as likely to die during a heat wave compared to the average for the city's residents.[85]

The effects of climate change also both reflect and accentuate inequality at the workplace. In the United States, Latinos make up a disproportionate share of the workforce in crop and livestock production and in construction, jobs where climate change directly affects working conditions.[86] California, with its large agricultural industry, is one of several states in which a mainly Latino immigrant labor force faces extreme risks from heat. In California, the vast majority of farmworkers are Latino and most were not born in the United States. A CDC study found especially high numbers of deaths between 2005 and 2014 in Arizona, California, and Texas.[87] Even with precautions, heat poses a threat. Marc Schenker, director of the Western Center for Agricultural Health and Safety, describes the risks: "But temperatures are getting hotter and more extreme, and that puts people doing hard labor outdoors particularly at risk."[88] Both farm workers and farmers who employed them remark at extreme temperatures. "For a grower, these heat waves are coming in at all the wrong times—not in the summer when you are expecting it," one Imperial Valley farmer explained. I was irrigating in December. I've never done that."[89]

Indigenous peoples across the world also face disproportionate risks and high costs from climate change. In the Arctic, rapid climate change threatens the very way of life of many native peoples. Melting sea ice, shorter winters, and wildfires reduce supplies of food, cut off transportation routes, and upend long-term customs. Indigenous peoples in Siberia can no longer use frozen rivers as travel routes for months on end. In the Yamal peninsula of northwest Siberia, the Nenets people have to delay crossing the Ob River with their reindeer herds to set up winter camps. "Our reindeer were hungry. There wasn't enough pasture," a Nenets reindeer herder explained. In varied regions of Siberia, tusk hunters are increasingly digging up the remains of Wooly Mammoths that become more exposed as permafrost melts. Climate change damages the ability of the reindeer themselves to find food. Spikes of winter warmth produce swathes of ice where snow would once have prevailed for far longer and thick sheets of ice make it harder for reindeer to eat lichen. Climate change similarly undermines the economy and culture of the Saami peoples of northern Scandinavia, Finland, and the Kola peninsula of Russia. Drought and fires further damage the food supply of reindeer.

In the tropics, climate change combines with extraction to damage the livelihoods of indigenous peoples. In Southeast Asia, burning down forest to clear land for oil palm cultivation destroys habitats, kills animals, and removes resources from indigenous peoples. Climate change aggravates the effects. On the island of Borneo, clearing land for new oil palm plantations decimated forests rich in diverse foods, reduced the habitat for animals and plants, and took away resources from the native Dayak people.[90] The destruction of forests, in turn creates feedback effects, by rapidly releasing stored carbon, and the spread of fire to peatland soils further enhances the release of carbon.[91] Far to the west along the vast Indonesian archipelago, clearing forests for oil plantations has led to similar damage for the Orang Rimba people in Sumatra. Human Rights Watch describes how "Many have been left homeless, live in plastic tents, and without livelihood support."[92]

In Australia, extreme heat and drought creates increasingly harsh living conditions for the Aboriginal population. Aboriginal people have adapted to living in the interior of Australia, but human-caused climate change brings new stress. Supplies of water are dwindling. The chief executive of the Central Land Council explained, "We're not getting the rain we used to, to recharge the aquifers. So as water is drawn out of the aquifers it becomes more saline and less potable."[93] On the Torres Straits islands, sea-level rise threatens low-lying communities.

In Africa, climate change threatens the foundations of the way of life of indigenous peoples who practice pastoralism in countries, including Sudan, Kenya, Tanzania, Somalia, Ethiopia, and Uganda. Pastoralists have shown resilience, turning to raising camels instead of cattle as in the case of the Borana pastoralists of southern Ethiopia as well as in northern Kenya.[94] Pastoralists have also tried to shift the range of livestock and modified fodder and employed irrigation.[95] However, prolonged and severe droughts in East Africa have reduced access to water and a reliable food supply for animals. Merely obtaining water has become an ever more time-consuming and arduous task, especially for women. The Turkana of Northern Kenya, for example, describe having to travel for 4.5 hours a day in high heat in search of water.[96] Women and girls, in particular, spend much of their time looking for and carrying water.

In South America as well, indigenous peoples face threats to their crops and water supply from climate change. Along the west coast and in the Andes, precipitation extremes have amplified El Niño to undermine cultivation of Quinoa. At the same time, melting of glaciers threatens water supplies in highlands. As temperatures rise, some farmers have moved cultivation of potato varieties to higher elevations. Indigenous farmers have also responded by diversifying crops, but the damage to the rural economy intensifies migration from small towns and villages to cities.[97]

East of the Andes, in the Amazon rainforest, indigenous people suffer from the combined effects of deforestation, extraction, and climate change. The head of an association of women in the state of Rondonia in central western Brazil describes the damage to fish: "the river frequently dries up, and the water that is left forms a sort of bay where the fish try to survive. When the rains return, the [thermal] shock of the cold rainwater mixing with the warm river water kills the small fish."[98] Longer dry spells reduce crop yields and intensify fires. The extensive fires, in turn, bring respiratory problems.

Taking lands and resources from native peoples accelerates and intensifies climate change. Land-use practices of indigenous peoples help to preserve forests and keep stored carbon stored.[99]

In Central America, indigenous communities count among those most affected by climate change. In Guatemala most children in indigenous communities already suffer from malnutrition. The malnutrition rate in the predominantly indigenous western highlands is around 65 percent.[100] In these already dire conditions, drought further reduces supplies of food and water. In the Guatemalan state of Alta Verapaz, the maize or corn crop has failed for indigenous farmers and supplies of water have dried up.

Unreliable precipitation and drought are similarly damaging the way of life for indigenous peoples in Mexican states, including Chiapas, Oaxaca, and Yucatan. Poor conditions for cultivating corn in the very valleys where ancestral forms of corn were first domesticated intersect with other causes to help drive migration of young adults who suffer from violence and can no longer make a go of farming.[101]

Within the borders of the United States, indigenous peoples similarly confront immense challenges from climate change. Many native peoples live in the expanding dry areas of the United States. Drought has increasingly damaged forage that the Navajo previously employed to feed their livestock. Water is increasingly in short supply for sheep. A Navajo described the challenges she faced in 2018: "The drought makes it difficult for us. . . . The biggest challenge is finding water for our livestock right now. We're the last few shepherds out here."[102]

Indigenous peoples who have previously responded to fluctuations in climate now confront rapid shifts. Climate change compounds the effects of competition for resources to reduce food supplies that tribal nations have long depended on in regions including the Pacific Northwest. The Quinault Indian Nation has suffered erosion of its chief village from the sea even as the once plentiful sockeye salmon have all but disappeared from rivers that are heating and dwindling after the demise of the former Anderson glacier in the Olympic Mountains that once helped to replenish and cooled the waters of the Quinault River. Warming waters in California and in the Pacific Northwest have intensified damage already caused by dams and diversion of water to salmon long fished by indigenous peoples.[103] Salmon populations have plummeted in the Salish Sea along the border of the United States and

Canada. The Suquamish nation and others have responded by seeking to restore and protect waterways.

On the Northern Great Plains, climate change increases risks for indigenous peoples. Extremes of precipitation have increased disruption of water supplies. Less reliable snowpack and earlier melt place more stress on livestock.[104] The Oglala Sioux nation has encountered extremes of precipitation with both massive precipitation events, including dangerous hail storms, and drought. In southern Montana on the Crow Reservation climate change affects the supply of local staples, including choke berries.[105]

Beyond the forty-eight contiguous states, a large proportion of tribal nations of the United States reside in Alaska, where they experience swift and intense climate change. Permafrost melt damages villages. The melt of sea ice makes it ever more dangerous to carry out hunting. Warming also threatens the ability to preserve and store food through smoking, fermentation, and cold storage.[106]

Climate change produces similar threats to First Nations in Canada. In the Canadian north, Inuit confront warming seas, melting permafrost, and shorter winters. Travel by sea ice and by ice bridges over rivers has become more difficult. At Dawson City on the Yukon River, residents can no longer count on the river to freeze sufficiently to build an ice bridge.[107] The melt of permafrost has broken infrastructure and damaged homes. Shifts in the range of game reduce access to traditional supplies of food. More broadly, residents of a rapidly warming region also see a profound threat to their identity.

Climate migration

Exacerbating inequality and disrupting supplies of food and water, climate change is also magnifying the force of important causes of displacement and migration. Displacement by climate change often intersects with other factors. In 2018, the UN General Assembly approved a Global Compact on Refugees that recognized that climate, environmental degradation, and natural disasters increasingly interact with the drivers of refugee movements, and similarly climate interacts with other factors that cause internal displacement.

Across North Africa, climate change in the early twenty-first century interacted with economic trends and conflict to help generate migration. The largest numbers move within Africa. Thus, Boko Haram and counterstrikes by the Nigerian government combined to displace more than 2 million people. International law classifies those who move within the borders of their country as internally displaced persons (IDPs). Smaller numbers of Africans travel all the way to Europe. They undertake this difficult and often dangerous journey for many reasons, but unreliable rains add to the reasons for venturing north from countries, such as Mali and Niger. In the Horn of Africa, severe drought has also heightened insecurity caused by war to boost the numbers of both migrants and the internally displaced.

The Lake Chad Basin in West Africa provides a striking example of how multiple factors intersect with climate change and environmental degradation to displace residents. Some 17 million of 38 million people in the Basin live in areas affected by the Boko Haram insurgency. Decreasing water supplies and arable land destabilized a region and displaced many who then suffered further displacement from violence.[108] Lake Chad shrank by some 90 percent over fifty years, with about half of that a result of climate change. Thus, many residents in search of resources moved and came into contact with Boko Haram. As one Nigerian refugee explained, "It took us to areas where we were deep into trouble with Boko Haram militants. Due to the fact that the river [a source of Lake Chad] was shrinking, we had to move to other places, and then from moving from one place to another, we met our fate. This has displaced us." The shrinking lake actually made it easier for Boko Haram to launch attacks: "As the Lake shrank," a refugee explained, "Boko Haram could move vehicles and machinery that could attack us."

Climate change has also interacted with other crises to push many residents of Central America to leave their homes and migrate. Extreme weather and drought in the region's dry corridor made more than 2 million people food insecure. Extremes of longer more intense droughts and bursts of heavy rain interacted with land degradation and the spread of invasive species to reduce crop yields. Numerous families sold tools and food to try to pay for food.[109] In El Salvador, Guatemala, and Honduras, climate change imperiled coffee farmers. Higher temperatures likely contributed to an outbreak of coffee leaf rust, a fungus that is highly damaging for coffee plants.[110]

In Guatemala, violence combines with climate change to drive migration by indigenous speakers of Q'anjobal/Kanjobal, Mam, and K'iché/Quiche. An indigenous farmer from the Quiche highlands described the desperate situation: "Corn, beans—when I was younger we often had full harvests . . . Now, nothing."[111]

In many regions climate change combines with other factors to create migration. It is difficult to quantify the exact numbers of people displaced across the globe by climate change, but the numbers are already immense. In 2017 alone, 68.5 million people were forcibly displaced, approximately one-third, or 22.5–24 million people, by sudden weather events.[112]

To describe someone such as a former resident of a region like Central America's dry corridor, the concept of a climate refugee makes intuitive sense as a person forced to leave and find a new home or place of safety, but here the reality of rapid climate change has outpaced the evolution of law. In international law, a refugee is a person who crosses an international border "owing to well-founded fear of being persecuted for reasons of race, religion, nationality, membership of a particular social group or political opinion." The world already has climate refugees in practice, before the creation of a legal term to refer to that group. Along with climate "refugees," a still larger number of people are displaced within national borders. Here, a rough approximate from existing terms would be that of an IDP.

Even without violence, famine, and war, climate change will continue to drive migration. A large proportion of the world's population lives near coastlines. Many residents may seek to stay in their communities and, when possible, take steps to defend them through adaptation, but large numbers of people will ultimately leave. One model for the United States projects net migration by the end of this century under a 1.8-meter sea-level rise scenario without adaptation as more than 2.5 million people move from

Miami-Fort Lauderdale-West Palm Beach, Florida, and some 500,000 out of New Orleans and Metairie, Louisiana.[113] With adaptation the projections drop to some 2 million and 373,000, respectively, for just two of the many affected metropolitan areas around the world.

CHAPTER EIGHT

Projections and uncertainties

- Climate projections and uncertainties
- Attribution analysis
- Energy choices
- Attacks on climate science
- Politics of climate change
- Climate agreements
- Economic interests
- Cognitive biases

Building on decades of work studying climate change, scientists have outlined possible future greenhouse gas emissions trajectories. We now possess a range of possible greenhouse gas concentration trajectories or pathways that cover the rest of the twenty-first century. On a shorter timescale, scientists have developed an increased ability to investigate the connections between climate change and specific extreme weather events in terms of the increased probability of such events that result from climate change. Based on these findings, states of the world have debated and created a series of international agreements to address and curb climate change. States of the world have debated how to do this through a series of climate agreements centering on emissions. As emissions have continued to rise and the effects of climate change have intensified, new climate movements have emerged, calling for additional strategies and also in many instances advocating for addressing climate justice or the unequal burdens caused by climate change.

At the same time, we still face uncertainties. With a truly minute number of exceptions, scientists have reached a consensus that human action is driving climate change, but complex variables will affect the precise degree and rate of change. Because use of fossil fuels has led to climate forcing, our energy choices will also greatly affect which emissions pathways we ultimately follow. Uncertainty also stems in large part from how we process and comprehend the findings of climate science and overwhelming evidence of climate change. Climate science has faced attacks, and economic interests, political orientation, and cognitive biases also all shape responses to climate change.

Climate projections and uncertainties

Climate scientists use an array of general circulation models, or global climate models (GCMs) to generate projections about how greenhouse gas emissions will alter climate. The GCMs simulate the movement of energy throughout the Earth system in response to a new starting condition, such as increased CO_2. Each grid contains a set of established mathematical equations based on the laws of physics and known biological processes, and grid sizes can be adjusted for global versus regional model runs. The first GCMs, dating back to the 1970s, had comparatively fewer parameters of the climate system, but models have grown in complexity and sophistication since that time, lending greater confidence in their projections. Despite some limitations of the earliest models, their projections have been consistent with those of more recent, advanced models.

To provide a reference point for climate modelers to evaluate future changes, a common set of climate change scenarios are used. The Intergovernmental Panel on Climate Change (IPCC) introduced the first set of scenarios in 1992, with updated scenarios published in 2000. In its 2014 report, the IPCC selected four "representative concentration pathways" (RCPs), which cover a range of possible emissions scenarios for this purpose. The RCPs include assumptions about population growth, energy use, and changes in land cover, among others. The most conservative pathway, RCP2.6, is based on ambitious strategies to curb climate change, while RCP8.5 assumes that our greenhouse gas emissions continue into the next century; RCPs 4.5 and 6 incorporate some strategies to reduce warming and therefore have emissions leveling off by 2100.

Climate projections for a future in which we continue to follow our current same trajectory are sobering at best. Even with aggressive climate policies to curb emissions, global temperatures will likely climb at least 1° C.[1] A high emissions trajectory would likely raise average temperatures by more than 3°C by the end of the twenty-first century, to levels above those ever experienced in human history.[2] These are not necessarily the high-end estimates because even the high emissions pathway could lead to more extreme outcomes produced either by climate feedbacks or by expanding energy production and consumption that does not simultaneously shift sharply away from carbon.

Continuing along our current path would also lead to greater increases in sea level. Previous projections forecast average global sea-level rise of up to 1 meter by 2100, but recent studies have doubled that estimate.[3] Growing knowledge about the effects of warmer oceans on ice sheets shifted projections toward these greater increases. Detailed research and modeling of melting both from above and below ice sheets have raised the prospect of greater instability and faster melting of ice in both Greenland and in Antarctica. The end result, in either case, would be faster sea-level rise. Research into possible feedback effects has yielded even more dire estimates for the possible maximum effects of a high emissions path. Thus, the famed climate scientist James Hansen and colleagues have noted possible sea-level rise on the order of several meters.[4] These forecasted changes would pose a major threat to many of the chief cities around the world. Cities like New York, Boston, Shanghai, and Dhaka would not immediately be flooded, but significant sections would become uninhabitable without investments in sea walls at an immense expense, and coastal cities would then confront still further increases in sea level.

Future precipitation patterns are more difficult to forecast, but the overall outlook calls for greater extremes. In the simplest sense, wet areas get wetter and dry regions get drier. Individual precipitation events are likely to become more extreme and more frequent.[5] Intensifying extremes of precipitation and drought would place poorer societies under the greatest pressure, boosting the likelihood of instability and migration.

Despite model improvements and growing confidence in future projections, areas of uncertainty remain. It is difficult for models to predict, for example, what clouds will be like in a warmer world. Clouds represent a source of uncertainty in part because they can act as either a negative or a positive feedback to global warming. An increase in low-level clouds would partly counteract the warming, providing a negative feedback—low clouds tend to reflect more sunlight compared to the heat energy that they absorb, resulting in a net cooling effect. High-level clouds, in contrast, result in net warming because they absorb more of Earth's heat energy relative to the amount of sunlight reflected back to space, and therefore produce a positive feedback. Current projections are for a net increase in high-level over low-level clouds, which would amplify our current warming trend. This remains an area of active research.

Other climate feedbacks are also likely to enhance global warming. The vast stores of carbon currently trapped in frozen Arctic soils—permafrost—represents just one example. The melting of permafrost initiates decay of ancient plant and animal remains that are currently frozen into the soil. The CO_2 or methane released during the decay would provide additional greenhouse gases, amplifying warming and creating a positive feedback. Estimating the effects of this feedback presents similar challenges to measuring the effects of clouds. In the case of permafrost, there is uncertainty regarding how much and the rate at which the carbon will be released. A recent analysis suggests a slow release of this stored carbon in the decades and centuries to come.[6] Most models used in the most recent IPCC report did not fully account for the permafrost feedback, so temperatures are likely to rise more than they are currently projected.

Such feedbacks in the climate system factor in to an estimate of climate sensitivity, the amount of warming that accompanies a doubling of CO_2 above preindustrial levels. Current estimates of climate sensitivity range from 2.5°C to 4°C, and climate scientists continue to work on narrowing this range. Uncertainties connected to climate feedbacks, as well as the response times of various parts of the Earth system—the ocean warms much more slowly than land for example—pose challenges to constraining these sensitivity estimates. Even using the lower end of the sensitive range continues us on a path of warming that intensifies climate-related risks. In its Sixth Assessment Report released in 2021, the IPCC stated that continued warming will lead to greater extremes including increased heat waves, heavy precipitation, and more intense droughts. Some changes, such as ocean warming, glacial melt, permafrost thaw, and sea level rise, may be irreversible over the next few centuries to millennia.[7]

Attribution analysis

On a shorter timescale, we are also increasingly able to identify which extreme weather events can be attributed to climate change. For much of the history of climate science, it has only been possible to describe extreme events as consistent with climate change. Thus, in many regions precipitation extremes of drought and heavy bursts of rain or (if it is cold enough) snow are consistent with projections for climate change.

Advances in the field of attribution analysis now make it possible to analyze the connections between climate change and particular weather events. We cannot identify a particular day's weather as the outcome of climate change, but statistical analysis makes it possible to determine the extent to which climate change contributes to extreme events such as prolonged droughts, intense extended heatwaves, and heavy precipitation.[8]

Attribution analysis issues findings about probability, or the extent to which climate change made a particular extreme event more likely. In an early example of attribution analysis, a scientific paper in 2004 analyzed the European summer heatwave of 2003 and found with a confidence level of > 90 percent "that human influence has at least doubled the risk of a heatwave exceeding this threshold magnitude."[9] A follow-up study in 2014 of summer heatwaves in Europe determined that "events that would occur twice a century in the early 2000s are now expected to occur twice a decade."[10]

Attribution analysis finds that climate change has increased the probability for some extreme weather events and large storms. Thus, a study of a stormy January 2018 in Europe did not find that climate change had increased the probability of large storms that struck Germany, the Netherlands, France, and

Switzerland.[11] However, attribution analysis is every year identifying multiple extreme weather events as more likely because of climate change. For 2018 alone, attribution analysis identified the imprint of climate change or the higher probability of events including drought in the southwestern United States, flooding in the mid-Atlantic, summer heatwave in Spain and Portugal, late spring drought in south China, summer heatwave in northeast Asia, and high temperatures in Japan in June.[12] Advances in attribution analysis will continue to increase our ability to connect extreme weather events to climate change.

Energy choices

Core principles of climate science date back to the nineteenth century, and a massive body of research in multiple connected fields provides us with a strong understanding of the causes and effects of climate change, but as the RCPs demonstrate, uncertainties still exist about whether we end up with an extreme or more moderate outcome for climate change. One of the uncertainties relates to human energy use—the more we maintain a path of continued heavy extraction and use of fossil fuels, the more likely we are to follow the higher RCP with more rapid warming, and greater potential for feedbacks.

The ingenuity of engineers and scientists in identifying and exploiting new sources of carbon places humans in an unprecedented position of having to choose whether or not to make use of vast stores of fossil fuel. In the late twentieth century, rising energy consumption led to some predictions of peak oil, the idea that the world had reached a point of maximum oil production. For a similar example of an early prediction of peak natural gas production, the Federal Power Commission staff concluded in 1975 that natural gas output had peaked in 1974.[13] However, the oil and gas industry has succeeded in vastly increasing reserves. At most points in human history, this would have seemed like a good thing, but in the twenty-first century it places people in a predicament, as the idea of a carbon budget makes clear. The carbon budget approach determines the amount of carbon that can be released without encountering the most extreme consequences of climate change—here again, holding warming to not more than 2°C is a frequently cited goal. In 2014, the Global Carbon Project, a group of researchers and scientists, estimated that the world had already used up two-thirds of the carbon allotment and that emissions were on track to exceed the budgeted amount within thirty years.[14]

Estimates of the carbon budget vary, though none is long. Most projections envision using up the carbon budget in a matter of decades. For holding temperature increases to below 1.5°C, estimates as of 2018 ranged from 0 to 15 years at current emission rates, and other studies produced estimates of up to 31 years.[15] In 2019, the Global Carbon Project declared that emissions would need to decline rapidly to net zero by 2050 (in other words, in less than three decades) to meet the targets of the Paris Climate Agreement. As of 2019 emissions growth had started to slow, but natural gas and oil use nonetheless continued to lead to modest overall carbon emissions growth.

A high rate of emissions would rapidly deplete and exceed the carbon budget, thereby pushing the world toward the high-end warming scenario RCP 8.5. This scenario has sometimes been referred to as a "business as usual scenario," but is better understood as a high emissions scenario. One analysis referred to RCP 8.5 as a scenario "with modest rates of technological change and energy intensity improvements, leading in the long term to high energy demand and GHG emissions in absence of climate change policies."[16] Climate scientists have debated which pathway would be most likely with our current emissions trends, with some seeing RCP 8.5 as a high-end estimate.

The chances of following a high emissions pathway depend in part on projections about the future role of coal. The highest emissions pathway might occur only with expansion of coal mining. Reliance on coal for producing power has already diminished in several countries; however, existing plans for extracting coal would still use up most of the carbon budget to hold warming to no more than 1.5°C.[17] The International Energy Agency calculates that all unabated coal-fired power plants would have to close by 2040 to hold to a temperature increase of 2°C.[18] Even as some countries have moved to retire such plants, others, including countries highly exposed to damage from climate change, continue to depend on coal-fired power plants. Bangladesh, home to a large and highly vulnerable population along its low-lying coasts and in the subsiding megacity of Dacca, had some twenty-nine new coal-fired power plants under construction or in planning as of 2019, but in 2021 scrapped plans for 10 plants.[19] India's National Electricity Plan foresees expanded coal capacity, though falling costs for renewables may speed a shift away from coal. China is a major coal producer and plans to expand oil and gas output. China's climate goals as of 2020 included a target of peak emissions in 2020, but the country's state grid and the China Electricity Council are pushing for building more coal-fired power plants.[20]

Australia is a country severely affected by climate change but also deeply invested in mining coal. In a country that saw uncontrolled wildfires in its summer of 2019–20, work on coal projects continues. Following federal Australian approval, the state of Queensland approved a new coal project in June 2019. Funded by the Adani group, an Indian coal company, the project would open a mine called the Carmichael mine in the Galilee Basin of Queensland. The mine would export coal to India, another country facing increasing risks from intensifying climate change. Critics point not only to the boost in carbon emissions but also to possible risks to the Great Barrier Reef created by coastal dredging. Despite the wildfires of 2019–20, work on the project continued into 2020.

Doubling down on extracting fossil fuels would push the world off of a low emissions pathway. A 2019 analysis, the Production Plan Report, calculated that plans for fossil fuel production would lead to carbon emissions in 2030 that are 53 percent higher than the estimated limit for an increase of no more than 2°C and 120 percent higher than consistent with an increase of no more than 1.5°C. The gaps were even larger for 2040 at 110 percent and 210 percent higher respectively.[21] The production gap, according to this report, was greatest for coal, though oil and gas production plans would also greatly overshoot carbon emissions consistent with a temperature increase of no more than 2°C.

A limited number of countries stand out for producing a large proportion of fossil fuels and for plans to increase production. Russia, for example, plans to expand natural gas production. Argentina plans to grow gas and oil production, and Nigeria and Iraq seek to nearly double oil output. Fossil fuel production is also projected to increase in Australia, Canada, and the United States.[22] Canada projects higher oil production.

The future trends for the use of natural gas will also influence emissions pathways. Debate has focused in particular on the role of hydraulic fracturing, or fracking, to increase production of natural gas, as well as on the role of nuclear power. Fracking involves pumping a mix of fluid and sand into a rock formation, such as shale, to split the rock, which allows for the extraction of oil and gas. The principle is not new, but advances in fracking succeeded in sharply boosting oil and gas production in the United States. Opponents of fracking have frequently voiced concerns that the chemicals injected into rock formations to force out oil and gas could also damage water supplies. The sheer volume of water required for fracking operations has also raised concerns. Proponents of fracking argue that increasing production of natural gas provides a bridge fuel toward a cleaner future. Natural gas, indeed, is a less carbon-intensive fossil fuel than oil and even more so than coal. Thus, in principle, replacing coal with natural gas reduced greenhouse gas emissions from a higher emissions pathway more reliant on coal.

On the other hand, methane released during the fracking process would contribute to moving toward a higher emissions pathway. Methane itself is a highly potent greenhouse gas; thus, methane leaked in the process of fracking could negate any potential benefits by shifting to a less carbon-intense fuel. Old gas lines also leak methane in many communities far removed from production sites. Reducing such leaks would require effective regulation, mechanisms for ensuring compliance by gas producers, and extensive investment in the natural gas delivery system. In the United States, the former Trump administration in 2020 eliminated a regulation requiring that oil and gas companies control methane leaks, a move rescinded by the Biden administration. In the United States and Canada, so-called orphan wells or oil and gas wells abandoned by insolvent companies leak methane and also threaten water supplies and soil. Orphan wells can date back a century in some cases, though the fracking boom increased the number in the twenty-first century.

Even if methane emissions could be controlled through strict drilling regulations and repairs to pipelines, expanded production of fracked natural gas, in some economic models, would delay a shift to a low-carbon future. If costs remained low, a high supply of fracked natural gas could contribute not only to the fall of coal production but could also slow the development of renewable energy on the scale necessary to avert even business-as-usual climate projections.

The effects of the coronavirus pandemic have accentuated questions about the economic basis for continuing development of natural gas. Even before the pandemic struck, some analysts had begun to question the business models of many natural gas producers. In part, the very supply of natural gas exceeded demand. Thus the number of natural gas drilling operations in Pennsylvania fell in 2019. Major oil companies wrote down the value of natural gas assets. In other words, they reported that these assets were worth less than previously estimated. Chevron and Exxon Mobil in late 2019 wrote down the value of natural gas assets, and Royal Dutch Shell also wrote down the value of oil and gas assets. New questions about the economics of natural gas were not, however, simply a product of success. Analysis, in particular, raised doubts about the actual level of natural gas reserves claimed by many natural gas producers. Many existing wells simply did not produce as much natural gas as drillers had forecast. A *Wall Street Journal* analysis of 2019 determined that "Two-thirds of projections made by the fracking companies between 2014 and 2017 in America's four hottest drilling regions appear to have been overly optimistic, according to the analysis of some 16,000 wells operated by 29 of the biggest producers in oil basins in Texas and North Dakota."[23] In addition, falling costs for renewable energy may call into question optimistic projections for future natural gas production.

Extraction of unconventional fossil fuels would also influence the future trajectory of emissions. These unconventional fossil fuels include tar sands in which bitumen, a very thick oil, is mixed with sand, clay, and water. The process of extracting such oil requires removing massive amounts of sand and clay, and that upgrading or purification process itself requires energy, thereby adding to the total emissions from using tar sands. The Canadian province of Alberta possesses extensive tar sands, and producers extract tar sands though large surface mines. The controversial Keystone XL pipeline was one of several pipelines planned to transport tar sands from the interior of Canada to the United States and global markets.

Emissions pathways depend both on the rate at which humans reduce use of fossil fuels and on how quickly we shift to new energy sources. Most plans for curbing human-caused climate change and dropping to a lower RCP call for a massive expansion of clean energy. In simplest terms, this would require expanding production of low-carbon or zero-carbon forms of energy. Solar and wind, along with hydropower, are the most obvious examples. Such energy sources are by no means all new: windmills have long existed, as have water wheels. But a world that consumes far more energy than at any point in human history will require deployment of clean energy sources on an unprecedented scale.

Costs for wind and solar power have been falling, and individual countries such as Denmark and Germany have made alternative energy into major power sources. Thus in 2020, Denmark produced more than 56 percent of its electric power from wind.[24] In Germany, the share of power produced by renewables increased from 6.6 percent in 2000 to nearly 45 percent in 2020, and on individual spring and summer days the total surpassed 70 percent and even reached as high as 85 percent. At the same time, accelerating and expanding the use of clean energy globally will require scaling up production and deployment on a massive level.

Wind and solar output have also increased dramatically in the UK. Energy use during the Covid-19 pandemic revealed the shift in power production in a country synonymous with the history of the Industrial Revolution. As power consumption fell, Britain's National Grid took remaining coal-fired power plants offline and as of June 2020, the country went two months without using any power produced from coal. The pandemic accentuated emerging trends: the previous year Britain went more than eighteen days in June 2019 without drawing on power generated from coal, and for ninety-one days in 2019 output from renewable sources exceeded output from fossil fuels.[25]

In South America, Uruguay stands out for expanding production of power from renewable sources. As of 2020, Uruguay generated close to 43 percent of its electricity from wind and solar (mostly from wind) and much of the rest of its electricity from hydropower. Chile as of 2020 generated 9.47 percent of electrical power from solar sources, while Brazil produced 84 percent of its electrical power from renewable energy, with some 64.58 percent coming from hydropower.[26]

The costs for renewable energy will influence future emissions. This is not a one-to-one relationship with lower renewable costs automatically reducing emissions, but lower renewable energy costs remove a major obstacle to realizing a lower emissions scenario. The levelized cost of energy or the estimated full cost of energy—for example, across the full lifespan of a power plant—has been dropping for solar, wind, and geothermal energy.[27]

Attacks on climate science

Despite the grim outcomes outlined in the high emissions projections, modern societies face major political, cognitive, and economic obstacles in processing and responding to climate change. Refusal in some quarters to accept the findings of climate science poses one such obstacle. It is difficult to address a problem if a significant minority claims that it does not exist or, in some cases, that rapid warming will actually be beneficial.

Climate science grew at a rapid pace in the late twentieth and early twenty-first centuries. Scientists from around the world expanded knowledge of almost every aspect of climate science and of the history of climate. They published findings both in existing scientific journals and in new journals dedicated to particular fields of climate science. To inform responses, the United Nations in 1988 established the Intergovernmental Panel on Climate Change. The IPCC produced its *First Assessment Report* in 1990, and has continued to create reports every five or six years, leading up to the *Fifth Assessment Report* in 2013. The *Sixth Assessment Report* is scheduled to be completed in 2022. In 2007, the IPCC, along with former US vice president Al Gore, received the Nobel Peace Prize for efforts to publicize the latest information about climate change.

At the same time, a strong backlash against climate science emerged in several countries in a variety of forms. Industries that would likely face adverse effects from regulation of carbon emissions in some cases supported efforts to cast doubt on climate science. Varied organizations and internet sites attacked or worked to discredit the findings of climate science. Several large broadcast, online, and print media platforms amplified these voices.

It is possible to see the effects of such mobilization in controversies over the "hockey stick," the Climate Research Unit in the United Kingdom, and the IPCC. The "hockey stick" refers to the curve of a graph first published by climate scientists Michael Mann, Raymond S. Bradley, and Malcolm Hughes in 1998.[28] The graph plotted global temperatures since 1400 CE, later extended to the past two millennia, and showed a sharp rise at the end toward the late twentieth century, a shape that resembled the outline of a hockey stick. In 2001, the IPCC placed the graph in its *Third Assessment Report*. In response, a campaign sought to discredit the scientists and the report by attacking the validity of the graph. These attacks gained public attention, but multiple scientific assessments of historical temperature trends confirmed the findings of a sharp temperature increase from the original hockey stick graph.

In another major attack on the scientific consensus that human activity is causing climate change and warming, still unidentified hackers broke into emails at the Climate Research Unit at the University of East Anglia in England, the major climate research center in the United Kingdom. Selective leaks of excerpts taken out of context aimed to create the impression that climate scientists were engaged in conspiracy, but multiple findings and inquiries found no evidence of any such improper action. Finally, a paragraph in the IPCC *Fourth Assessment Report* on the future pace of melting of Himalayan glaciers was found to be faulty, leading to additional claims of conspiracy by skeptics.

Some of the climate attacks stem from a lack of understanding of the science itself. The term "global warming" conjures perceptions that each day will be warmer than the previous; in this view, a cold or snowy day is enough to dismiss the idea of a warming trend. A more sophisticated argument, but along the same line of thinking, then emerged with the idea that global warming had slowed. Even as CO_2 continued to increase, the temperature record did not appear to keep pace during the early 2000s.

Skeptics seized on this apparent hiatus in the temperature rise as evidence against a warming trend. Climate scientists, in contrast, pointed out that the warming continued—some have noted that much of the added energy is showing up in the ocean rather than in the atmosphere,[29] while others argue that there was, in fact, no hiatus at all.[30]

Starting in 2014, a series of years with record or near record warmth followed. Most temperature data sets showed 2014 to be the warmest on record. All of the major global surface temperature data sets then found that 2015 set a new record for warmth, and most subsequently determined that 2016 set yet another new record. The major data sets kept by NASA, NOAA, the Climate Research Unit, the Japanese Meteorological Agency, and now by other research organizations, including Berkeley Earth and Copernicus, the European Union's Earth Observation program, vary slightly but all showed consistent high temperatures during the period from 2014 to 2020. Indeed, despite a La Niña event, the year 2020 was either the warmest or close to the warmest year, depending on the exact data set. Overall, most of the warmest years on record have occurred since 2000; by NASA's analysis, nineteen of the warmest years have occurred since 2000.[31]

In no case did any of these campaigns or controversies alter the fundamental scientific consensus that human action is leading to climate change, but mobilization against climate science helped heighten differences in perceptions of climate change in different countries. A public opinion survey released in 2014, for example, revealed major differences between countries in the percentage of respondents who answered that climate change is "largely the result of human activity." In the United States 54 percent of respondents took that position, a total lower than the 71 percent in Canada, and far lower than in many other countries such as 79 percent in Brazil, 80 percent in France, 84 percent in Italy, and 93 percent in China.[32] Survey results vary over time, and different questions produce different results, but respondents from the United States continue to view climate change with less urgency than do respondents in most other regions. Another survey released by Pew in 2015 found that 41 percent in the United States agreed that "climate change is harming people now," a lower proportion than the 48 percent in Asia and the Pacific, the 52 percent in Africa, the 60 percent in Europe, and the 77 percent in Latin America. Only respondents from the Middle East, at 26 percent, answered yes less often.[33]

Politics of climate change

Politics also influences responses to climate science and climate change. Green parties that in many cases predate widespread concern about climate change have in several countries identified themselves with efforts to curb climate change. Several European countries have a well-established tradition of political parties with roots in environmental issues. These preceded widespread public recognition of climate change as a key issue. Green parties in Europe have since made addressing climate change a central priority. The Green Party in Germany, for example, was founded in 1980 and emerged from opposition to nuclear power and antiwar movements. The German Greens call for going to all renewable power by 2030 and for an end to the age of coal. Greens today have reached the political mainstream in Germany, and, as of recent years, their results in polls and elections suggested that the German Greens might be on the verge of moving past the German Social Democratic Party as one of Germany's two most powerful parties, along with the Christian Democratic Union (CDU). Green parties, or Greens, have achieved their strongest results primarily in northern and Central Europe in countries including Finland and Austria, and in 2020, the Greens in Ireland also joined a coalition government.

The level of political division over climate science varies greatly by country, but in the United States, significant numbers of elected political leaders publicly denounce or question the validity of climate science. The strongest political opposition to climate science in the early twenty-first century was found in the Republican Party and in regions with major fossil fuel industries. Public opinion surveys showed sharp differences along party lines in the percentage of Americans who accept that the Earth has been warming or who thought that it was an important issue. Scattered political leaders have also expressed skepticism about climate science in countries such as Australia and Canada.

Concern over climate change has increased in the United States but remains sharply polarized along political lines. Thus, 59 percent of respondents to a Pew survey in 2020 said that climate change was affecting where they lived, with the share who identified it as a serious problem rising from 45 percent in 2015 to 53 percent in 2020. There was a greater left-right divide on climate change in the United States than in any other country surveyed. The gap between left and right on seeing climate change as affecting where they currently live was 51 percent in the United States compared to 34 percent in Australia, the next-most polarized country.[34]

In the United States and other countries, identity and political attitudes affect how people process information even in an age of extreme weather events that in some cases can now be attributed to climate change. Australia, for example, suffered intense wildfires during the Southern Hemisphere's summer of 2019–20, but not all processed the disasters as evidence of climate change. Climate change made the extent of fire more probable. An early attribution analysis found that "The probability of a Fire Weather Index this high has increased by at least 30% since 1900."[35] However, some observers and even witnesses insisted on seeing the fires as the result of individual human actions such as arson. Most such acts were careless rather than actual arson, and fires would not have spread to the extent that they did or caused comparable damage in the absence of climate change.

Climate agreements

On the international level, the United Nations has addressed climate change through a series of conferences. This effort began in earnest with the United Nations Framework Convention on Climate Change of 1992, which entered into force in 1994 with the goal of stabilizing greenhouse gas emissions "at a level that would prevent dangerous anthropogenic (human induced) interference with the climate system."[36] The first major attempt to meet this goal took place with the Kyoto Protocol of 1997. Under the terms of the Kyoto Protocol, the parties or negotiating states agreed to make cuts in carbon emissions measured from 1990 levels. These cuts varied—the EU, for example, pledged cuts of 8 percent and the United States 7 percent. Cuts averaged 5.2 percent. Many states ultimately ratified the Kyoto Protocol, but the United States ceased the attempt to ratify in 2001, and Canada withdrew from the Protocol in 2011. The Protocol also did not include mandatory emissions cuts for developing countries such as China and India—they had, up to the point of the onset of negotiations, historically produced less carbon emissions than had industrialized countries. Through a combination of emissions by nonparticipating countries and, in some cases, failure to meet Kyoto emissions cuts, carbon emissions actually rose substantially from the baseline date of 1990 by almost 50 percent by 2010.[37]

Such efforts to craft global agreements to curb emissions confront the question of how to set a target for limits to warming. Holding the global temperature increase to no more than 2°C compared to the preindustrial era emerged as a goal in global warming negotiations and agreements. Discussion of 2°C as a limit can be traced as far back as the 1970s, and by the 1990s European discussion of climate policy began to establish it as a goal. The Copenhagen talks in 2009 identified 2°C as an international goal, but that limit did not make it into the final document. A follow-up agreement at Cancun in 2010 committed governments to "hold the increase in global average temperature below 2°C above pre-industrial levels."[38] In 2015, the G-7 leaders, at the urging of host Angela Merkel, chancellor of Germany, agreed to the 2°C target.

The product of long negotiations, the 2°C goal has itself given rise to debate. Some scientists have argued that it is too conservative and may give rise to a false sense of security. Given the many feedback mechanisms created by warming, the 2°C goal may not be sufficient to prevent major challenges to human societies as well as to many species. James Hansen, a pioneering climate scientist, wrote that a 2°C rise would "subject young people, future generations and nature to irreparable harm."[39] In contrast, some critics of the goal contended that the 2°C goal is too simple and called instead for focusing on a set of "vital signs," which would include such factors as concentration of CO_2 in the atmosphere, ocean temperatures, and high-latitude temperatures. In response, several leading climate scientists have argued for the value of setting goals, the difficulty of selecting useful vital signs, and the danger that abolishing a temperature goal would provide an excuse for further delay and inaction. Hans Joachim Schellnhuber, a climate adviser to German chancellor Merkel and to Pope Francis, noted both the value and potential risks of the 2°C goal. The goal at least gave governments something to aim at: "Two degrees Celsius is a compromise, but it is at least a tangible and feasible target, so this is something." However, he added that it might not provide safety. "But the two-degree guardrail is somewhere around or above the tipping point. So two degrees is not a good compromise! It is the dividing line between dangerous and catastrophic climate change."[40] In 2015, the Paris Climate Agreement retained the goal of limiting temperature increases to no more than 2°C, but added the aim "to limit the temperature increase to 1.5°C above pre-industrial levels, recognizing that this would significantly reduce the risks and impacts of climate change."[41]

The protracted negotiations over climate result not simply from any failings of international and national political institutions and leaders but also from real disputes over how to apportion costs and benefits. Continued human-induced warming would create broad challenges for all human societies, but some are likely to bear the highest costs early on. A climate change vulnerability index produced in 2013 included such countries as South Sudan, Haiti, Sierra Leone, Guinea Bissau, and Bangladesh as most at risk, though these countries vary enormously in their level of economic development. In general, poorer populations are least able to insulate themselves from the early effects of climate change. This imbalance in risks stems both from major gaps in resources and in some cases from the likely regional effects of change. Thus, poorer societies have fewer resources to deploy to meet new challenges posed by climate change. Guinea Bissau, for example, is a low-lying country with a population of only 1.6 million in West Africa, but with a GDP of below $1,000 per person, the country could devote few resources to adaptation. Its carbon emissions are barely measurable as a percentage of total human greenhouse gas emissions, and there are many countries in a similar position. At the same time, some of the countries least able to devote massive resources to addressing climate changes may see major changes in precipitation. Thus, the increasing frequency of extreme precipitation and drought is likely to prove especially damaging to Haiti, the country with the lowest per capita GDP in the Western Hemisphere.

On a global level, these differences in risk and resources influence responses to climate change because some of the most affected populations have the least power to shape and influence agreements. Simultaneously, residents of wealthier countries, who have collectively contributed more

CO_2 but may feel themselves to be less immediately affected, may therefore be less supportive of action. Such disparities in level of risk can be found even on the regional level, where residents of poor neighborhoods in urban areas face more severe consequences from warming than do residents of wealthier neighborhoods. Extreme heat waves that become more likely with climate change lead to higher mortality, but the risk of death from heat is not evenly spread. Thus extreme heat waves, whether in Pakistan, the United States, or elsewhere, are most likely to lead to deaths among the poor.

Along with apportioning costs and benefits between more and less economically developed countries, balancing costs between past, present, and future emitters of greenhouse gases poses another major challenge for addressing climate change. Over the course of much of the Industrial Revolution, the early industrial powers have emitted by far the most carbon. The United States, for example, emitted the largest total amount of carbon in the twentieth century. From 1850 through 2002, the United States contributed to 29.3 percent of carbon emissions, more than any other single country. The United Kingdom, the first industrialized country and the world's leading emitter in 1850, accounted for 6.3 percent of emissions, below Germany at 7.3 percent and above Japan at 4.1 percent. Taking into account this past pattern, the Kyoto Accords set emissions cuts only for developed countries. Under the terms of Kyoto, those countries that over the course of the Industrial Revolution had led the world in emissions and had gained the largest economic benefits would cut their emissions first.

As of the early twenty-first century, the United States and its counterparts among the early industrial countries still contributed substantially to carbon emissions, but since the late twentieth century emissions have increased in multiple regions of the world. As of 2018, the United States was still the top cumulative emitter of CO_2 since 1850, but that percentage had declined to 25 percent. New industrial powers in Asia raised carbon emissions, and China surpassed the United States as the world's top emitter annually of total carbon. In 2018 top carbon emitters in the twenty-first century included China, the United States, India, and Russia along with the European Union States.[42]

As emissions in countries such as China and India have increased, debate over how to curb emissions has focused in large part on how to balance emissions cuts between new leaders in carbon emissions and historic leaders. The historic industrial powers clearly contributed the most over time to the climate change already taking place, but surging emissions around the world will generate further climate change. On both sides, the shifting balance in emissions can serve as a convenient political excuse for delay. Political leaders in newer industrial countries could argue, with reason, that the older industrial powers should go first. At the same time, some politicians from the older industrial countries called for delaying action so long as not all countries joined in. A plausible path toward limiting climate change will require global action by all major emitters, but even a good-faith effort will require agreement on how to apportion or share cuts.

Over many decades, developed countries used carbon more efficiently than developing nations. Emissions intensity, or the level of greenhouse gas emissions per GDP, is higher in both Russia and China, as well as Indonesia and Canada, than in the United States.[43] However, measuring carbon emissions per person still places the United States far above countries including China and Indonesia, though the United Arab Emirates, Saudi Arabia, and Australia have outpaced the United States in per capita emissions. The idea of reducing emissions intensity provided one way to try to bring developing countries into a treaty to reduce emissions overall. Thus in 2015, China promised to reduce carbon intensity of GDP by 60 to 65 percent below 2005 levels by 2030. However, this would still lead to rising total emissions in China, now the world's largest emitter of carbon.

After years of multiple rounds of negotiations, the Paris Climate Change Conference of 2015 yielded a new agreement in which individual countries submitted their own plans to cut carbon emissions. Under the terms of the agreement, countries will have to submit new and more stringent plans every five years. The Paris Agreement revealed the continued difficulty of finding a way to ensure emissions cuts, but it did break new ground: in contrast to the Kyoto Accords that focused on cuts by established developed countries, the Paris Agreement included most of the world's countries.

In November 2016, the results of the US presidential election created a new set of questions about how to address climate change. In his campaign, Donald Trump pledged support for coal and promised to take the United States out of the Paris Climate Agreement, and in 2017 announced that the United States would withdraw from the agreement. He also opposed regulations to reduce carbon emissions from power plants. A reduced commitment by the United States to national and international action to curb greenhouse gas emissions would further hamper efforts to hold down overall warming. In the 2020 US presidential election, former vice president Joe Biden defeated Trump and pledged to reenter the Paris Climate Agreement.

Economic interests

Delay in confronting climate change stems not simply from denial or rejection of scientific findings but also more broadly from perception of climate change. Simply recognizing that humans are causing climate change does not necessarily equate to understanding the full dimensions of the threat or to being prepared to support action to try to curb climate change. Economic interests and cognition also influence understanding of climate change and climate science.

The very growth of CO_2 levels in Earth's atmosphere to levels unprecedented in human history stems from multiple waves of industrialization and revolutions in transportation powered mainly by fossil fuels. The composition of the world's largest corporations reflects this reality. Metrics for compiling lists of the world's largest companies vary, but by any measure, energy companies that specialize in producing oil and gas make up a large share of the world's largest corporations as determined by revenue. Such companies made up eight of the world's twenty-five largest companies by revenue in 2015, with another company heavily involved in mining.[44] Powerful economic sectors would have much to lose from a shift away from fossil fuels, and countries and states highly dependent on these companies also have economic motives to view such a shift with concern. Persian Gulf states, including Saudi Arabia, and Russia count among the world's largest oil and gas producers. So too do Canada, Brazil, and Mexico. The hydraulic fracturing, or "fracking," revolution made the United States the world's largest producer of natural gas, and US crude oil production after a long period of decline also increased between 2005 and 2015. Among countries with large coal reserves, China, the United States, Australia, India, Indonesia, and Russia count among the top producers. Simply possessing fossil fuel reserves does not automatically force a country to obstruct efforts to curb global warming, but climate negotiators have criticized Saudi Arabia for seeking to water down language and minimize goals in international climate talks. At the same time, even countries with greener international reputations may find it difficult to turn away from major and long-established industries. Thus, Norway in 2016 issued drilling licenses for the exploration of new oil reserves in the Arctic.

A shift away from fossil fuels would reduce employment in oil, gas, and mining companies. Energy production is not a zero-sum game; thus, employment in renewable sectors is already growing at a faster rate than employment in established fossil fuel production in many regions, but sharply reducing extraction of fossil fuel raises the question of what measures to take to assist displaced workers.

The sheer density of infrastructure designed for the heavy use of fossil fuels also poses an obstacle to change. Over many generations, we have cumulatively made enormous fixed investments in infrastructure based on the use of fossil fuels to the point where such infrastructure has become the norm. In some cases, new energy sources can be inserted into existing infrastructure, but economists also refer to sunken costs, or to costs that cannot be recovered. The United States, for example, has made far greater investments in roads than in railways. In a society in which the large majority commutes in a personal motor vehicle, it is easy to see the costs associated with building and maintaining roads as the norm, but these also represent a sunken cost, parts of which could have been invested elsewhere, such as in railways.

Assumptions about costs also create obstacles toward shifting away from fossil fuels. We tend to differentiate between the immediate upfront cost to the consumer and costs to be paid at some later time. Relying extensively on fossil fuels, we count the cost of producing and buying oil, gas, and coal. In the most obvious example, any motorist at a pump probably knows how much gas costs on a given day. However, we do not make anyone pay upfront for the associated costs both of diseases associated with pollution and of damage caused by climate change. So while the consumer at a gas station is acutely aware of the price of gas, we can still, individually and collectively, pollute for free. Fossil fuels remain artificially cheap so long as no one has to pay the full cost upfront.

Cognitive biases

Cognition and, more specifically, the way in which humans identify and respond to potential threats also affects readiness to undertake major actions. Evolution itself may slow our response to climate change. Human evolution provides us with the means to identify and respond to certain threats, but we do not always have the capacity to fully evaluate risks. A normal activity such as driving, for instance, can seem safer than something unusual—being bitten by a poisonous snake—even though, statistically speaking, driving poses a greater risk. We also tend to be good at sensing immediate physical risks, but much worse at identifying and responding to more complex long-term dangers such as climate change. Humans have the capacity to analyze complex problems and to craft long-term plans, but we are also quick to respond to what we perceive as immediate threats. We may both overestimate the likelihood of such dangers and move comparatively quickly when we sense ourselves at immediate risk. A complex, multicausal problem unfolding over decades is easier for us to ignore. We are readier, in other words, to take immediate action to try to evade a large carnivorous animal (though vanishingly few of us will ever meet one hunting us) than to make changes to reduce even grave risks that unfold over the course of our lifetime.

Cognitive bias distorts our perception of climate change. Research suggests we are more ready to accept greater risks to avert possible losses than to achieve potential gains.[45] We may value what we already have more than what we can gain. Psychologists refer to this as loss aversion or the caring more about losses than gains. In the case of taking action on climate change, we may fear the loss of a lifestyle that has brought many comforts. Opponents of action on climate change repeatedly seek to prompt feelings of loss aversion with claims of things that climate action will take away.

Emerging fields of research address these issues by applying behavioral science to try to better understand human perceptions of climate change. One debate centers on the question of whether it is more effective to present people with a wider and potentially grim picture or to take a more optimistic approach that informs them of discrete individual actions that they can take. However, unless such individual actions are tied to a much broader shift in norms, they will be insufficient to avert the most extreme scenarios for climate change.

Paradoxically for those who recognize the scale of challenge posed by climate change, the very enormity of the problem can also undermine action. Anxiety can lead to fear and create a sense of helplessness. Alternatively, the most dire projections for the future of climate change can support apathy, a belief that action is not possible and will not matter. Thus, some commentators describe a future of inevitable social collapse.

CHAPTER NINE

Responses to climate change

- New climate movements
- Decarbonization
- Technical approaches and geoengineering
- Adaptation
- Individual action
- Climate economics
- Into the Anthropocene

Climate change and human history have become so closely connected that it is not possible to separate one from the other. Complex societies over millennia and centuries have become more resilient to climate shifts, but we have now reached the end of any trend toward decoupling between the fortunes of human societies and climate change. Because human activity has become a prime agent of forcing, current and future energy use will shape future climate change. In any reasonable scenario we now confront momentous choices about what future we wish to live in.

We have multiple options for seeking to reduce greenhouse gas emissions and curb damage from climate change. After more than two centuries of industrialization, dependence on fossil fuels is built into virtually all sectors of the economy. Plans to mitigate climate change by reducing greenhouse gas emissions therefore focus on decarbonization of multiple sectors ranging from power to transportation and agriculture. In some cases, proposals focus on reducing emissions of other greenhouse gases, in particular methane. A different approach, geoengineering, suggests trying to remake Earth's climate through technical approaches carried out at a large scale. Even with successful efforts to reduce emissions there will be significant further climate change, and in worst-case scenarios that will be extreme, so another approach stresses adaptation or how to try to live with the effects of climate change. The global scale of climate change means that any effective response will have to be collective, but many people, asking how to personally respond to climate change, have also looked to individual actions. Finally, as climate change and human history become ever more closely connected, determining our response to climate change will reveal how we see our economies and societies, and how we seek to define our relationship with the only planet we know of that supports a vast array of life.

New climate movements

As climate change has intensified, new organizations and movements have emerged to press for action. Long-standing environmental organizations have publicized climate change and pressed for policy changes, but the years preceding 2020 saw the rise of new grassroots movements committed to drawing on new groups of supporters and leaders and to employing new strategies. Many of these new movements focus both on reducing greenhouse gas emissions and on climate justice or on addressing the unequal effects of climate change.

Young people have taken an increasingly visible role in leading climate movements, especially Fridays for Future founded in 2018. The starting point for Fridays for Future can be traced to the actions of one person, then-fifteen-year-old Greta Thunberg of Stockholm Sweden, who in August 2018 began a school strike for climate. Others soon joined her in striking to call for the Swedish government to enact policies to hold to the targets of the Paris Climate Agreement. The group sought to build moral pressure for action, and in 2019 created a declaration to hold global warming to no more than 1.5°C above preindustrial levels, to ensure climate justice and to listen to the best science. Thunberg herself has frequently stated that politicians must pay attention to science. In 2019, she traveled to and from the Americas by sailboat in order to speak to the UN General Assembly. The group held strikes around much of the world in November 2019, and resumed strikes in September 2020 (Figure 9.1).

Fridays for Future has gained the greatest traction in Europe. In Germany, for example, the resumption of school strikes in September 2020, after interruption by the Covid-19 pandemic, was a national news story. The movement has spread to the United States but started later, in 2019.

Thunberg herself gained much praise but also sometimes faced vitriolic attacks, claiming that she was a pawn and sometimes criticizing her in personal terms. At the same time, expressions of gratitude for youth action appear to have left her underwhelmed. She has made the point that youth action is necessary only because of previous failure by adults. At the United Nations, she told delegates, “How dare you? You have stolen my dreams and my childhood.”

FIGURE 9.1 *Fridays for Future demonstration in Hamburg. Source: Wikimedia Commons, https://commons.wikimedia.org/wiki/File:Fridays_For_Future_Hamburg_Greta_Thunberg_2019-03-01.JPG.*

New climate movements have also called for greater diversity and a greater voice and more respect for people of color, indigenous peoples, and for grassroots organizers. Numerous initiatives in the United States and elsewhere have called for greater diversity in climate movements and for recognizing the work that diverse activists and experts already carry out. Thus, in Britain, a group called Climate Reframe, aims to highlight the work of diverse experts and climate activists. Many opinion pieces have called for greater diversity and for broader representation. In 2019 Peter Wang Hjemdahl observed that "'our global movement for environmental conservation is predominantly led by white people' even though . . . 'it's poor people of color who end up bearing the largest share of climate change's burden."[1]

Vanessa Nakate, a climate activist from Uganda, pointed out how the Associated Press cropped her out of a photo taken with Greta Thunberg. Climate activists elsewhere have decried similar treatment even by climate organizations.

Indigenous peoples do not share a single position on how to respond to climate change but have taken on leading roles in several campaigns. Indigenous peoples in Canada and the United States helped to mobilize opposition to oil pipeline projects, including the Keystone XL pipeline that aimed to transport oil produced from tar sands in Alberta Canada. They focused not just on the effects of climate change but on possible threats to water supply and to sites of cultural and historic importance.

In North Dakota, thousands, including indigenous people, gathered at an encampment in 2016 and 2017 to protest the construction of the Dakota Access Pipeline designed to transport oil from shale fields in North Dakota. In early 2017, remaining protesters were removed by National Guard and law enforcement agencies.

After the May 25, 2020, killing of George Floyd, the record of climate movements on race gained greater attention in the United States. Elizabeth Yeampierre, co-chair of the Climate Justice Alliance, indicated that those who built a lot of environmental institutions focused on conservation, wildlife, and open space "but didn't care about black people. There is a long history of racism in those movements."[2]

Along with identifying new leaders and drawing on more diverse support, another new approach to climate advocacy calls for a greater emphasis on methods such as civil disobedience. In 2018, a protest in London that announced a "Declaration of Rebellion against the UK Government" led to further protests and to the formation of an organization called Extinction Rebellion, or XR. The word *rebellion* refers to a conscious shift from what XR terms "traditional strategies" to nonviolent civil disobedience. XR points to the government response to the Paris Climate Agreement as cause for grassroots action through what the group calls "Citizens Assemblies" to lead governments.

Extinction Rebellion in the UK has blocked bridges and public sites in London, and similar actions have taken place in European countries, including France, Germany, Spain, and the Netherlands; in Australia and New Zealand, and on a smaller scale in Canada and the United States. It has encouraged mass arrests as a tactic, though the group also indicates that it is not necessary to be arrested. After interruption by the Covid-19 pandemic, protests resumed in the fall of 2020 with actions such as a march in Hobart Australia to bring attention to the death of kelp forests off the island of Tasmania. In London members sought to block roads leading to printing presses of newspapers the group charged with downplaying climate change.

Extinction Rebellion consciously seeks to disrupt on the grounds that previous campaigns and protests have not led to decisive action to curb climate change, but that very disruption has met with public criticism. At the same time, some observers suggest that the group's approach succeeds in gaining attention for its cause and that even critics may then take pains to establish their credentials on climate change.[3]

New climate movements have also emerged outside of Europe. With origins in climate marches of 2014, the group Climate Mobilization pressures governments at all levels to declare a Climate Emergency. The group calls for transforming the US economy to reach zero emissions, and to rapidly draw down carbon. The group in particular calls for speed. Campaigns for declarations of a Climate Emergency have spread widely with the governments of Scotland, Wales, and the UK declaring climate emergencies in 2019. In the United States, cities, including Austin, Miami, and New York, have issued such declarations, as have many smaller communities.

In the United States, new climate movements are taking on a more active part in politics. Founded in 2017, the Sunrise Movement, for example, expressly seeks to support candidates at all levels of government and endorses and works on behalf of candidates. The Sunrise Movement also backs the Green New Deal.

In the United States, the Green New Deal, a plan introduced in 2018 by Representative Alexandria Ocasio-Cortez and Senator Ed Markey, reflected rising attention to climate justice. The resolution observed that climate change along with pollution "exacerbated systemic racial, regional, social, environmental, and economic injustices (referred to in this preamble as 'systemic injustices') by disproportionately affecting indigenous peoples, communities of color, migrant communities, deindustrialized communities, depopulated rural communities, the poor, low-income workers, women, the elderly, the unhoused, people with disabilities, and youth." As of 2020, the Green New Deal was far from passing, though supporters spoke of their efforts to influence other plans.

Other emerging forms of climate activism focus on economic and legal action, including divestment. The opposite of investment, divestment means selling or withdrawing investments to try to advance social or political goals. Modeled on the divestment movement of the 1970s and 1980s that pressured institutions to withdraw investments in South Africa or in companies that did business in South Africa during the Apartheid era of state-sanctioned racial segregation and discrimination, student groups have pressured universities to divest their endowments from fossil fuel companies.

Divestment from fossil fuel campaigns began in 2010 and 2011 and have pressed colleges and universities chiefly in the United States and the United Kingdom to withdraw investments from fossil fuels. University and college presidents have often responded by stressing their financial responsibilities. As of 2020 most colleges and university endowments had not changed their investment portfolios to withdraw from fossil fuels, but a growing number have declared plans to divest or to reduce fossil fuel holdings. Even as opponents of divestment stressed financial considerations, some stressed the growing risk of investments in fossil fuels as a reason to divest. Thus, the University of California system in 2019 decided to drop investments in fossil fuels from its investment on financial grounds. The system's chief investment officer and the chair of the Board of Regents pointed to the "financial risk" created by maintaining holdings in fossil fuels.

Divestment also emerged as an issue for foundations, religious institutions, and for local governments and pension funds. With some prominent exceptions, foundations have generally been slow to divest. Some have taken partial steps, such as refraining from investment in coal. Pension funds have begun to consider divestment. In Britain, NEST, a pension fund for workers saving under an auto-enrollment plan, moved in 2020 to begin partial divestment. For pension funds, as with colleges and universities, the debate over divestment has focused on financial risks as well as other effects of climate change. Thus, a large Swedish pension fund that in 2020 announced an end to fossil fuel investments pointed to "a substantial uncertainty for companies involved in coal, oil and natural gas activities."[4]

Climate activism has also emphasized the costs and damage of fossil fuel extraction through lawsuits. Such lawsuits seek to recoup damages for costs created by climate change. A 2020 report put the number of lawsuits related to climate at more than 1500.[5] Most have focused on regulations or permitting, but more recent lawsuits have taken a far broader approach. Some, filed on behalf of young people, point to the damage caused to future generations. Others argue that fossil fuel extraction is linked to human rights violations. Lawsuits have targeted major fossil fuel companies, or "carbon majors." These have called for payment of damages resulting from climate change. Some, taking a financial approach, charge that fossil fuel companies knowingly hid risks, and associated costs, from investors. To date, fossil fuel companies have held their own in court in the United States: Exxon Mobil won a 2019 verdict in a case brought by the state of New York, but the number of cases is growing and also spreading globally.

Climate activists have also moved to target carbon emissions at their source through campaigns to keep carbon in the ground. Among several reasons for objecting to the Keystone XL project, climate movements called for keeping carbon in the ground. Opponents of large coal mining projects in Australia have also made the same arguments.

Decarbonization

Because human activity is leading to climate change, and energy consumption is responsible for nearly three-quarters of greenhouse gas emissions, almost all proposals for curbing climate change call for replacing fossil fuels with renewable energy.[6] Germany has described this shift as an Energiewende, or energy transition, though in Germany the concept is linked both to replacing fossil fuels with renewables and to a shift away from nuclear power. Many countries have described the goal of replacing energy sources as moving to zero or net-zero emissions, though plans differ in whether they allow so-called carbon credits, which allow the purchase of credits for emissions that, in principle, are tied to reducing emissions elsewhere, to count toward net-zero emissions. Finland in 2019 pledged to become carbon neutral by 2035. The Finnish plan in contrast to some others does not include purchasing carbon credits, though that is subject to review in 2025. The European Council of the European Union in December 2020 endorsed a reduction in emissions to 55 percent of 1990 levels by 2030 to then "reach net-Zero by 2050." The transition or shift in energy has also been described as decarbonization or deep decarbonization to decouple modern economies from carbon emissions.

Plans for a continuing energy transition call for increasing energy production from renewable sources, including solar and wind. Costs for solar energy have plummeted. Wind turbines both on land and offshore produce electricity. Advances in wind turbines have led to prototypes with a wind turbine height of more than 800 feet and a diameter of more than 700 feet. The International Energy Agency forecast in 2020 that wind and solar capacity could exceed that of gas by 2023 and coal by 2024.

Countries such as Germany and Switzerland envision an energy transition without nuclear power, but advocates of nuclear energy point out that it contributes no greenhouse gas emissions from power production. However, accidents, most recently at the Fukushima nuclear plant in Japan in 2011, have increased concerns about the safety of nuclear power. Building conventional nuclear power plants is also extremely expensive with initial construction costs of many billions of dollars, and storage of radioactive waste creates still unresolved problems. To date, there is no consensus about how to store the most highly radioactive waste. In the United States, a plan to store such waste in a repository under Yucca Mountain in Nevada came to an end after more than thirty years of discussion, design, construction, and debate. Ongoing research continues into alternative designs for delivering nuclear power, including thorium reactors and breeder reactors. Aside from questions of cost, concerns remain about safety and the possibility of nuclear proliferation.

Rapidly reducing carbon emissions would also require decarbonizing the transportation sector, which contributes around 16 percent of global emissions. Interest in diesel as a more energy-efficient alternative to gas has rapidly faded, in large part because of a scandal in which major automobile manufacturers falsified data to cheat on pollution emissions tests. Decarbonizing transportation has centered ever more on electric vehicles. Such electric vehicles still account for only a small portion of motor vehicle sales, but by 2020 market capitalization for Tesla vastly exceeded that of the world's largest automakers measured by sales. Leaving aside the question of how much Tesla or any company should be valued, the high valuation reflects investors' expectations about the future. Indeed, numerous other automobile manufacturers are increasing investment in electric vehicles. In 2021, General Motors announced it would stop selling gasoline-powered cars and light trucks by 2035. Whatever the manufacturer, the source of electricity for electric vehicles will determine how much their adoption decarbonizes transportation. To achieve reduction in carbon emissions, a shift toward electric vehicles will also require decarbonizing electricity generation.

Decarbonizing transportation would extend far beyond personal motor vehicles to trucks, trains, shipping, and aviation. Electrification of trucks and busses has started, but costs and distance of charge may slow adoption. For railways, diesel-locomotives are widespread, but electric locomotives are also well established, so reducing emissions from railways would require the investment necessary to convert rail lines to electricity. In the United States, most rail lines rely on diesel power. However, many of the world's high-speed trains already run on electrical power.

It is easy for humans as a terrestrial species to overlook the role of shipping in decarbonization. Many cargo ships employ bunker fuel, an extremely low-grade fuel widely employed in shipping. For cargo ships, a shift from bunker fuel to diesel would lower air pollution. Development of zero-emissions cargo ships is still largely in the concept stage with a number of designs for prototypes. These incorporate varied energy sources. Maersk, the world's largest container shipping company, joined several other large corporations in 2020 in an initiative dubbed Transform to Net Zero.[7]

Aviation faces particular challenges in decarbonizing. An electric aircraft has to draw on sufficient power to account for the weight of batteries, and, unlike a traditional aircraft, which ends a flight with less fuel than when it starts, an electric-powered airplane does not lose weight during flight. Making airplanes more fuel-efficient would reduce their carbon emission, but aside from small electrical-powered aircraft, plans for zero-emissions commercial aircraft are also in the concept phase.

Many, but not all, of the plans for decarbonization rely on shifting from fossil fuels to electricity generated by renewable sources. An energy transition therefore would involve not only vastly increasing renewable energy production and replacing vehicles and trains powered by fossil fuels with versions that draw on electricity but would also mean upgrading the electrical power grid and focusing on increasing storage capacity. Transportation and power systems that rely increasingly on electricity will require continuing improvements in batteries, energy storage, and recycling of batteries. At the same time, concepts for decarbonizing some sectors, namely, shipping and aviation, also propose drawing on other sources of energy, including hydrogen and "zero-emission" fuels.

Hydrogen fuel cells have a long history dating back to the nineteenth century. Fuel cells combine oxygen and hydrogen to produce water and electricity. NASA has used liquid hydrogen as a fuel for rockets and has employed fuel cells in its space program, including in the Apollo Command Module. Fuel cells provide another zero-emissions power source, but to date the costs, difficulty refueling, and problems storing a flammable fuel have slowed widespread commercial development.

Beyond transportation, decarbonization would also have to extend to manufacturing and agriculture. Concrete and steel, key materials for modern construction, both rely heavily on fossil fuels. The very production of cement, a critical ingredient in concrete, creates CO_2 emissions. Indeed, if cement were a country, its production as of 2019 would have made it the third largest global emitter.[8] Zero-emissions concrete would therefore involve either capturing CO_2 during production or dispensing with concrete altogether in favor of alternatives. Steel production is another carbon-intensive industry. To convert iron ore into steel, producers use fossil fuels, typically coal. Initiatives for reducing the carbon intensity of steel production include increasing recycling, employing hydrogen, carbon capture, and piloting new metallurgical techniques.

Among other major industries, virtually the entire plastic sector is based on fossil fuels. The raw materials for plastics production derive from petroleum and other fossil fuels. A 2019 report explained, "Nearly every piece of plastic begins as a fossil fuel, and greenhouse gases are emitted at each of each stage of the plastic lifecycle." Plastic is a major source of pollution and also contributes substantially to global warming. The report found that "At present rates, these greenhouse gas emissions from the plastic lifecycle threaten the ability of the global community to meet carbon emissions targets."[9] Decarbonizing this sector would involve cutting plastic production, using renewable energy for production, and employing bioplastics or plastics made from renewable biomass.

Decarbonization would also affect how we live and how we eat. Home energy audits that provide tips on conserving energy are already available in many regions, and energy conservation in buildings in turn reduces costs for heating and cooling. Green building is a rapidly growing sector for construction that focuses on carbon emissions along with air quality and pollution. Materials as well as design and energy sources can reduce carbon emissions. For example, treated blends of wood provide a possible alternative to steel. Air-conditioning on a heating planet with growing middle classes would also have to be part of an energy transition, not only by using energy from renewable sources to power air-conditioning but also by dispensing with coolants that are themselves highly powerful greenhouse gases. Hydrofluorocarbons, or HFCs, absorb heat far more efficiently than CO_2, and chemical companies are developing a number of alternatives.

Carbon emissions for food are complex to measure and vary vastly. Producing meat, in particular beef, leads to high greenhouse gas emissions including methane, whereas most vegetables, grains, and beans generate far lower greenhouse gas emissions. Producers are trying different feeds and supplements to reduce methane emissions from raising cattle, but in general a world with lower greenhouse emissions would also see lower rates of meat consumption.

Technical approaches and geoengineering

In a world with massive demand for energy, it makes sense to focus on reducing greenhouse gas emissions produced from the energy sector, but a different set of approaches focus on enhancing natural carbon storage or on using technology to store or capture carbon. Other approaches attempt to nullify the effects of climate change. Just as degradation of forests can increase carbon emissions, building up natural carbon sinks is one possible means of addressing rising CO_2 levels in Earth's atmosphere. Planting trees and adjusting farming practices can in principle boost carbon storage. However, simply planting trees by itself would not remove the need to implement other policies, and efforts to build up natural carbon sinks come with possible trade-offs.

Given the role of rising CO_2 concentrations in climate forcing, some researchers have looked at ways to reduce emissions during energy production or to remove CO_2 directly from the atmosphere. The phrase *clean coal* refers not to coal itself but to the idea of capturing CO_2 produced at coal-fired power plants. The technology for capturing CO_2 exists, but to date it has proven extremely difficult to do so at the necessary scale to run a coal-fired power plant or to do so in a cost-effective way. Any such plan must also store captured carbon. One possible use: employing captured carbon for fracking or oil production would boost output of oil and gas through enhanced oil recovery, or EOR.

Numerous companies have investigated and in some cases developed systems for carbon capture. Price and scale and the effects of EOR remain challenges. EOR amounts to removing carbon to enhance carbon emissions. In principle, EOR might reduce the total carbon emissions from oil productions, but calculations of the net effect of EOR on carbon emissions vary. Captured CO_2 could also be employed to substitute for carbon dioxide from other sources in products, ranging from concrete to plastics, and even beverages. Without subsidies or policy interventions, none of these uses would currently create products with competitive prices, but companies are engaged in using captured carbon for building materials and varied products.

Instead of reusing or recycling carbon, carbon capture could also take the form of burying carbon. Removing and storing carbon dioxide on an extremely large scale would likely require storing much captured carbon underground, most likely in salt formations. The period in which carbon would be removed or sequestered is also an issue. Removing or sequestering carbon for a short period of time would bring few benefits. Employing carbon in building materials, in contrast, would sequester it for much longer time spans.[10]

Another intriguing idea calls for removing CO_2 directly from the atmosphere. Again, this is possible on a very modest level, but on nowhere near the scale necessary to counteract greenhouse gas emissions. A demonstration project in Iceland succeeded in converting CO_2 into calcite, and research into this option will likely continue.[11] Research is ongoing into carbon mineralization or using chemical reactions to store carbon in minerals. Companies in Iceland and Switzerland have joined in a project to store carbon through mineralization in basalt. Geologists have also piloted storing carbon in rock waste from a diamond mine in northern Canada. Cost and preventing leaching of any associated pollutants pose challenges. Companies are investigating the use of carbon mineralization in construction products[12] such as synthetic limestone.

Another geoengineering scenario for removing CO_2 from the atmosphere focuses on increasing carbon uptake in the ocean by algae. Similar to the idea of planting more trees on land, fertilizing ocean waters to increase photosynthesis by algae would effectively move CO_2 from the atmosphere into the ocean. When algae die and their remains sink to the depths of the ocean, they theoretically carry that carbon with them for long-term storage on the seafloor. Research into fertilizing ocean waters with iron in order to increase photosynthesis in the ocean has been going on for decades; the idea is based on research that showed greater algal growth during glacial periods, when winds delivered more iron to the ocean, acting as a fertilizer. While recent studies on iron fertilization demonstrated an overall increase in algal growth with the added iron, that did not always translate into sequestering the carbon into the deep ocean. Long-term iron fertilizing could, according to some models, actually reduce the ability of ocean waters to absorb CO_2.[13] Proponents of ocean iron fertilization suggest that it could benefit fisheries. Increasing levels of phytoplankton could, however, have unintended and unpredictable consequences for ocean life. One possible negative outcome would be to reduce oxygen levels, which in turn could expand marine dead zones.

A different set of proposals for addressing climate change would not necessarily capture carbon at all but would seek to shield the Earth from the effects of climate change in other ways. Proposals for geoengineering of the atmosphere conventionally focus on reducing the amount of solar energy to reach the Earth's surface by blocking it through some mechanism, such as mirrors, balloons, or aerosols of some kind pumped into the atmosphere. In a sense, these proposals would seek to mimic the effects produced at times by volcanic eruptions in which particles ejected into the atmosphere temporarily reduce insolation. Pilot projects to test the feasibility of solar geoengineering, or of dimming sunlight reaching the earth, are also proceeding. One, tentatively scheduled for 2021, would test the flight of a balloon that would disperse a small quantity of calcium carbonate and then observe the effects.[14]

The feasibility of any of these plans for geoengineering remains unproven. Moreover, even if it were possible to reduce warming through some kind of mechanical geoengineering, that would not curb many of the effects of climate change. Shielding the Earth would have no effect on the rapid acidification of the Earth's oceans, which in itself poses a threat to many forms of life. Also, any breakdown of a system of geoengineering could lead to abrupt, sharp, and potentially catastrophic spikes in warming.

Adaptation

By the late twentieth and the early twenty-first centuries anthropogenic global warming had already started to reverse a trend in which increasingly complex societies had gained more ability to withstand climate fluctuations of the Holocene. Over many centuries, technological improvement, scientific gains, fast transportation, and effective administration reduced the dangers of drought and provided more capacity to withstand other fluctuations with minimal harm. However, climate extremes started to become far more damaging, leading to a renewed focus on the idea of adaptation.

Local leaders and activists have forged a broad movement to make communities more resilient to the effects of climate change. As international efforts to curb global warming have lagged, a host of communities around the world have explored measures for responding and adapting to a rapidly shifting climate. A meeting of representatives of local governments from around the world produced a charter in Durban, South Africa, in 2011 that called for "mainstreaming adaptation as a key informant of all local government development planning," and regular meetings between civic leaders have continued at the international level.[15] In the United States, local elected officials created Resilient Communities for America in 2013.

Drought has encouraged and indeed forced communities across the world to consider water-saving measures with varying effectiveness. In California the drought that began in 2011 grew so severe that some Californians cheered when they saw raindrops. The state introduced stringent water conservation measures in the summer of 2014 but struggled to meet the water reduction targets over several months. In Brazil, government authorities were slow to respond to the depletion of water in reservoirs that supply massive cities like São Paulo, but eventually reduced water pressure to decrease flow and gave discounts to those who cut water use. In Australia, water restrictions aimed to curb urban consumption of water during recent droughts.

Major drought-stricken areas imposed some water restrictions, but to a significant degree, the main response in many regions consisted of hoping that rains would return to sufficient levels. Drought-affected regions from California, to the American Southwest, to Brazil, and to Australia face the prospect of a new normal with overall drying and long periods of time with limited water. Adaptations for the long term include systematic measures to encourage the use of graywater, or employing water used for showering, bathing, or washing clothes for other purposes, such as watering plants and trees. Cities in arid regions are also turning to programs to encourage homeowners to take out lawns and replace nonnative grass with desert landscapes.

Residents of large metropolitan areas face the prospect of living with unaccustomed heat. Phoenix, Arizona, for example, grew rapidly as a city in the sunbelt known for warm temperatures, but climate change has brought prolonged intense heat. In 2020, Phoenix experienced fourteen days with temperatures above 115°F, fifty-three days with high temperatures of 110 or above Fahrenheit, and 145 with high temperatures above 100. This level of heat has forced many residents to change their habits and to shift outdoor activities to cooler parts of the day.[16]

Options for adaptation to rising seas range from wetlands restoration to lifting up buildings, to creating massive storm barriers. The Netherlands, with a wealth of experience in living at or below sea level, has provided a vast store of information and data. On a local level, some homeowners place their houses on pilings or stilts or try to protect coastal properties with large rocks. Adaptation has entered building design: instead of placing electrical and mechanical equipment in the basement in new offices or apartment buildings, taking account of climate change means locating them higher up in new buildings and creating ground floors as open space that can be exposed to water without causing damage.

Cities around the world are exploring projects to try to provide protection from rising seas: the East Coast of the United States alone provides multiple examples. Staten Island in New York City obtained funding to build a five-mile-long sea wall intended to protect against damage of the kind suffered from Hurricane Sandy in 2012. To protect New York City from future storm surges, the Army Corps of Engineers investigated building a six-mile-long barrier at an estimated cost of $119 billion. The plan was "indefinitely postponed" in 2020, but the challenge posed by rising seas in a built-up area with massive investment remains.

Charleston, South Carolina, another community at immediate risk from rising seas, is considering a massive seawall project. Charleston is built on marshlands and old landfill, and the combination of subsidence and rising seas is already leading to repeated flooding under what are now normal conditions. In 2020, Charleston officials backed a plan to build an eight-mile-long sea wall to protect the downtown at an estimated cost of $2 billion.[17]

Communities in southern Florida are also implementing projects to try to deal with rising waters. Miami Beach has built new pumps and sewers, but this is an adaptation without a projected end point for the change that requires adaptation. Still, Miami Beach's mayor spoke of hoping to buy the city fifty years' time.

On the coast of the Gulf of Mexico, Louisiana has developed a Coastal Master Plan at an estimated cost of $50 billion to fund numerous projects to seek to build up and protect land. One project calls for cutting through a federal levee south of New Orleans to let loose flows of sediment to build up land to the south. The plan carries immediate risks of destroying fish and crustaceans dependent on brackish water.[18]

Leaving out the questions of damage to plants and animals, total cost, future funding, and the plan's ability to keep pace with continued sea-level rise and more powerful storms, the 2017 version of the Louisiana's Coastal Master Plan made very little mention of climate change. The plan also severed the state's massive oil and gas industry from the root causes of the problem it purported to address, simply treating the oil and gas extraction and processing industry as an asset. "Louisiana's wetlands," it stated, "protect valuable infrastructure from storm surge and flooding. These assets include fisheries, oil and gas pipelines, petroleum reserves, and the

Henry Hub (a national distribution point for natural gas)." Maintaining the existing oil, gas, and petrochemical industries counted simply as a benefit, or, as the 2017 plan declared: "The master plan benefits all communities identified as being associated with the oil and gas industry."[19]

In the face of high costs for staving off rising seas and storm surges, another response is to build new towns. In Louisiana, the small town of Isle de Jean Charles has already lost most of its land because of engineering projects designed to prevent floods that diverted sediment and also because of climate change. Oil drilling has also increased coastal erosion in the Mississippi Delta. A plan to build a new town with a federal grant of $48 million would move the residents of Isle de Jean Charles some 40 miles north. Other towns, such as Jean Lafitte, now regularly suffer heavy financial losses after major storms. In eight years after Hurricane Katrina, the federal flood insurance payments for a town of 2,000 residents amounted to $9.3 million.[20]

As these examples show, it is possible to find plan after plan to build seawalls or undertake other projects intended to protect towns and cities, but the sheer number and cost cast doubt on these as a plausible long-term solution for the United States, let alone for the world. Just to take the example of one small island, the costs of a project to build protective seawalls and jetties at Tangier Island in Chesapeake Bay run into the millions of dollars, and more ambitious projects would easily cost tens of millions of dollars: none, of course, would actually curb rising sea levels. Perhaps, some wealthy areas will be able to fund such projects, but many lack the resources to do so. A 2019 study by the Center for Climate Integrity set the overall cost for building sea walls for the vulnerable US cities and infrastructure at $400 billion. The overall costs of these projects are difficult to determine, but by any calculation immense. A dike for Houston alone might cost $32 billion.

Similar projects and proposals, some on a vastly greater scale, can be found around the world. In 2020, scientists outlined plans for creating barriers across the entire North Sea in Europe. One dam would cross from Scotland to Norway—the other from England to France. The estimated cost: $250–550 billion. The authors outlined the plan not to advocate for the barrier but to emphasize the threat posed by climate change.

The Netherlands, a country built as its name indicates, is engaged in raising the level of existing dikes to try to keep pace with rising waters. Drawing on long-experience building below sea level and providing spillways for water, the country's engineering firms and consultants are now expanding their work globally.

In China, cities, including Shanghai, are constructing seawalls and reengineering urban areas. Shanghai has built extensive seawalls across Hangzhou Bay and gates to protect against overflowing rivers. Shanghai is also one of China's sponge cities: in 2014, China opened a sponge city initiative. The idea is to make cities function more like sponges or better able to absorb water, by using permeable pavement and by creating ponds, wetlands, and tanks and tunnels to store water.

In Southeast Asia, Jakarta, the largest city and, at present, the capital of Indonesia with a population of 10 million and 30 million in the metropolitan area, provides a dramatic example of the challenges of adaptation. Jakarta was built close to the ocean in a region with thirteen rivers. Rising sea levels are rapidly swamping the city because it is simultaneously sinking due to extreme depletion of groundwater and lax regulations. The city lacks sufficient water, so residents extract water from limited aquifers below the city. Buildings in north Jakarta have already been abandoned. Authorities considered a proposed network of dikes called the Great Garuda, named after a mythical bird.[21] The project would have cost an estimated $40 billion, and critics warned that it would trap polluted water in new lagoons. It was canceled, though other seawalls have been constructed in Jakarta Bay, and a new dike is planned to better defend Jakarta from flooding. Authorities have turned to relocation as an alternative or complementary response. In April 2019, Indonesia's president instead announced that Indonesia would build a new capital.

Around the world, the sheer cost of building sea walls and retractable barriers with costs in the realm of tens of billions of dollars for individual projects would further separate rich and poor communities because towns and cities with less wealth will be less likely to afford such barriers.

Seawalls raise controversy for other reasons besides cost. They may redirect sand, create environmental problems by potentially trapping sewage, and, of course, they do not prevent further sea-level rise.

Other proposals for adaptation call for building up natural defenses that can cushion, at least for a time, against the effects of rising seas. Marshes and wetlands offer protection against storm surges. Restoring wetland and inland areas can also offer some protection against flooding from heavy precipitation. Protecting and restoring ecosystems could also help to store carbon and boost biodiversity.

Financial markets also create incentives to adapt, though this can lead to shifting costs from private to public entities. Flood insurance provides an example. In the United States, controversy over changes to the federal flood insurance map dramatized some of the likely challenges of adaptation. In 2012 the US Congress authorized the Federal Emergency Management Agency to draw new flood maps because the US national flood insurance program was in debt. FEMA then issued revised maps that marked an expanded area as prone to floods. As a result, homeowners in many regions confronted soaring insurance bills that increased by thousands of dollars. Constituents contacted their elected representatives, and in 2014 Congress reversed most of the changes to the flood insurance program. The financial shock to homeowners was significant, but the new flood risk maps did not even take into account the effects of rising seas. The entire episode suggested that the public is not ready to come to terms with the real cost of attempting to adapt to climate change.

In the United States alone, over $1 trillion of coastal real estate is at risk, and potential losses are changing lending and borrowing patterns in coastal areas. A large proportion of homeowners in the United States borrow thirty-year mortgages, but with sea-level rise and more powerful storms, an increasing number of houses close to coasts face the risk of major damage. Over the next thirty years the proportion of homes in coastal areas facing such damage will only increase. To reduce their risk, banks are increasing the proportion of costs that buyers must put down as down payments. Financial institutions are also offloading the risk by transferring mortgages in coastal areas to Fannie Mae and to Freddie Mac, enterprises created and chartered by the US Congress that buy and can package mortgages into mortgage-backed securities.[22]

Insurance companies are also trying to reduce their exposure in areas affected by increased risk of wildfires. After the 2016 Fort McMurray wildfires in Alberta, Canada, insurance companies allowed many customers to take cash payments to rebuild further away from nearby homes, and other homes incorporated new features to provide greater protection from fire. In California, enormous losses from wildfires have increased losses from insurance companies that responded by raising premiums and canceling policies. As insurance companies started to withdraw from issuing policies for homeowners in fire-prone areas, California in 2019 instituted a one-year moratorium on the practice. However, the financial challenges will likely grow. The Camp Fire that burned through much of the town of Paradise California in November 2018 dramatized both the danger to life and the enormous costs. The wildfire killed eighty-five people, injured many more, including firefighters, and resulted in costs of $16.5 billion, making it the most expensive disaster in terms of insurance in the world in 2018. Sparks from an electrical tower likely started the fire, but heat and drought contributed to its spread: the number of days with prime conditions for wildfires has increased from eight to ten in Paradise since the 1970s.[23]

Increased risk and higher costs from wildfires are also affecting the insurance industry in Australia. Along with California, the insurance giant Munich Re identified the southwestern United States and southeastern Australia as regions most affected by the increased insurance costs created by the connection between climate change and wildfires. Higher costs could make insurance unaffordable for many customers. Ernst Rauch, chief climatologist for Munich Re, explained in 2019 that "Affordability is so critical [because] some people on low and average incomes in some regions will no longer be able to buy insurance."[24]

Along with cost, plans for adaptation face a problem of defining an end point. Storm surge barriers, erected at great expense, cannot protect all coastal areas, even in affluent societies, and plans to hoist buildings then raise the question of what to do in the future as sea levels continue to rise. Adaptation projects require projections of future sea-level rise, but at present with a continuing rise in carbon emissions, it is impossible to pinpoint a reasonable estimate for that process. Expensive projects that might protect against a sea-level increase of one foot would have to be replaced by still further projects.

Growing numbers of cities and towns now make resilience in the face of climate change part of their planning process, but all of these plans, however well crafted, face a common problem: How can planners calculate the end point for change? How can communities adapt to extremes that exceed the projections for any given time period? Local and state politics have also challenged the effort. In North Carolina, for example, a law passed in 2012 banned the state from taking into account scientific projections of sea-level rise when conducting coastal planning.

Adaptation is also taking the form of exploiting new opportunities. On balance, accelerating climate change will severely disrupt human societies, but as many cities face rising seas and key agricultural regions suffer from extremes of drought and flooding, some regions may become more hospitable for growing crops. Indeed, in the public discussion over policy responses, some voices rejecting intervention tout the marvels that they claim warming will bring to the world. On a very small scale, the gains for farming can be seen in Greenland, where yields of potatoes and vegetables have increased. In Britain, winegrowers are exploring the possibility of boosting future production. On a much larger scale, farmers in countries such as Canada are growing grain further north and increasing the production of corn. Climate change is bringing significant disruption to Russia with unprecedented heat waves and higher risks of wildfire, but may also increase yields for agriculture in Siberia. Climate change to date has shifted patterns of yields for crops, including corn and soybeans. In Russia, the melting of permafrost strengthens feedbacks, but for now it has also provided new areas for cultivating soybeans.

Warming is unleashing efforts to exploit resources from areas that are becoming more accessible for shipping as well as for mining and other extractive industries. The search for trade routes in far northern waters is old. Before Europeans had fully determined the geography of northern North America, they sought an elusive northwest passage, a route that would take them north around North America, en route to Asia. In the 1570s Martin Frobisher, an English explorer, led expeditions in pursuit of the passage. Other explorers followed. Henry Hudson sailed into Hudson Bay—he was last seen when his crew mutinied and placed him with his son and a few other crewmen into a small boat. There were other attempts, but the first successful crossing took place only in 1906 when the Norwegian explorer Roald Amundsen completed a three-year expedition.

The melting Arctic has raised new interest in a possible future northwest passage viable for shipping. North of Eurasia, shipping interest in the Russian North Sea Route is also increasing. Russia has carried out naval exercises in the area, and a container ship took the route.

Energy and mining companies have also sought oil, gas, and mineral deposits in the Arctic. Conditions are still severe enough to create setbacks:

on December 31, 2012, Royal Dutch Shell's immense *Kulluk* drilling rig ran aground in Alaska. Shell began drilling again in the summer of 2015, but high costs and disappointing findings prompted the company to halt its operations. Norway in 2014 opened up the southeast Barents Sea for exploration. The country's future as an oil and gas producer depended on more finds: "For Norway to continue to be a long-term reliable supplier of oil and gas it is important to explore for and develop," the deputy oil and energy minister explained.[25] In early 2015 Norway offered new oil leases in the Arctic. In 2020 Norway awarded offshore production licenses in "mature" or previously exploited offshore areas and outlined plans for new oil exploration areas, mostly in the Barents Sea in the Arctic. Norway's minister of petroleum and energy stated that "A steady supply of new acreage is crucial in order to maintain activity on Norway's continental shelf."[26] Such drilling creates the potential to magnify the feedback effects of rapid warming at high latitudes: more production of oil will generate still more CO_2.

Individual action

The long list of possible responses to climate change often features collective action, but many individuals seek to respond to climate change on a personal level. This can take the form of re-examining household energy use and sources, trying to reduce unnecessary energy consumption, and, where the individual or family has a choice, choosing to adopt renewable energy, such as installing solar panels or purchasing power produced by solar or wind. A smaller house or dwelling might reduce energy use. Efficient appliances, and yes, light bulbs, also lower energy use. Individuals may also choose to replace vehicles with more fuel-efficient replacements or with electric vehicles.

Action at the individual level can also mean reconsidering shopping habits. Fast fashion, for instance, contributes to greenhouse gas emissions, so the individual might decide to make more use of used clothing. Diet enters into the list of choices because some food items, such as meat, contribute disproportionately to greenhouse gas emissions. The individual also confronts choices for vacations. Trips closer to home and less air travel would reduce greenhouse gas emissions.

By undertaking these (or some of these as well as many other possible actions) the individual can build a connection to a broader effort. Personal action may also yield wider effects by influencing others. Thus, an individual who undertakes a change to try to reduce their carbon footprint on a personal level knows that this will have no measurable effect on overall global greenhouse gas emissions. Nonetheless, by making changes individuals can serve as a model for others and start to spread particular shifts in consumption more widely.

On the individual level, it is difficult to avoid questions of inequity and possible hypocrisy. Some changes in consumption involve shifting to more expensive products or goods out of reach, for now, of many. Also, dependence on fossil fuels is so deeply built into our current systems that reducing greenhouse gas emissions on an individual level can mean giving up treasured family pastimes, losing connections with friends and relatives, and living with constrained horizons. Furthermore, climate activists find themselves open to attacks that they display hypocrisy. Some climate experts have argued that shifting focus from fossil fuel companies and governments to the individual can serve as a pretext to divert attention from the sectors most responsible for greenhouse gas emissions. Thus, the climate scientist Michael Mann describes "'deflection campaigns' aimed to divert attention from big polluters and place the burden on individuals," though noting that individual action is important.[27]

Climate economics

Individual choices about how to reduce personal carbon footprint raise questions about consumption, spending, and lifestyle, and at the collective level, the need to make sharp emissions cuts leads to questions, not just about individual economic sectors, types of industry, and modes of transportation, but about the very organization of the world economy. In one vision of the future world economy, markets can provide a mechanism to lower emissions. This would have to go beyond merely proclaiming a product "green" to win brand recognition. If policy makers can succeed in creating a real cost for carbon emissions, companies, in this scenario, will compete to come up with solutions.

A couple of market-based approaches that could help reduce greenhouse gas emissions and therefore mitigate climate change include a carbon tax and an emissions trading scheme, both of which essentially put a price on carbon and begin to incorporate the life-cycle cost of emissions. Such a market-based approach aims to make consumers consider the full societal cost of carbon-intensive goods. It also provides motivation to alter consumption patterns in a way that reduces carbon emissions overall.

Implementing a carbon tax, which sets a standard price per ton of carbon emissions, encourages reductions in emissions as the taxpayer attempts to reduce cost. Under one variant of the plan, the revenue collected by the tax would be redistributed so that the actual tax would be revenue-neutral: this is sometimes referred to as "tax and dividend." Neither form of carbon tax would create an absolute cap on emissions. Opposition to introducing or raising taxes provides a chief obstacle to this approach.

Cap and trade, or emissions trading, works by placing an overall limit on emissions from most, if not all, sectors of the economy, and then distributes permits for emissions. The different sectors either receive a certain number of emission allowances or allocations, or they obtain allocations through a bidding process. Once the various companies have their allowances, they can then sell or trade them. The carbon market would drive the prices of these allowances. Key challenges for cap and trade include determining the initial cap and the prices for allocation and allowances. Thus, a cap-and-trade plan introduced in the European Union in 2005 provided too many allowances, which kept prices and revenue very low. Cap-and-trade systems have also tended to focus on particular sectors rather than on entire economies.

A very different reading of the path forward counters that capitalism itself as currently organized has fostered the climate crisis through a focus on growth and corporate profits. This approach calls for remaking the predominant economic system to address climate change.[28] Such an approach would likely require rethinking measures of growth. GDP, the main traditional measure of growth, calculates the monetary value of all goods and services produced in a particular period of time but leaves out possible effects of such output. Economists have proposed alternatives that would take into account the environment as well as health, inequality, and people's satisfaction with work.

The clash between economic visions in itself dramatizes the enormity of the choices humans face looking toward the future. The future of human societies will be closely connected with climate change. The timescale for significant projected climate effects is now far shorter than an average human lifespan. Whatever choices we make, we know that enormous numbers of people who are living today will experience the consequences.

Into the Anthropocene

Human prehistory makes up both a sliver of geologic time but also a vast expanse of time from our current perspective. During a period of several million years human ancestors evolved during an overall period of cooling, and over several hundred thousands of years our own species, *Homo sapiens*, lived as hunter-gatherers through glacial maxima and through sometimes severe climate fluctuations. Climate change revealed human vulnerability, but early humans also showed great resilience. Thus, many lineages died out, from reasons that could have included climate change, but *Homo sapiens* survived glacial maxima and developed the capacity to find food in highly varied biomes.

After the last glacial maximum, *Homo sapiens* flourished during the Holocene. We predate the Holocene, but all human civilizations to date emerged during this period of comparatively warm and stable climate. Fluctuations within the Holocene interacted with other factors to influence human history in many ways; some societies benefited from favorable conditions for agriculture or for pastoralism, while others suffered from cold spells or prolonged drought. Gradually complex societies developed greater resiliency and the ability to insulate themselves from the worst consequences of climate fluctuations.

With the Industrial Revolution, the relationship between human history and climate changed. At first, greenhouse gas emissions from nascent industry were small enough as to have little effect on climate, but as industrialization deepened and spread and as human energy use soared, humans became the chief cause of climate forcing and caused pronounced and unprecedented warming along with many other severe effects of rapid climate change.

Human forcing has created such striking change that we are beginning to see climatic changes that surpass Holocene fluctuations as a result of human activity. In light of this, some scientists contend that we have entered a new geological epoch, the Anthropocene[29] and have proposed formal adoption of the term to the International Commission on Stratigraphy (ICS). This has triggered debate among geologists: Does the human impact on the environment warrant its own epoch, and if so, when did it begin and how is it identified in the rock record? Some suggest that the beginning of the Industrial Revolution marks a logical beginning of the Anthropocene, while others suggest an earlier start when humans began transforming the planet for agriculture thousands of years ago.[30] The Dutch chemist Paul Crutzen, who won the Nobel Prize for research on ozone-depleting chemicals and helped popularize the term *Anthropocene*, suggested along with a colleague that the term was not just a question of stratigraphy: "But that shouldn't stop us from seeing and learning what it means to live in this new Anthropocene epoch, on a planet that is being anthroposized at high speed."[31] This notion of being anthroposized stressed transformation by human action and at speed unknown in all of human history and prehistory. The Anthropogenic Working Group of the ICS, citing anthropogenic signatures such as the observed increase in greenhouse gases, expansive use of fertilizers for agriculture, habitat loss and species extinction, increased erosion, and accumulation of plastics, voted in 2019 to approve this new epoch to succeed the Holocene, beginning around 1950.[32]

Still a subject of discussion, it is uncertain whether or not the Anthropocene will be certified as a formal epoch as part of the geological timescale, but the term has nevertheless gained popularity in academic settings as well as with the general public. It is now common to see scientists reference the Anthropocene in peer-reviewed articles, and several academic publications are named for the era. Referencing the Anthropocene acknowledges that humans have become a primary agent of global climate change. Indeed, use and discussion of the term have spread beyond geology, climate science, and even science to the humanities and to art, literature, and music. Such use has also prompted critiques on varied grounds: that the term shows a kind of "hubris" in naming an era after ourselves or that it blames humanity as a whole for climate change, when many have little responsibility.[33]

We cannot foretell if the term *Anthropocene* will stick and gain wide acceptance, but even if it does not, another word may emerge to describe the new era that human activity has created because the change is real and marked. Immense work has gone into and will continue to go into climate science, modeling, and into guidance on how to curb and respond to climate change, but the core findings are stunningly apparent: a temperature graph that rises strikingly and abruptly and keeps rising, glaciers that melt so rapidly that they become markedly smaller within less than a normal lifetime, marine species that disappear from waters in which they long-lived, gardeners who notice a decidedly earlier spring and later first frost—the list goes on. Humans, with a record-high population now live in a time when human-caused climate change places new pressures and stress on societies and the world we share with other forms of life on earth. At the same time, our very knowledge of these trends and considerable technological capacity also create at least the chance for people to shape future climate to try to reduce the magnitude of the break with the climate in which civilizations have flourished.

GLOSSARY

8.2 k or 8 k event: an abrupt cooling event that occurred around 8200 years ago, and lasted for approximately 160y

Abanaki: Algonquian speakers of northern New England and Quebec

African Humid Period: period 9000 to 6000y ago when the Sahara had abundant grass and shrubs due to increased precipitation –see Green Sahara

Alps: high mountain range of Europe

Anatolia: peninsula in western Asia that makes up most of modern day Turkey

Andes: mountain range of western South America

Angkor: city of Khmer Empire with temple complex of Angkor Wat

Antarctic Cold Reversal: a period of Southern Hemisphere cooling between approximately 14,700 and 13,000y ago that interrupted the overall warming period between 20,000 and 10,000y ago

Anthropocene: proposed new geologic epoch marked by human transformation of planet earth and its climate

Assyria: powerful state of northern Mesopotamia–engaged in late phase of expansion from 900-612 BCE during the Iron Age

Attribution analysis: scientific analysis to identify whether climate change contributes to particular weather events

Australopiths: early extinct species related to humans

Balkans: peninsula of southeastern Eruope

Bantu: group of African languages found today from Central to Southern Africa

Bering Strait: narrow body of water that today separates Siberia and Alaska and was previously the site for a land bridge

Biome: life zone made up of region and its plants and animals

Bronze Age: period identified with the use of bronze for tools and weapons–depending on region began between 3300 and 1900 BCE

Bubonic plague: disease caused by plague bacteria that has caused several pandemics, including the Black Death of the mid-14th Century CE in Europe and Western Asia

Byzantine Empire: name given for the Eastern or Later Roman Empire that survived the end of the Western Empire and endured though with much less territory, until defeat by the Ottoman Turks in 1453 CE

Cahokia: complex site of Mississippi culture with population peak around 1200 CE

Carolingian: family that became rulers of the Franks

Chaco Canyon: site in New Mexico with development of great houses between 800 and 1150-1200 CE

Climate feedbacks: processes in the climate system that either amplify or counteract an initial change in climate

climate forcings: factors that cause climate to change

Climate Research Unit (CRU): climate research center at University of East Anglia in Britain

Clovis: culture of North America from some 13,000 to 11,000 years ago identified with tools

Cognitive Bias: distortions in judgement or errors in thinking–studied in psychology

Columbian Exchange: process of exchange of populations plants, animals, pathogens between Eurasia and Africa and the Americas

Coronavirus: RNA virus–transmission of a coronavirus started the global pandemic that spread globally in 2020

Dansgaard-Oeschger events: abrupt warming events followed by gradual cooling that occurred during the last glacial period

Decarbonization: reducing carbon emissions in our economies and societies

***Denisovans*:** extinct human species that had contact with modern humans.

Divestment: withdrawing assets to press for political and policy change

Eccentricity: shape of Earth's orbit

El Nino-Southern Oscillation (ENSO): climate phenomenon associated with changes in atmospheric pressure across the equatorial Pacific Ocean. This climate oscillation causes worldwide changes in weather.

Export-led growth: an economic approach of pursuing economic growth by emphasizing exports

Extinction Rebellion: climate movement that employs civil disobedience

Fertile Crescent: crescent shaped region in the Middle East from the Tigris and Euphrates River to the Nile-an early center for farming and for civilizations

First Intermediate Period: 2160-2055 BCE Period of decentralization in Egypt after the Old Kingdom followed by revived central power of Middle Kingdom

Fordism: economic model combining mass production of consumer durables with mass consumption

Fracking (hydraulic fracturing): process of injecting fluids into rock formations to extract oil and gas

Fridays for Future: Youth climate movement founded in 2018

Geoengineering: intervening in earth's environment, often through technology, to try to modify climate

Globalization: increasing communication, commerce, travel and interaction on a global scale that builds connections around the world

Grasslands: biome that contains mostly grass, often found in climates that have a distinct dry season

Great Rift Valley: large valley in East Africa formed by tectonic plates moving apart

Green New Deal: climate change policy proposal in US that focuses on clean energy, job creation, and inequality

Green Sahara: periods of increased precipitation in the Sahara caused by an expansion of the monsoon region that supported more abundant vegetation and plants and animals

Greenhouse Effect: process that keeps Earth warmer than it otherwise would be as a result of gases in the atmosphere that absorb heat

Greens: political parties identified in particular with a focus on the environment

Grindelwald fluctuation: a cooler period within the Little Ice Age that occured between 1560 and 1630

Han Dynasty: imperial dynasty of China from 206 BCE–220 CE–established pattern of long-lasting dynastic rule

Heinrich events: cold periods during the last glacial periods during which cold polar ocean waters expanded equatorward

Himalayas: highest mountain range of world between South Asia and Tibetan Plateau

Hockey Stick: in context of climate change, graph of temperature over time that shows sharp recent rise because of human activity

Holocene: the current geological epoch that began approximately 11,700y ago

Hominin: humans are the only surviving species–also includes extinct human ancestors and related species.

***Homo erectus*:** extinct human species that emerged around 2 million years ago and dispersed widely from Africa into Eurasia.

***Homo sapiens*:** species of modern humans who emerged in Africa and dispersed globally

Huns: nomadic people who migrated west and entered Roman Empire

Hypoxia: low oxygen conditions

Indus River Civilization: Bronze Age civilization of Indus River with cities in what is now Pakistan

Industrialization: process of developing industry and manufacturing in a region or country–often associated with urbanization or widespread growth of cities and shift of population away from rural areas

Insolation: incoming solar radiation

Internal colonization: expansion of settlement within a region for example within Europe during the Middle Ages

IPCC (Intergovernmental Panel on Climate Change): body established by United Nations to compile information about climate change

Iron Age: period identified by use of iron for tools and weapons–began between 1200 and 600 BCE

ITCZ: intertropical convergence zone – region near the equator where trade winds meet, characterized by warm temperatures and abundant rainfall

James Watt: inventor of improved steam engine

Keystone XL: pipeline project to transport oil from tar sands in Alberta

Khmer Empire: state established in southeast Asia from 9th to 15th centuries CE–known especially for Angkor Wat

Kyoto Protocol: International climate agreement of 1997

Laki: volcanic fissure in Iceland

LALIA: Late Antique Little Ice Age - a cool period that occurred 536 to 660 CE

Lascaux: caves from paleolithic or old stone age in southwestern France with numerous paintings

Laurentide ice sheet: Large ice sheet that stretched from the Arctic into the northern United States during the Last Glacial Maximum.

Levant: region of the eastern Mediterranean, including coastal areas of what are now Turkey, Syria, and Lebanon

LGM: Last glacial maximum, the most recent glacial period around 20,000y ago

LIA: Little Ice Age, a relatively cool period that occurred between ca. 1300-1850.

Loess: accumulation of windblown sediment

Magdalenian: culture of the late stone age found in Europe

Mapungubwe: kingdom of 10th -14th century CE in southern Africa

Mauna Loa Observatory: Atmospheric research observatory, located on the top of Mauna Loa, Hawaii, that has been collecting carbon dioxide measurements since 1958.

Maya: complex society of central America with numerous cities during classic period of 200-900 CE

MCA: Medieval Climate Anomaly, also referred to as the Medieval Warm Period, a period of relative warmth around 900 to 1200.

Megadrought: severe and prolonged (at least 20y) drought conditions

Megafauna: large animals—most became extinct after the Last Glacial Maximum

Meiji Restoration: restoration of imperial rule in Japan–associated with modernization and industrialization

Mesopotomia: land of the Tigris and Euphrates rivers–home to early cities and Bronze Age civilizations

Microfossils: microscopic fossils, typically plankton shells, that can be used to trace past climate changes

Middle Ages (Medieval period): historical period associated with Europe from end of Western Roman Empire in 476 CE until roughly the 14th-15th centuries CE

Milankovitch cycles: changes in Earth's orbit and tilt that influences the amount of solar radiation Earth receives (see precession, eccentricity, and obliquity)

Ming Dynasty: Chinese Dynasty after Mongol rule–ended in 1644

Minoan: Bronze Age civilization of the island of Crete

Mongols: Mainly semi-nomadic people of steppes who under Ghengis Khan created an empire and under his successor conquered China. Mongols created states in the 13th century across a wide swathe of Eurasia

Monsoon: a seasonal shift of winds that results in a rainy summer season and dry winter season in certain tropical regions

Mycenae: Bronze Age civilization of southern Greece

NASA (National Aeronautics and Space Administration): United States space agency that also provides information about climate change on Earth

Natufians: culture of the stone age with sites found in modern day Israel, Jordan, Lebanon, Syria and Palestine–hunters who built permanent settlements

Nazca: coastal plain in southern Peru

***Neanderthals*:** extinct human species that lived in Europe and western Asia and had contact with *Homo sapiens*

Near East: region that includes Middle East and Ottoman lands in Europe

Neolithic or New Stone Age: final period of the stone age starting approximately 12,000 years ago

Newcomen steam engine: early steam engine

NOAA (National Oceanic and Atmospheric Administration): United States federal scientific agency that focus on oceans, waterways, and atmosphere

Normans: descendants of Norse who settled in Normandy in northern France who took power in England in 1066 CE

North Atlantic Oscillation (NAO): climate variation caused by changes in atmospheric pressure between the northern and central Atlantic, which influences the intensity and location of the jet stream.

Obliquity: tilt of Earth's rotational axis

Ocean acidification: the increase in acidity (decreased pH) observed as a result of the ocean's uptake of carbon dioxide

Old Kingdom: sequence of Egyptian dynasties from c. 2686–2150 BC known for the great pyramids

Ottoman Empire: established by Ottoman Turks who conquered the remnants of the Byzantine Empire and expanded their rule into southeastern Europe,

North Africa, and parts of the Middle East and the Arabian peninsula–ended in 1922

Paleolithic: old stone age–period in which humans developed stone tools starting approximately 2.58 or even 3.3 million years ago

Paris Agreement: international climate agreement of 2015

People's Republic of China: Chinese state founded in 1949 after the Chinese Civil War of 1945-1949

Permafrost: permanently (at least for two years) frozen soil

Pilgrims: Puritan English settlers who arrived in what is now Massachusetts on Mayflower

Plankton: Organisms, mostly microscopic, that passively float in salt or freshwater. Their remains can be used in climate research to investigate what their environment was like (e.g., water temperature) while they were living

Pleistocene: geological epoch preceding the Holocene, between 2.6 million and 11,700 years ago.

Pluvials: periods of increased rainfall

Precession: change in the direction of Earth's rotational axis caused by the wobble of Earth's rotation and its orbital plane

Qing Dynasty: Dynasty established by Manchus after the collapse of the Ming Dynasty.

Quaternary: the geological period that began approximately 2.6 million years ago, and includes both the Pleistocene and Holocene epochs

Rain Forest: dense forests with high and consistent year-round rainfall

Rainshadow: arid region on the downwind side of a mountain

RCPs: representative concentration pathways, various scenarios used by the IPCC to project future changes in temperature as a result of greenhouse gas emissions

Refugia: areas in which humans other animals and plants survived during glacial maximum

Roman Climate Optimum: period of comparatively stable climate from 400 BC to 200 CE that benefitted societies, including Rome

Roman Empire: large empire of antiquity–founded as Republic, expanded around Mediterranean and held lands in Europe, North Africa, and Near East before end of the Western Roman Empire in 476 CE

Sapropels: organic-rich mud layers deposited in the Mediterranean Sea that are typically indicative of bottom waters with little to no oxygen.

Savanna: extensive grasslands with scattered trees

Scandinavia: region of northern Europe, including peninsula that now contains Norway and Sweden and adjacent areas, including Denmark

Seventeenth century crisis: period of warfare, conflict and instability in the 17th century CE

Shamal: hot, dry wind that blows in the Persian Gulf region that can cause sandstorms

Speleothem: deposit of calcium carbonate in caves

Tambora: volcano in Sumbawa in Indonesia

Tang Dynasty: 618-907 CE resumed period of powerful dynastic rule in China

Tar sands: petroleum deposits mixed with sand, clay, and water

Tectonic plates: pieces of Earth's crust and upper mantle

Titicaca: Lake at high altitude on borders of modern-day Bolivia and Peru

Tundra: biome characteristic of the Arctic, with low temperatures, short growing seasons, and permafrost

Uplift: upward vertical movement of Earth's surface in response to plate tectonics

Vikings: Norse people who migrated and traveled by sea and water from Scandinavia during the period from before 800 CE to around 1100 CE. Vikings engaged in raids, traded, and created settlements.

Weathering: process of breaking down rocks and minerals at Earth's surface

Younger Dryas: return to near glacial conditions around 12,000y ago as the Earth was warming from the last glacial maximum

Yuan Dynasty: established by Mongols in China in the 13th century and ended in 1368 CE

Zoonotic diseases: diseases that cross from animals to humans

NOTES

Introduction

1 Jarred Diamond, *Collapse: How Societies Choose to Succeed or Fail* (New York: Penguin, 2005).

Chapter One

1 J. R. Toggweiler and H. Bjornsson, “Drake Passage and Palaeoclimate,” *Journal of Quaternary Science* 15 (2000): 319–28.

2 R. M. DeConto and D. Pollard, “Rapid Cenozoic Glaciation of Antarctica Induced by Declining Atmospheric CO_2,” *Nature* 421 (2000): 245–9; R. M. DeConto and D. Pollard, “A Coupled Climate-Ice Sheet Modeling Approach to the Early Cenozoic History of the Antarctic Ice Sheet,” *Palaeogeography, Palaeoclimatology, Palaeoecology* 198 (2003): 39–52.

3 M. Pagani, M. Huber, Z. Liu, S Bohaty, J. Henderiks, W. Sijp, S. Krishnan, and R. DeConto, “The Role of Carbon Dioxide during the Onset of Antarctic Glaciation,” *Science* 334 (2011): 1261–4.

4 P. Pearson, G. Foster, and B. Wade, “Atmospheric Carbon Dioxide through the Eocene-Oligocene Climate Transition,” *Nature* 461 (2009): 1110–14.

5 M. E. Raymo, W. R. Ruddiman, and P. N. Froelich, “Influence of Late Cenozoic Mountain Building on Ocean Geochemical Cycles,” *Geology* 16 (1998): 649–53; M.E. Raymo and W.F. Ruddiman, “Tectonic Forcing of Late Cenozoic Climate,” *Nature* 359 (1992): 117–22.

6 L. Keigwin, “Isotopic Paleoceanography of the Caribbean and East Pacific: Role of Panama Uplift in Late Neogene Time,” *Science* 217 (1982): 350–3.

7 G. H. Haug and R. Tiedemann, “Effect of the Formation of the Isthmus of Panama on Atlantic Ocean Thermohaline Circulation,” *Nature* 393 (1998): 673–6.

8 P. J. Polissar et al., “Synchronous Rise of African C4 Ecosystems 10 Million Years Ago in the Absence of Aridification,” *Nature Geoscience* 12 (2019): 657–60.

9 J. J. Imbrie Hays and N. Shackleton, "Variations in the Earth's Orbit: Pacemaker of the Ice Ages," *Science* 194 (1976): 1121–32.

10 P. B. Demenocal, "African Climate Change and Faunal Evolution during the Pliocene–Pleistocene," *Earth and Planetary Science Letters* 220, no. 1–2 (2004): 3–24; and Axel Timmerman and Tobias Friedrich, "Late Pleistocene Climate Drivers of Early Human Migration," *Nature* (2016): 92–5.

11 J. E. Kutzbach, "J. E. Monsoon Climate of the Early Holocene: Climate Experiment with Earth's Orbital Parameters for 9000 Years Ago," *Science* 214 (1981): 59–61.

12 M. Rossingnol-Strick, W. Nesteroff, P. Olive, and C. Vergnaud-Grazzini, "After the Deluge: Mediterranean Stagnation and Sapropel Formation," *Nature* 295 (1982): 105–10.

13 "Humans Migrated Out of Africa to Escape Drying Climate, New Study Says," *Science News*, October 17, 2017, http://www.sci-news.com/geology/humans-o ut-of-africa-escape-drying-climate-05331.html.

14 Timmerman and Friedrich, "Late Pleistocene Climate Drivers of Early Human Migration."

15 Eugene I. Smith et al., "Humans Thrived in South Africa through the Toba Eruption about 74,000 Years Ago," *Nature* 555 (2018): 511–4.

16 Swpan Mallick et al., "The Simons Genome Diversity Project: 300 Genomes from 142 Diverse Populations," *Nature* 538 (2016): 201–6.

17 Chris Clarkson et al., "Human Occupation of Northern Australia by 65,000 Years Ago," *Nature* 547 (2017): 306–10.

18 Ray Tobler, Adam Rohrlach, Julien Soubrier et al., "Aboriginal Mitogenomes Reveal 50,000 Years of Regionalism in Australia," *Nature* 544 (2017): 180–4; Luca Pagani, Daniel John Lawson, Evelyn Jagoda, Alexander Mörseburg, Anders Eriksson, Mario Mitt, Florian Clemente et al. "Genomic Analyses Inform on Migration Events during the Peopling of Eurasia," *Nature* 538, no. 7624 (2016): 238–42; and James M. Bowler, Harvey Johnston, Jon M. Olley, John R. Prescott, Richard G. Roberts, Wilfred Shawcross, and Nigel A. Spooner. "New Ages for Human Occupation and Climatic Change at Lake Mungo, Australia," *Nature* 421, no. 6925 (2003): 837–40.

19 Gifford H. Miller, Marilyn L. Fogel, John W. Magee, Michael K. Gagan, Simon J. Clarke, and Beverly J. Johnson, "Ecosystem Collapse in Pleistocene Australia and a Human Role in Megafaunal Extinction," *Science* 309, no. 5732 (2005): 287–90.

20 S.A. Hocknull et al., "Extinction of Eastern Sahul Megafauna Coincides with Sustained Environmental Deterioration," *Nature Communications* 11 (2020): 2250, https://doi.org/10.1038/s41467-020-15785-w.

21 F. Saltré et al., "Climate-Human Interaction Associated with Southeast Australian Megafauna Extinction Patterns," *Nature Communications* 10 (2019): 5311, https://doi.org/10.1038/s41467-019-13277-0.

22 Tsutsumi Takashi, "MIS3 Edge-Ground Axes and the Arrival of the First *Homo Sapiens* in the Japanese Archipelago," *Quaternary International* 248 (2012): 70–8.

23 Akira Iwase et al. "Timing of Megafaunal Extinction in the Late Pleistocene on the Japanese Archipelago," *Quaternary International* 255 (2012): 114–24.

24 Chris Stringer, *Lone Survivors: How We Came to be the Only Humans on Earth* (New York: Times Books, Henry Holt and Company, 2012), 225 and 227–8.

25 Benjamin Vernot et al., "Excavating Neandertal and Denisovan DNA from the Genomes of Melanesian Individuals," *Science* 352, no. 6282 (2016): 235–9.

26 Pasquale Raia et al., "Past Extinctions of Homo Species Coincided with Increased Vulnerability to Climatic Change," *One Earth* 3 (2020): 480–90.

27 Loukas, P. Barton, Jeffrey Brantingham, and Duxue Ji, "Late Pleistocene Climate Change and Paleolithic Cultural Evolution in Northern China: Implications from the Last Glacial Maximum," *Developments in Quaternary Sciences* 9 (2007): 105–28.

28 J. R. Jones et al., "Adaptability, Resilience and Environmental Buffering in European Refugia during the Late Pleistocene: Insights from La Riera Cave (Asturias, Cantabria, Spain)," *Science Reports* 10, no. 1217 (2020). https://doi.org/10.1038/s41598-020-57715-2.

29 Yaroslav V. Kuzmin, "Extinction of the Woolly Mammoth (Mammuthus Primigenius) and Woolly Rhinoceros (Coelodonta Antiquitatis) in Eurasia: Review of Chronological and Environmental Issues," *Boreas* 39, no. 2 (2010): 247–61.

30 R. Nick, E. Barton, Roger M. Jacobi, Dick Stapert, and Martin J. Street, "The Late-Glacial Reoccupation of the British Isles and the Creswellian," *Journal of Quaternary Science* 18, no. 7 (2003): 631–43; Maria Pala et al., "Mitochondrial DNA Signals of Late Glacial Recolonization of Europe from Near Eastern Refugia," *The American Journal of Human Genetics* 90, no. 5 (2012): 915–24.

31 Xiaoyun Cai et al., "Human Migration through Bottlenecks from Southeast Asia into East Asia during Last Glacial Maximum Revealed by Y Chromosomes," *PLoS One* 6, no. 8 (2011): e24282; and Min-Sheng Peng et al., "Inland Post-Glacial Dispersal in East Asia Revealed by Mitochondrial Haplogroup M9a'b," *BMC Biology* 9, no. 1 (2011): 1; and Ian Buvit et al., "Last Glacial Maximum Human Occupation of the Transbaikal, Siberia," *PaleoAmerica* 1 (2015): 374–6.

32 Barton, Brantingham, and Ji, "Late Pleistocene Climate Change and Paleolithic Cultural Evolution in Northern China," 105–28.

33 Marcus J. Hamilton and Briggs Buchanan, "Archaeological Support for the Three-Stage Expansion of Modern Humans across Northeastern Eurasia and into the Americas," *PloS one* 5, no. 8 (2010): e12472.

34 J. M. Broughton and E. M. Weitzel, "Population Reconstructions for Humans and Megafauna Suggest Mixed Causes for North American Pleistocene Extinctions," *Nature Communications* 9, 5441 (2018). https://doi.org/10.1038/s41467-018-07897-1.

35 Tobias Anderman et al., "The Past and Future Human Impact on Mammalian Diversity," *Science Advances* 6 (2020). DOI:10.1126/sciadv.abb2313.

Chapter Two

1 Barbara Fraser, "The First South Americans: Extreme Living," http://www.nature.com/news/the-first-south-americans-extreme-living-1.16038; and Lamont Doherty Earth Observatory, "Climate in the Peruvian Andes: From Early Humans to Modern Challenges," http://www.ldeo.columbia.edu/news-events/climate-peruvian-andes-early-humans-modern-challenges.

2 Jessica L. Metcalf, Chris Turney, Ross Barnett, Fabiana Martin, Sarah C. Bray, Julia T. Vilstrup et al., "Synergistic Roles of Climate Warming and Human Occupation in Patagonian Megafaunal Extinctions During the Last Deglaciation," *Science Advances* 2 (June 17, 2016). DOI: 10.1126/sciadv.1501682.

3 Ibid.

4 Philippe Crombé, Philippe De Smedt, Neil S. Davies, Vanessa Gelorini, Ann Zwertvaegher, Roger Langohr, Dirk Van Damme et al., "Hunter-Gatherer Responses to the Changing Environment of the Moervaart Palaeolake (Nw Belgium) during the Late Glacial and Early Holocene," *Quaternary International* 308 (2013): 162–77.

5 Eline D. Lorenzen, David Nogués-Bravo, Ludovic Orlando, Jaco Weinstock, Jonas Binladen, Katharine A. Marske, Andrew Ugan et al., "Species-Specific Responses of Late Quaternary Megafauna to Climate and Humans," *Nature* 479, no. 7373 (2011): 359–64.

6 Ehud Weiss, Wilma Wetterstrom, Dani Nadel, and Ofer Bar-Yosef, "The Broad Spectrum Revisited: Evidence from Plant Remains," *Proceedings of the National Academy of Sciences of the United States of America* 101, no. 26 (2004): 9551–5.

7 Ofer Bar-Yosef, "The Natufian Culture in the Levant, Threshold to the Origins of Agriculture." *Evolutionary Anthropology: Issues, News, and Reviews* 6, no. 5 (1998): 159–77.

8 Li Liu, Sheahan Bestel, Jinming Shi, Yanhua Song, and Xingcan Chen, "Paleolithic Human Exploitation of Plant Foods during the Last Glacial Maximum in North China," *Proceedings of the National Academy of Sciences* 110, no. 14 (2013): 5380–5.

9 Jade d'Alpoim Guedes, Jacqueline Austermann, and Jerry X. Mitrovica, "Lost Foraging Opportunities for East Asian Hunter-Gatherers Due to Rising Sea Level Since the Last Glacial Maximum," *Geoarchaeology* 31 (2016): 255–66.

10 Hong Shi, Xuebin Qi, Hua Zhong, Yi Peng, Xiaoming Zhang, Runlin Z. Ma, and Bing Su, "Genetic Evidence of an East Asian Origin and Paleolithic Northward Migration of Y-Chromosome Haplogroup N," *PloS one* 8, no. 6 (2013): e66102.

11 Wallace S. Broecker, James P. Kennett, Benjamin P. Flower, James T. Teller, Sue Trumbore, Georges Bonani, and Willy Wolf, "Routing of Meltwater from the Laurentide Ice Sheet during the Younger Dryas Cold Episode," *Nature* 341 (1989): 318–21.

12 Julian B. Murton, Mark D. Bateman, Scott R. Dallimore, James T. Teller, and Zhirong Yang, "Identification of Younger Dryas Outburst Flood Path from Lake Agassiz to the Arctic Ocean," *Nature* 464 (2010): 740–3.

13 For review, see Annelies van Hoesel, Wim Z. Hoek, Gillian M. Pennock, and Martyn R. Drur, "The Younger Dryas Impact Hypothesis: A Critical Review," *Quaternary Science Reviews*, 83 (2014): 95–114.

14 A. E. Carlson, "The Younger Dryas Climate Event," in S.A. Elias ed., *The Encyclopedia of Quaternary Science*, vol. 3 (Amsterdam: Elsevier, 2013), 126–34.

15 David G. Anderson, Kirk Maasch, and Daniel H. Sandweiss, eds.,*Climate Change and Cultural Dynamics: A Global Perspective on mid-Holocene Transitions* (Amsterdam: Elsevier/Academic Press, 2011); and David G. Anderson, Albert C. Goodyear, James Kennett, and Allen West, "Multiple Lines of Evidence for Possible Human Population Decline/Settlement Reorganization during the Early Younger Dryas," *Quaternary International*, 242 (2011): 570–83.

16 Metin I. Eren, "On Younger Dryas Climate Change as a Causal Determinant of Prehistoric Hunter-Gatherer Cultural Change," in Metin I. Erin, ed. *Hunter-Gatherer Behavior: Human Response during the Younger Dryas* (Walnut Creek: Left Coast Press, 2012), 11–20; and Dean R. Snow, *Archaeology of Native North America* (London and New York: Routledge, 2016), 46.

17 Stella M. Blockley and Clive S. Gamble, "Europe in the Younger Dryas: Animal Resources, Settlement and Funerary Behavior," in Eren, *Younger Dryas*, 190.

18 Bar-Yosef, "The Natufian Culture in the Levant, Threshold to the Origins of Agriculture," and Ofer Bar-Yosef "Climatic Fluctuations and Early Farming in West and East Asia," *Current Anthropology* 52 (2011): S175–93.

19 Gordon Hillman, Robert Hedges, Andrew Moore, Susan Colledge, and Paul Pettitt, "New Evidence of Lateglacial Cereal Cultivation at Abu Hureyra on the Euphrates," *The Holocene* 11, no. 4 (2001): 383–93.

20 Arlene M. Rosen and Isabel Rivera-Collazo, "Climate Change, Adaptive Cycles, and the Persistence of Foraging Economies during the Late Pleistocene/Holocene Transition in the Levant," *Proceedings of the National Academy of Sciences* 109, no. 10 (2012): 3640–5.

21 Chantel E. White and Cheryl A. Makarewicz, "Harvesting Practices and Early Neolithic Barley Cultivation at el-Hemmeh, Jordan," *Vegetation History and Archaeobotany* 21, no. 2 (2012): 85–94; and George Willcox, George, Sandra Fornite, and Linda Herveux, "Early Holocene Cultivation before Domestication in Northern Syria," *Vegetation History and Archaeobotany* 17, no. 3 (2008): 313–25.

22 Sylvi Haldorsen, Hasan Akan, Bahattin Çelik, and Manfred Heun, "The Climate of the Younger Dryas as a Boundary for Einkorn Domestication," *Vegetation History and Archaeobotany* 20 (2011): 305–18.

23 Dorian Q. Fuller, George Willcox, and Robin G. Allaby, "Cultivation and Domestication Had Multiple Origins: Arguments against the Core Area Hypothesis for the Origins of Agriculture in the Near East," *World Archaeology* 43 (2011): 628–52.

24 Liu et al., "Paleolithic Human Exploitation of Plant Foods during the Last Glacial Maximum in North China."

25 Bar-Yosef, "Climatic Fluctuations and Early Farming in West and East Asia."

26 Tracey L.D. Lu, "The Occurrence of Cereal Cultivation in China," *Asian Perspectives* 45 (2006): 129–58.

27 Mark Nathan Cohen, *The Food Crisis in Prehistory: Overpopulation and the Origins of Agriculture* (New Haven: Yale University Press, 1977).

28 Hong-Xiang Zheng, Shi Yan, Zhen-Dong Qin, Yi Wang, Jing-Ze Tan, Hui Li, and Li Jin, "Major Population Expansion of East Asians Began before Neolithic Time: Evidence of mtDNA Genomes," *PLoS One* 6, no. 10 (2011): e25835.

29 Bar-Yosef, "Climatic Fluctuations and Early Farming in West and East Asia."

30 Shahal Abbo, Simcha Lev-Yadun, Manfred Heun, and Avi Gopher, "On the 'lost' crops of the Neolithic Near East," *Journal of Experimental Botany* 64 (2013): 815–22.

31 X. Yang, W. Wu, L. Perry et al., "Critical Role of Climate Change in Plant Selection and Millet Domestication in North China," *Science Reports* 8, no. 7855 (2018). https://doi.org/10.1038/s41598-018-26218-6.

32 Li Liu, Gyoung-Ah Lee, Leping Jiang, and Juzhong Zhang, "Evidence for the Early Beginning (c. 9000 cal. BP) of Rice Domestication in China: A Response," *The Holocene* 17 (2007): 1059–68.

33 Yunfei Zheng, Gary W. Crawford, Leping Jiang, and Xugao Chen, "Rice Domestication Revealed by Reduced Shattering of Archaeological Rice from the Lower Yangtze Valley," *Scientific Reports* 6 (2016); and Zhenhua Deng, Ling Qin, Yu Gao, Alison Ruth Weisskopf, Chi Zhang, and Dorian Q. Fuller, "From Early Domesticated Rice of the Middle Yangtze Basin to Millet, Rice and Wheat Agriculture: Archaeobotanical Macro-Remains from Baligang, Nanyang Basin, Central China (6700–500 BC)," *PloS one* 10, no. 10 (2015): e0139885.

34 Fuler, "The Occurrence of Cereal Cultivation in China."

35 Oris I. Sanjur, Dolores R. Piperno, Thomas C. Andres, and Linda Wessel-Beaver, "Phylogenetic Relationships Among Domesticated and Wild Species of Cucurbita (Cucurbitaceae) Inferred from a Mitochondrial Gene: Implications for Crop Plant Evolution and Areas of Origin," *Proceedings of the National Academy of Sciences* 99 (2002): 535–40; and Anthony J. Ranere, Dolores R. Piperno, Irene Holst, Ruth Dickau, and José Iriarte, "The Cultural and Chronological Context of Early Holocene Maize and Squash Domestication in the Central Balsas River Valley, Mexico,"*Proceedings of the National Academy of Sciences* 106 (2009): 5014–18.

36 Huai Wang, Tina Nussbaum-Wagler, Bailin Li, Qiong Zhao, Yves Vigouroux, Marianna Faller, Kirsten Bomblies, Lewis Lukens, and John F. Doebley, "The Origin of the Naked Grains of Maize," *Nature* 436, no. 7051 (2005): 714–9; and Ranere et al., "Cultural and Chronological Context of Early Holocene Maize and Squash Domestication," and http://www.eurekalert.org/pub_releases/2008-06/asop-eoo062308.php.

37 Ranere et al., "Cultural and Chronological Context of Early Holocene Maize and Squash Domestication."

38 Jonathan Haas, Winifred Creamer, Luis Huamán Mesía, David Goldstein, Karl Reinhard, and Cindy Vergel Rodríguez, "Evidence for Maize (Zea Mays)

in the Late Archaic (3000–1800 BC) in the Norte Chico Region of Peru," *Proceedings of the National Academy of Sciences* 110 (2013): 4945–9.

39 Charles R. Ortloff, "Hydraulic Engineering in Ancient Peru and Bolivia," in Helaine Selin, ed., *Encyclopaedia of the History of Science, Technology, and Medicine in Non-Western Cultures* (Dordrecht: Springer, 2016) https://doi.org /10.1007/978-94-007-7747-7_10076.

40 Daniel A. Contreras, "(Re) Constructing the Sacred: Landscape Geoarchaeology at Chavín de Huántar, Peru," *Archaeological and Anthropological Sciences* 9 (2017): 1045–57.

41 Bruce D. Smith, "Eastern North America as an Independent Center of Plant Domestication," *Proceedings of the National Academy of Sciences* 103 no. 33(2006): 12223–8. doi:1073/pnas.0604335103.

42 Judith Ann Carney, and Richard Nicholas Rosomoff, *In the Shadow of Slavery: Africa's Botanical Legacy in the Atlantic World* (Berkeley: University of California Press, 2009).

43 Tim Denham, "Early Agriculture and Plant Domestication in New Guinea and Island Southeast Asia," *Current Anthropology* 52 (2011): 379–95; Simon G. Haberle, Carol Lentfer, Shawn O'Donnell, and Tim Denham, "The Palaeoenvironments of Kuk Swamp from the Beginnings of Agriculture in the Highlands of Papua New Guinea," *Quaternary International* 249 (2012): 129–39.

44 Bill Gammage, *The Biggest Estate on Earth: How Aborigines Made Australia* (Sydney: Allen & Unwin, 2011).

45 Christopher R. Gignoux, Brenna M. Henn, Joanna L. Mountain., and Ofer Bar-Yosef, "Rapid, Global Demographic Expansions after the Origins of Agriculture," *Proceedings of the National Academy of Sciences of the United States of America* 108, no. 15 (2011): 6044–9. http://www.jstor.org/stable /41126604; and Wolfgang Haak, Peter Forster, Barbara Bramanti, Shuichi Matsumura, Guido Brandt, Marc Tänzer, Richard Villems et al., "Ancient DNA from the First European Farmers in 7500-Year-Old Neolithic Sites," *Science* 310 (2005): 1016–18.

46 Zuzana Hofmanová, Susanne Kreutzer, Garrett Hellenthal, Christian Sell, Yoan Diekmann, David Díez-del-Molino, Lucy van Dorp et al., "Early Farmers from across Europe Directly Descended from Neolithic Aegeans," *Proceedings of the National Academy of Sciences* (2016): 201523951.

47 Pontus Skoglund, Helena Malmström, Ayça Omrak, Maanasa Raghavan, Cristina Valdiosera, Torsten Günther, Per Hall et al., "Genomic Diversity and Admixture Differs for Stone-Age Scandinavian Foragers and Farmers," *Science* 344, no. 6185 (2014): 747–50.

48 Andrew Curry, "Archaeology: The Milk Revolution," *Nature* (2013), http:// www.nature.com/news/archaeology-the-milk-revolution-1.13471.

49 Lucas Bueno, Adriana Schmidt Dias, and James Steele, "The Late Pleistocene/ Early Holocene Archaeological Record in Brazil: A Geo-Referenced Database," *Quaternary International* 301 (2013): 74–93; and Charles R. Clement, William M. Denevan, Michael J. Heckenberger, André Braga Junqueira, Eduardo G. Neves, Wenceslau G. Teixeira, and William I. Woods,

"The Domestication of Amazonia before European Conquest," *Proceedings of the Royal Society B* 282, no. 1812 (2015): 20150813.

50 Gignoux et al., "Rapid, Global Demographic Expansions after the Origins of Agriculture."

51 Ibid.

52 Christoph Zielhofer, Lee Clare, Gary Rollefson, Stephan Wächter, Dirk Hoffmeister, Georg Bareth, Christopher Roettig et al., "The Decline of the Early Neolithic Population Center of' Ain Ghazal and Corresponding Earth-Surface Processes, Jordan Rift Valley," *Quaternary Research* 78 (2012): 427–41.

53 Lee Clare, "Pastoral Clashes: Conflict Risk and Mitigation at the Pottery Neolithic Transition in the Southern Levant," *Neo-Lithics* 1 (2010): 1331; and Anna Belfer-Cohen and A. Nigel Goring-Morris, "Becoming Farmers," *Current Anthropology* 52 (2011): S209–20.

54 R. B. Alley and A. M. Agustsdottir, "The 8 K Event: Cause and Consequences of a Major Holocene Abrupt Climate Change," *Quaternary Science Reviews* 24, no. 10 (2005): 1123–49.

55 Nicolás E. Young, Jason P. Briner, Dylan H. Rood, and Robert C. Finkel, "Glacier Extent during the Younger Dryas and 8.2-ka Event on Baffin Island, Arctic Canada," *Science* 337, no. 6100 (2012): 1330–3.

56 Jean-François Berger, Laurent Lespez, Catherine Kuzucuoglu, Arthur Glais, Fuad Hourani, Adrien Barra, and Jean Guilaine, "Interactions between Climate Change and Human Activities During the Early to mid-Holocene in the Eastern Mediterranean Basins," *Climates of the Past* 12 (2016): 1847–77.

57 P.M.M.G. Akkermans, J. van der Plicht, O. P. Nieuwenhuyse, A. Russell, A. Kaneda, and H. Buitenhuis, "Weathering Climate Change in the Near East: Dating and Neolithic Adaptations 8200 Years Ago," *Antiquity* 84 (2010): 71–85.

58 Jean-François Berger, and Jean Guilaine, "The 8200calBP Abrupt Environmental Change and the Neolithic Transition: A Mediterranean Perspective," *Quaternary International* 200, no. 1 (2009): 31–49.

59 Bernhard Weninger, Eva Alram-Stern, Eva Bauer, Lee Clare, Uwe Danzeglocke, Olaf Jöris, Claudia Kubatzki, Gary Rollefson, Henrieta Todorova, and Tjeerd van Andel, "Climate Forcing Due to the 8200 cal yr BP Event Observed at Early Neolithic Sites in the Eastern Mediterranean," *Quaternary Research* 66, no. 3 (2006): 401–20.

60 Mélanie Roffet-Salque et al., "Evidence for the Impact of the 8.2-kyBP Climate Event on Near Eastern Early Farmers," *Proceedings of the National Academy of Sciences* 115 (2018): 8705–9. doi:10.1073/pnas.1803607115.

61 Pascal Flohr, Dominik Fleitmann, Roger Matthews, Wendy Matthews, and Stuart Black, "Evidence of Resilience to Past Climate Change in Southwest Asia: Early Farming Communities and the 9.2 and 8.2 ka Events," *Quaternary Science Reviews* 136 (2016): 28.

62 Karen Wicks and Steven Mithen, "The Impact of the Abrupt 8.2 ka Cold Event on the Mesolithic Population of Western Scotland: A Bayesian Chronological Analysis Using 'Activity Events' as a Population Proxy," *Journal of Archaeological Science* 45 (2014): 240–69.

63 J.E. Kutzbach, "Monsoon Climate of the Early Holocene: Climate Experiment with the Earth's Orbital Parameters for 9000 Years Ago," *Science* 214 (1981): 59–61.

64 Paul C. Sereno, Elena A. A. Garcea, Hélène Jousse, Christopher M. Stojanowski, Jean-François Saliège, Abdoulaye Maga, Oumarou A. Ide et al., "Lakeside Cemeteries in the Sahara: 5000 Years of Holocene Population and Environmental Change," *PLoS One* 3 (2008): e2995; Katie Manning and Adrian Timpson, "The Demographic Response to Holocene Climate Change in the Sahara," *Quaternary Science Reviews* 101 (2014): 28–35; and Sibel Barut Kusimba, *African Foragers: Environment, Technology, Interactions*, Vol. 4 (Walnut Creek: Rowman Altamira, 2003), 52.

65 Christopher M. Stojanowski and Kelly J. Knudson, "Changing Patterns of Mobility as a Response to Climatic Deterioration and Aridification in the Middle Holocene Southern Sahara," *American Journal of Physical Anthropology* 154 (2014): 79–93.

66 Stefanie Kahlheber and Katharina Neumann, "The Development Of Plant Cultivation in Semi-Arid West Africa," in T.P. Denham, J. Iriarte and L. Vrydaghs eds., *Rethinking Agriculture: Archaeological and Ethnoarchaeological Perspectives* (Walnut Creek: Left Coast Press, 2007), 331, 337.

67 Cécile L. Blanchet, Martin Frank, and Stefan Schouten, "Asynchronous Changes in Vegetation, Runoff and Erosion in the Nile River Watershed During the Holocene," *PloS one* 9 (2014): e115958; and Julie Dunne, Richard P. Evershed, Mé Salque, Lucy Cramp, Silvia Bruni, Kathleen Ryan, Stefano Biagetti, and Savino di Lernia, "First Dairying in Green Saharan Africa in the Fifth Millennium BC," *Nature* 486 (2012): 390–4.

68 Jörg Linstädter and Stefan Kröpelin, "Wadi Bakht Revisited: Holocene Climate Change and Prehistoric Occupation in the Gilf Kebir Region of the Eastern Sahara, SW Egypt," *Geoarchaeology* 19 (2004): 753–78.

69 Douglas Post Park, "Climate Change, Human Response and the Origins of Urbanism at Timbuktu: Archaeological Investigations into the Prehistoric Urbanism of the Timbuktu Region on the Niger Bend, Mali, West Africa," *Azania: Archaeological Research in Africa* 4 (2012): 246–7.

70 Sen Li, Carina Schlebusch, and Mattias Jakobsson, "Genetic Variation Reveals Large-Scale Population Expansion and Migration during the Expansion of Bantu-Speaking Peoples," *Proceedings of the Royal Society B* 281, no. 1793(2014): 20141448.

71 Stefanie Kahlheber, Koen Bostoen, and Katharina Neumann, "Early Plant Cultivation in the Central African Rain Forest: First Millennium BC Pearl Millet from South Cameroon," *Journal of African Archaeology* 7 (2009): 253–72.

72 Richard Oslisly, Richard, Lee White, Ilham Bentaleb, Charly Favier, Michel Fontugne, Jean-François Gillet, and David Sebag, "Climatic and Cultural Changes in the West Congo Basin Forests Over the Past 5000 Years," *Philosophical Transactions of the Royal Society of London B: Biological Sciences* 368 (2013): 20120304; and Rebecca Grollemun, Simon Branford, Koen Bostoen, Andrew Meade, Chris Venditti, and Mark Pagel, "Bantu Expansion Shows That Habitat Alters the Route and Pace of Human

Dispersals," *Proceedings of the National Academy of Sciences* 112 (2015): 13296–301.

73 Terry M. Brncic, Katherine J. Willis, David J. Harris, and Richard Washington, "Culture or Climate? The Relative Influences of Past Processes on the Composition of the Lowland Congo Rainforest," *Philosophical Transactions of the Royal Society of London B: Biological Sciences* 362 (2007): 229–42; Ulrich Salzmann and Philipp Hoelzmann, "The Dahomey Gap: An Abrupt Climatically Induced Rain Forest Fragmentation in West Africa during the Late Holocene," *The Holocene* 15 (2005): 190–9, http://www.jsg.utexas.edu/research_symposium/files/kim_h_2013.pdf.

74 Jean Maley, "Elaeis Guineensis Jacq.(Oil Palm) Fluctuations in Central Africa during the Late Holocene: Climate or Human Driving Forces for This Pioneering Species?," *Vegetation History and Archaeobotany* 10 (2001): 117–20; and Germain Bayon, Bernard Dennielou, Joël Etoubleau, Emmanuel Ponzevera, Samuel Toucanne, and Sylvain Bermell, "Intensifying Weathering and Land Use in Iron Age Central Africa," *Science* 335 (2012): 1219–22.

75 Florian Thevenon, David Williamson, Annie Vincens, Maurice Taieb, Ouassila Merdaci, Michel Decobert, and Guillaume Buchet, "A Late-Holocene Charcoal Record from Lake Masoko, SW Tanzania: Climatic and Anthropologic Implications," *The Holocene* 13 (2003): 785–92.

76 Max Engel and Helmut Brückner, "Holocene Climate Variability of Mesopotamia and Its Impact on the History of Civilisation," *EarthArXiv* (2018). doi:10.31223/osf.io/s2aqt.

77 Neil Roberts, Warren J. Eastwood, Catherine Kuzucuoğlu, Girolamo Fiorentino, and Valentina Caracuta, "Climatic, Vegetation and Cultural Change in the Eastern Mediterranean during the mid-Holocene Environmental Transition," *The Holocene* 21, no. 1 (2011): 147–62.

78 Dan Lawrence, Graham Philip, Hannah Hunt, Lisa Snape-Kennedy, and T.J. Wilkinson, "Long Term Population, City Size and Climate Trends in the Fertile Crescent: A First Approximation," *PloS one* 11, no. 3 (2016): e0152563.

79 Jared Diamond, "The Worst Mistake in the History of the Human Race. (Adoption of Agriculture)," *Discover* (May 1987): 64–6.

80 Matthias Merker, Camille Blin, Stefano Mona, Nicolas Duforet-Frebourg, Sophie Lecher, Eve Willery, Michael GB Blum et al., "Evolutionary History and Global Spread of the Mycobacterium Tuberculosis Beijing Lineage," *Nature Genetics* 47 (2015): 242–9.

81 Kent V. Flannery and Joyce Marcus, *The Creation of Inequality: How our Prehistoric Ancestors Set the Stage for Monarchy, Slavery, and Empire* (Cambridge, MA: Harvard University Press, 2012).

Chapter Three

1 Emmanuel Le Roy Ladurie, *Times of Feast, Times of Famine* (New York: Doubleday, 1971).

2 Jared Diamond, *Collapse: How Societies Choose to Fail or Succeed* (New York: Viking: 2005).

3 Patricia A. McAnany and Norman Yoffee, eds., *Questioning Collapse: Human Resilience, Ecological Vulnerability, and the Aftermath of Empire* (New York: Cambridge University Press, 2010); Ronald K. Faulseit, ed., *Beyond Collapse: Archaeological Perspectives on Resilience, Revitalization, and Transformation in Complex Societies* (Carbondale: SIU Press, 2015).

4 Camilo Ponton et al., "Holocene Aridification of India," *Geophysical Research Letters*, 39 (2012): L03704 doi:10.1029/2011GL050722.

5 Gwen Robbins et al., "Infection, Disease, and Biosocial Processes at the end of the Indus Civilization," *PloS one* 8, no. 12 (2013): e84814.

6 Fenggui Liu and Zhaodong Feng, "A Dramatic Climatic Transition at~ 4000 cal. yr BP and its Cultural Responses in Chinese Cultural Domains," *The Holocene* 22 (2012): 1181–97; Xiaoping Yang et al., "Groundwater Sapping as the Cause of Irreversible Desertification of Hunshandake Sandy Lands, Inner Mongolia, Northern China," *Proceedings of the National Academy of Sciences* 112 (2015): 702–6.

7 Glenn M. Schwartz and Naomi F. Miller, "The 'Crisis' of the Late Third Millennium B.C: Ecofactual and Artificial Evidence from Umm el-Marra and the Jabbul Plain,"; Walther Sallaberger, "From Urban Culture To nomadism: A History of Upper Mesopotamia in the Late Third Millennium," in Catherine Kuzucuoğlu and Catherine Marro, eds., *Socie ́tés Humaines et changement Climatique à la fin du troisiè me millénaire: une crise a-t -elle eu lieu en Haute-M é s opotamie? Actes du colloque de Lyon, 5–8 de ́cembre 2005* (Istanbul: Institut français d'é tudes anatoliennes-Georges Dumézil, 2007), 199 and 417–19.

8 Takaaki K Watanabe et al., "Oman Corals Suggest That a Stronger Winter Shamal Season Caused the Akkadian Empire (Mesopotamia) Collapse," *Geology* 47 (2019): 1141–5. doi:10.1130/G46604.1.

9 Stacy A. Carolin et al., "Precise Timing of Abrupt Increase in Dust Activity in the Middle East Coincident with 4.2 ka Social Change," *PNAS* 116 (2019): 67–72; doi:10.1073/pnas.1808103115; Evangeline Cookson, Daniel J. Hill, and Dan Lawrence, "Impacts of Long Term Climate Change During the Collapse of the Akkadian Empire," *Journal of Archaeological Science* 106 (2019): 1–9.

10 Roger J. Flower et al., "Environmental Changes at the Desert Margin: An Assessment of Recent Paleolimnological Records in Lake Qarun, Middle Egypt," *Journal of Paleolimnology* 35 (2006): 1–24.

11 J.-D. Stanley et al., "Nile Flow Failure at the End of the Old Kingdom, Egypt: Strontium Isotopic and Petrologic Evidence," *Geoarchaeology* 18 (2003): 395–402, doi:10.1002/ gea.10065.

12 Ellen Morris, "'Lo, Nobles Lament, Poor Rejoice': State Formation in the Wake of Social Flux," in Glenn M. Schwartz and John J. Nichols, eds., *After Collapse: The Regeneration of Complex Societies* (Tucson: University of Arizona Press, 2006), 58–71.

13 Y. Zong et al., "Environmental Change and Neolithic Settlement Movement in the Lower Yangtze Wetlands of China," *The Holocene* 22 (2012): 659–73; H. Kajita et al., "Extraordinary Cold Episodes During the mid-Holocene in

the Yangtze Delta: Interruption of the Earliest Rice Cultivating Civilization," *Quaternary Science Reviews* 201 (2018): 418–28.

14 Haiwei Zhang et al., "Hydroclimatic Variations in Southeastern China during the 4.2 ka Event Reflected by Stalagmite Records," *Climate of the Past* 14 (2018): 1805–17.

15 Michael Williams, "Dark Ages and Dark Areas: Global Deforestation in the Deep Past," *Journal of Historical Geography* 26 (2000): 28–46; Jessie Woodbridge et al., "The Impact of the Neolithic Agricultural Transition in Britain: A Comparison of Pollen-Based Land-Cover and Archaeological 14 C Date-Inferred Population Change," *Journal of Archaeological Science* 51 (2014): 216–24.

16 W.F. Ruddiman, "The Anthropogenic Greenhouse Era Began Thousands of Years Ago," *Climate Change,* 61 (2003): 261–93; W.F. Ruddiman, "The Early Anthropogenic Hypothesis: Challenges and Responses," *Review of Geophysics* 45 (2007): RG4001, doi:10.1029/2006RG000207.

17 Dorian Q. Fuller et al., "The Contribution of Rice Agriculture and Livestock Pastoralism to Prehistoric Methane Levels: An Archaeological Assessment," *The Holocene* 21, no. 5 (2011): 743–59.

18 W.F. Ruddiman et al., "Late Holocene Climate: Natural or Anthropogenic?," *Review of Geophysics*, 54 (2016): 93–118.

19 David Kaniewski et al., "Middle East Coastal Ecosystem Response to Middle-To-Late Holocene Abrupt Climate Changes," *Proceedings of the National Academy of Sciences* 105, no. 37 (2008): 13941–6.

20 James Henry Breasted and William Rainey Harper, *Ancient Records* 2. Series, vol. 4 (Chicago: University of Chicago Press, 1906), 45.

21 Brandon L. Drake, "The Influence of Climatic Change on the Late Bronze Age Collapse and the Greek Dark Ages," *Journal of Archaeological Science* 39 (2012): 1862–70.

22 Christopher Bernhardt, Benjamin Horton, and Jean-Daniel Stanley, "Nile Delta Vegetation Response to Holocene Climate Variability," *Geology* 40 (2012): 615–18.

23 A. Bernard Knapp, and Sturt W. Manning, "Crisis in Context: The End of the Late Bronze Age in the Eastern Mediterranean," *American Journal of Archaeology* 120 (2016): 99–149. doi:10.3764/aja.120.1.0099.

24 Erika Weiberg et al., "Mediterranean Land Use Systems from Prehistory to Antiquity: A Case Study from Peloponnese (Greece)," *Journal of Land Use Science* 14 (2019): 1–20.

25 David Kaniewski et al., "Environmental Roots of the Late Bronze Age Crisis," *PLoS One* 8(2013): e71004.

26 Israel Finkelstein et al., "Egyptian Imperial Economy in Canaan: Reaction to the Climate Crisis at the End of the Late Bronze Age," *Ägypten und Levante* 27 (2017): 249–60. doi:10.1553/AEundL27s249.

27 Ian Armit et al., "Rapid Climate Change Did Not Cause Population Collapse at the End of the European Bronze Age," *Proceedings of the National Academy of Sciences* 111 (2014): 17045–9.

28 Katherine A. Adelsberger and Tristram R. Kidder, "Climate Change, Landscape Evolution, and Human Settlement in the Lower Mississippi Valley, 5 500-2 400 Cal B.P," in Lucy Wilson, Pam Dickinson, and Jason Jeandron, eds., *Reconstructing Human-Landscape Interactions* (Newcastle: Cambridge Scholars Publishing, 2007), 91–2; Kenneth E. Sassaman, *The Eastern Archaic, Historicized* (Lanham: Rowman Altamira, 2010), 190–5.

29 Tristram R. Kidder, "Trend, Tradition, and Transition at the End of the Archaic," in David Hurst Thomas and Matthew C. Sanger, eds., *Trend, Tradition, and Turmoil: What Happened to the Southeastern Archaic* (New York: American Museum of Natural History, 2010), 23–32.

30 Adam W. Schneider and Selim F. Adalı, "'No Harvest Was Reaped': Demographic and Climatic Factors in the Decline of the Neo-Assyrian Empire," *Climatic Change* 127 (2014): 435–46. For a Dissenting View, see Arkadiusz Sołtysiak, "Drought and the Fall of Assyria: Quite Another Story," *Climatic Change* 136, no. 3–4 (2016): 389–94.

31 Ashish Sinha, "Role of Climate in the Rise and Fall of the Neo-Assyrian Empire," *Science Advances* 5 (2019): eaax6656.

32 Graeme T. Swindles, Gill Plunkett, and Helen M. Roe, "A Delayed Climatic Response to Solar Forcing at 2800 Cal. BP: Multiproxy Evidence from Three Irish Peatlands," *The Holocene* 17, no. 2 (2007): 177–82.

33 B. van Geel et al., "Climate Change and the Expansion of the Scythian Culture after 850 BC: A Hypothesis," *Journal of Archaeological Science* 31, no. 12 (2004): 1735–42.

34 Rebecca W. Wendelken, "Horses and Gold: The Scythians of the Eurasian Steppes," in Andrew Bell-Fialkoff, ed., *The Role of Migration in the History of the Eurasian Steppe* (New York: St. Martin's Press, 2000), 199.

35 Saskia Hin, *The Demography of Roman Italy: Population Dynamics in an Ancient Conquest Society (201 BCE-14 CE).* (Cambridge: Cambridge University Press, 2013), 86.

36 Michael McCormick et al., "Climate Change during and after the Roman Empire: Reconstructing the Past from Scientific and Historical Evidence," *Journal of Interdisciplinary History* 43 (2012): 169–220.

37 Elio Sgreccia, Vincenza Mele, and Gonzalo Miranda, *Le Radici della bioetica: atti del Congresso internazionale, (Roma, 15-17 febbraio 1996),* vol. 1 (Milano: Vita e pensiero, 1998), 153.

38 Marco Moriondo et al., "Olive Trees as Bio-Indicators of Climate Evolution in the Mediterranean Basin," *Global Ecology and Biogeography* 22 (2013): 818–33; Richard Hoffmann, *An Environmental History of Medieval Europe* (Cambridge: Cambridge University Press, 2014), 42.

39 Hin, *Demography of Roman Italy*, 91 and 95–6.

40 Palmyrena: City, Hinterland and Caravan Trade between Orient and Occident, http://www.org.uib.no/palmyrena/.

41 Graeme Barker, "A Tale of Two Deserts: Contrasting Desertification Histories on Rome's Desert Frontiers," *World Archaeology* 33 (2002): 488–507.

42 J. G. Manning, F. Ludlow, A. R. Stine et al., "Volcanic Suppression of Nile Summer Flooding Triggers Revolt and Constrains Interstate Conflict in Ancient Egypt," .*Nature Communications* 8 (2017). https://doi.org/10.1038/s41467-017-00957-y.

43 Joseph R. McConnell, Michael Sigl, Gill Plunkett, et al., "Extreme Climate after Massive Eruption of Alaska's Okmok Volcano in 43 BCE and Effects on the Late Roman Republic and Ptolemaic Kingdom," *PNAS* 117 (2020): 202002722.

44 Charles William Eliot, *Voyages and Travels: Ancient and Modern* (New York: Collier, 1938), 95 and 97.

45 Dio Cassius, *Roman History*, vol. 9: bks. 71– 80, trans. Earnest Cary and Herbert B. Foster (Cambridge, MA: Harvard University Press, 1927), 265.

46 Procopius, *History of the Wars*, vol. 5 (Cambridge, MA: Harvard University Press, 1978), 265.

47 Internet Encyclopedia of Philosophy, s.v., "Confucius (551–479 B.C.E)," http://www.iep.utm.edu/confuciu/.

48 Chen Bo and Gideon Shelach, "Fortified Settlements and the Settlement System in the Northern Zone of the Han Empire," *Antiquity* 88 (2014): 222–40.

49 Randolph Ford, "Barbaricum Depictum: Images of the Germani and Xiongnu in the Works of Tacitus and Sima Qian," *Sino-Platonic Papers* 207 (2010): 5.

50 Tristram R. Kidder and Haiwang Liu, "Bridging Theoretical Gaps in Geoarchaeology: Archaeology, Geoarchaeology, and History in the Yellow River Valley, China," *Archaeological and Anthropological Sciences* (2014), doi:10.1007/ s12520-014-0184-5.

51 Ruddiman et al., Late Holocene Climate."

52 Einhard, *Life of Charlemagne*, trans. S. E. Turner (New York: Harper and Brothers, 1880), 62.

53 Xunming Wang et al., "Climate, Desertification, and the Rise and Collapse of China's Historical Dynasties," *Human Ecology* 38 (2010): 157–72.

54 Zhudeng Wei, Xiuqi Fang, and Yun Su, "Climate Change and Fiscal Balance in China over the Past Two Millennia," *The Holocene* 24 (2014): 1771–84; Yun Su, XiuQi Fang, and Jun Yin, "Impact of Climate Change on Fluctuations of Grain Harvests in China from the Western Han Dynasty to the Five Dynasties (206 BC–960 AD)," *Science China Earth Sciences* 57(2014): 1701–12.

55 B. J. Dermody, R. Van Beek, Elijah Meeks, K. Klein Goldewijk, Walter Scheidel, Y. van Der Velde, M. Bierkens, M. J. Wassen, and S. C. Dekker, "A Virtual Water Network of the Roman World," *Hydrology and Earth System Sciences* 18 (2014): 5025–40.

56 Paul Erdkamp, "War, Food, Climate Change, and the Decline of the Roman World," *Journal of Late Antiquity* 12 (2019): 422–65.

57 Kyle Harper, *The Fate of Rome: Climate Disease and the End of an Empire* (Princeton: Princeton University Press, 2018), 131–6.

58 McCormick et al., "Climate Change during and after the Roman Empire," 189.

59 Stuart W. Manning, "The Roman World and Climate: Context, Relevance of Climate Change, and Some Issues," in W. V. Harris, ed., *The Ancient*

*Mediterranean Environment between Science and History (*Leiden and Boston: Brill, 2013), 103–70.

60 Adam Izdebski, "Why Did Agriculture Flourish in the Late Antique East? The Role of Climate Fluctuations in the Development and Contraction of Agriculture in Asia Minor and the Middle East from the 4th Till the 7th c. AD," *Milllenium, Jahrbuchzu Kultur und Geschichte des ersten Jahrtausends n. Chr* 8 (2011): 291–312.

61 Harper, *Fate of Rome*, 163, 192.

62 Edward R. Cook, "Megadroughts, ENSO, and the Invasion of Late-Roman Europe by the Huns and Avars," in Harris, ed., *The Ancient Mediterranean Environment between Science and History*, 89–102; McCormick et al., "Climate Change during and after the Roman Empire."

63 Procopius, *History of the Wars*, 246–7.

64 Stephen Williams and J. G. P. Friell, *Theodosius: The Empire at Bay* (New Haven: Yale University Press, 1995), 8.

65 Peter Heather, *The Fall of the Roman Empire: A New History of Rome and the Barbarians* (London: Macmillan, 2005), 286–7, 338–9, 369, and 374.

66 B. Lee Drake, "Changes in North Atlantic Oscillation Drove Population Migrations and the Collapse of the Western Roman Empire," *Science Reports* 7 (2017). https://doi.org/10.1038/s41598-017-01289-z.

67 Olga V. Churakova (Sidorova) et al., "A Cluster of Stratospheric Volcanic Eruptions in the AD 530s Recorded in Siberian Tree Rings," *Global and Planetary Change* 122 (2014): 140–50.

68 Ulf Büntgen et al., "Cooling and Societal Change during the Late Antique Little Ice Age from 536 to around 660 AD," *Nature Geoscience* 9 (2016): 231–6.

69 Harper, *Fate of Rome,* 234, 253–7.

70 Kristina Sessa, "The New Environmental History of the Fall of Rome: A Methodological Consideration," *Journal of Late Antiquity* 12 (2019): 211–55; and Paula Kouki, "Problems of Relating Environmental History to Human Settlement in the Classical and Late Classical Periods-the Example of Southern Jordan," in William V. Harris, ed., *The Ancient Mediterranean Environment between Science and History (*Netherlands: Brill, 2013), 197–211.

71 Guy Bar-Oz et al., "Ancient Trash Mounds Unravel Urban Collapse a Century before the End of Byzantine Hegemony in the Southern Levant," *PNAS* 116 (2019): 8239–48. doi:10.1073/pnas.1900233116.

72 John Haldon et al., "The Climate and Environment of Byzantine Anatolia: Integrating Science, History, and Archaeology," *Journal of Interdisciplinary History*, 45 (2014): 113–61; and John Haldon et al., "History Meets Palaeoscience: Consilience and Collaboration in Studying Past Societal Responses to Environmental Change," *PNAS* 115 (2018): 3210–18.

73 Fredric L. Cheyette, "The Disappearance of the Ancient Landscape and the Climatic Anomaly of the Early Middle Ages: A Question to Be Pursued," *Early Medieval Europe* 16 (2008), 133, 140, 143–4, 153.

74 Cheyette, "The Disappearance of the Ancient Landscape," 160–2; Hoffmann, *An Environmental History of Medieval Europe*, 68.

75 A. Volkmann, "Climate Change, Environment and Migration: A GIS-Based Study of the Roman Iron Age to the Early Middle Ages in the River Oder Region," *Post-Classical Archaeologies* 5 (2015): 69–94.

76 N.D. Brayshaw Roberts, C. Kuzucuoğlu, R. Perez, and L. Sadori, "The mid-Holocene Climatic Transition in the Mediterranean: Causes and Consequences," *The Holocene* 21, no. 1 (2011): 3–13.

Chapter Four

1 H.H. Lamb, "The Early Medieval Warm Epoch and Its Sequel," *Palaeogeography, Palaeoclimatology and Palaeoecology* 1, no. 1 (1965): 13–37.

2 Moinuddin Ahmed et al., "Continental-Scale Temperature Variability during the Past Two Millennia," *Nature Geos*cience 6 (2013): 339–46.

3 S. Stine, "Extreme and Persistent Drought in California and Patagonia during Medieval Time," *Nature* 369 (1994): 546–9.

4 Richard Seager et al., "Blueprints for Medieval Hydroclimate," *Quaternary Science Reviews* 26 (2007): 2322–36.

5 Thomas J. Crowley, "Causes of Climate Change over the Past 1000 Years," *Science* 289 (2000): 270–7.

6 Michael E. Mann et al., "Global Signatures and Dynamical Origins of the Little Ice Age and Medieval Climate Anomaly," *Science* 326 (2009): 1256–60.

7 C. A. Woodhouse, J. L. Russell, and E. R. Cook, "Two Modes of North American Drought from Instrumental and Paleoclimatic Data," *Journal of Climate* 22 (2009): 4336–47.

8 Fekri A. Hassan, "Extreme Nile Floods and Famines in Medieval Egypt (AD 930–1500) and Their Climatic Implications," *Quaternary Internationa*l 173–174 (2007): 101–12; Michael M. Santoro et al., "An Aggregated Climate Teleconnection Index Linked to Historical Egyptian Famines of the Last Thousand Years," *The Holocene* 25 (2015): 872–9.

9 Barry Cunliffe, *Britain Begins* (Oxford: Oxford University Press, 2013), 447.

10 Jared Diamond, *Collapse: How Societies Choose to Fail or Succeed* (New York: Viking, 2005), 198–201.

11 William J. D'Andrea et al., "Abrupt Holocene Climate Change as an Important Factor for Human Migration in West Greenland," *Proceedings of the National Academy of Sciences* 108 (2011): 9765–9.

12 Nicolás E. Young, Avriel D. Schweinsberg, Jason P. Briner, and Joerg M. Schaefer, "Glacier Maxima in Baffin Bay during the Medieval Warm Period Coeval with Norse Settlement," *Science Advances* 1 (2015): e1500806.

13 T. Kobashi et al., "High Variability of Greenland Surface Temperature Over the Past 4000 Years Estimated from Trapped Air in an Ice Core," *Geophysical Research Letters* 38 (2011): L21501.

14 Max T. and Charles D. Arnold, "The Timing of the Thule Migration: New Dates from the Western Canadian Arctic," *American Antiquity* 73 (2008): 527–38.

15 H. H. Lamb, *Climate, History, and the Modern World* (London: Routledge, 1995), 179 and 195.

16 Knut Helle, E. I. Kouri, and Jens E. Oleson, *The Cambridge History of Scandinavia* (Cambridge: Cambridge University Press, 2003), 1, 257.

17 Robert Rees Davies, *The First English Empire: Power and Identities in the British Isles 1093–1343* (Oxford: Oxford University Press, 2000), 151 and 153–5.

18 Robert Rees Davies, *The Age of Conquest: Wales, 1063–1415* (Oxford, Oxford University Press, 1987), 97–100.

19 Robert Bartlett, *The Making of Europe: Conquest, Colonization and Cultural Change 950–1350* (Princeton: Princeton University Press, 1994), 114 and 155; Richard Hoffmann, *An Environmental History of Medieval Europe* (Cambridge: Cambridge University Press, 2014), 136–7, 139, 167; John Aberth, *An Environmental History of the Middle Ages: The Crucible of Nature* (London: Routledge, 2013), 34.

20 Giles Constable, "The Place of the Magdeburg Charter of 1107/ 08 in the History of Eastern Germany and of the Crusades," in Kaspar Elm, Franz J. Felten, Nikolas Jaspert, and Stephanie Haarländer, eds., *Vita Religiosa im Mittelalter: Festschriftf ü r Kaspar Elm zum 70. Geburtstag* (Berlin: Duncker &Humblot, 1999), 298–9.

21 Len Scales, *The Shaping of German Identity: Authority and Crisis, 1245–1414* (Cambridge: Cambridge University Press, 2012), 50.

22 J. Preiser-Kapeller, "A Climate for Crusades: Weather, Climate, and Armed Pilgrimage to the Holy Land (11th to 14th Century)," *Medievalists.net*, http://www.medievalists.net/2013/12/a-climate-for-crusades-weather-climate-andarmed-pilgrimage-to-the-holy-land-11th-14th-century/.

23 Ronnie Ellenblum, *The Collapse of the Eastern Mediterranean: Climate Change and the Decline of the East, 950–1072* (New York: Cambridge University Press, 2012), 342.

24 Richard W. Bulliet, *Cotton, Climate, and Camels in Early Islamic Iran: A Moment in World History*. (New York: Columbia University Press, 2009), 69.

25 J. Preiser-Kapeller, "A Collapse of the Eastern Mediterranean? New Results and Theories on the Interplay between Climate and Societies in Byzantium and the Near East, ca. 1000–1200 AD," *Jahrbuch der Österreichischen Byzantinistik* 65 (2015): 195–242.

26 Ellenblum, *The Collapse of the Eastern Mediterranean*, 3.

27 Ibid, 29.

28 Ibid, 53, 129, and 193.

29 Bulliet, *Cotton, Climate, and Camels in Early Islamic Iran*, 69, 86, and 96.

30 Ellenblum, *The Collapse of the Eastern Mediterranean*, 75–6; Bulliet, *Cotton, Climate, and Camels in Early Islamic Iran*, 1.

31 Ellenblum, *The Collapse of the Eastern Mediterranean*, 94–105.

32 Yan, Q., Z. Zhang, H. Wang, and D. Jiang, "Simulated Warm Periods of Climate over China during the Last Two Millennia: The Sui-Tang Warm

Period Versus the Song-Yuan Warm Period," *Journal of Geophysical Research: Atmospheres* 120 (2015): 2229–41, doi:10.1002/2014JD022941.

33 Xunming Wang et al., "Climate, Desertification, and the Rise and Collapse of China's Historical Dynasties," *Human Ecology* 38 (2010): 159, 164.

34 Pingzhong Zhang et al., "A Test of Climate, Sun, and Culture Relationships from an 1810-Year Chinese Cave Record," *Science* 322, no. 5903 (2008): 940–2.

35 Gergana Yancheva et al., "Influence of the Intertropical Convergence Zone on the East Asian Monsoon," *Nature* 445 (2007): 74–7.

36 Y. Su, L. Liu, X. Q. Fang, and Y. N. Ma, "The Relationship between Climate Change and Wars Waged between Nomadic and Farming Groups from the Western Han Dynasty to the Tang Dynasty Period," *Climate of the Past* 12, no. 1 (2016): 137–50.

37 Nicola Di Cosmo, Clive Oppenheimer, and Ulf Büntgen, "Onterplay of Environmental and Socio-Political Factors in the Downfall of the Eastern Türk Empire in 630 CE," *Climatic Change* 145 (2017): 383–95. https://doi.org/10.1 007/s10584-017-2111-0; and Jie Fei, Jie Zhou, and Yongjian Hou, "Circa A.D. 626 Volcanic Eruption, Climatic Cooling, and the Collapse of the Eastern Turkic Empire," *Climatic Change* 81 (2007): 469–75. doi:10.1007/s10584-006-9199-y.

38 Wang et al., "Climate, Desertification, and the Rise and Collapse of China's Historical Dynasties," 162.

39 Rosanne D'Arrigo et al., "1738 Years of Mongolian Temperature Variability Inferred from a Tree-Ring Width Chronology of Siberian Pine," *Geophysical Research Letters* 28 (2001): 543–6.

40 Neil Pederson et al., "Pluvials, Droughts, the Mongol Empire, and Modern Mongolia," *Proceedings of the National Academy of Sciences* 111 (2014): 4375–9.

41 Victor Lieberman and Brendan Buckley, "The Impact of Climate on Southeast Asia, Circa 950–1820: New Findings," *Modern Asian Studies* 46, no. 5 (2012): 1056–9; Victor B. Lieberman, *Strange Parallels: Southeast Asia in Global Context, c 800-1830* (New York: Cambridge University Press, 2003), 1: 104–8 and 459.

42 Lieberman and Buckley, "The Impact of Climate on Southeast Asia, Circa 950–1820," 1061–2 and 1065.

43 Ibid., 1062.

44 Lieberman, *Strange Parallels*, 97; Lieberman and Buckley, "The Impact of Climate on Southeast Asia, Circa 950–1820," 1063.

45 Lieberman and Buckley, "The Impact of Climate on Southeast Asia, Circa 950–1820," 1065 and 1067.

46 Marc A. Abramiuk et al., "Linking Past and Present: A Preliminary Paleoethnobotanical Study of Maya Nutritional and Medicinal Plant Use and Sustainable Cultivation in the Southern Maya Mountains, Belize," *Ethnobotany Research and Applications* 9 (2011): 257–73.

47 Billie L. Turner and Jeremy A. Sabloff, "Classic Period Collapse of the Central Maya Lowlands: Insights about Human–Environment Relationships for

Sustainability," *Proceedings of the National Academy of Sciences* 109 (2012): 13908–14.

48 David A. Hodell, Jason H. Curtis, and Mark Brenner, "Possible Role of Climate in the Collapse of Classic Maya Civilization," *Nature* 375 (1995): 391–4; Jason H. Curtis, David A. Hodell, and Mark Brenner, "Climate Variability on the Yucatan Peninsula (Mexico) during the Past 3500 Years, and Implications for Maya Cultural Evolution," *Quaternary Research* 46 (1996): 37–47.

49 Peter M.J. Douglas et al., "Drought, Agricultural Adaptation, and Sociopolitical Collapse in the Maya Lowlands," *Proceedings of the National Academy of Sciences* 112, no. 18 (2015): 5607–12.

50 R. B. Gill, *The Great Maya Droughts: Water, Life, and Death* (Albuquerque, NM: University of New Mexico Press, 2000); Martín Medina-Elizalde and Eelco J. Rohling, "Collapse of Classic Maya Civilization Related to Modest Reduction in Precipitation," *Science* 335, no. 6071 (2012): 956–9.

51 Martín Medina-Elizalde et al., "High Resolution Stalagmite Climate Record from the Yucatán Peninsula Spanning the Maya Terminal Classic Period," *Earth and Planetary Science Letters* 298 (2010): 255–62.

52 David A. Hodell, Mark Brenner, and Jason H. Curtis, "Terminal Classic Drought in the Northern Maya Lowlands Inferred from Multiple Sediment Cores in Lake Chichancanab (Mexico)," *Quaternary Science Reviews* 24 (2005): 1413–27.

53 W. Christopher Carleton, David Campbell, and Mark Collard, "A Reassessment of the Impact of Drought Cycles on the Classic Maya," *Quaternary Science Reviews* 105 (2014): 151–61.

54 Chad S. Lane, Sally P. Horn, and Matthew T. Kerr, "Beyond the Mayan Lowlands: Impacts of the Terminal Classic Drought in the Caribbean Antilles," *Quaternary Science Reviews* 86 (2014): 89–98.

55 Zachary P. Taylor, Sally P. Horn, and David B. Finkelstein, "Pre-Hispanic Agricultural Decline Prior to the Spanish Conquest in Southern Central America," *Quaternary Science Reviews* 73 (2013): 196–200.

56 Tripti Bhattacharya et al., "Cultural Implications of Late Holocene Climate Change in the Cuenca Oriental, Mexico," *Proceedings of the National Academy of Sciences* 112, no. 6 (2015): 1693–8.

57 John Reed Swanton, *Early History of the Creek Indians and Their Neighbors* (Washington, DC: GPO, 1922), 168; Edward Gaylord Bourne et al., *Narratives of the Career of Hernando de Soto in the Conquest of Florida, as Told by A knight of Elvas, and in a Relation by Luys Hernández de Biedma, Factor of the Expedition*, vol. 2 (New York: Allerton, 1922), 89, 101, and 139.

58 William C. Foster, *Climate and Culture Change in North America AD 900–1600* (Austin: University of Texas Press, 2012), 32, 46, 51; David G. Anderson, "Climate and Culture Change in Prehistoric and Early Historic Eastern North America," *Archaeology of Eastern North America* 29 (2001): 166.

59 Charles R. Cobb and Brian M. Butler, "The Vacant Quarter Revisited: Late Mississippian Abandonment of the Lower Ohio Valley," *American Antiquity* 67 (2002): 625–41.

60 Larry V. Benson, Timothy R. Pauketat, and Edward R. Cook, "Cahokia's Boom and Bust in the Context of Climate Change," *American Antiquity* 74 (2009): 467–83.

61 Samuel E. Munoz et al., "Reply to Baires et al.: Shifts in Mississippi River Flood Regime Remain a Contributing Factor to Cahokia's Emergence and Decline," *Proceedings of the National Academy of Sciences* 112, no. 29 (2015): E3754.

62 John P. Hart, John P. Nass, and Bernard K. Means, "Monongahela Subsistence-Settlement Change?" *Midcontinental Journal of Archaeology* 30 (2005): 356–7.

63 Sharon Hull, Mostafa Fayek, F. Joan Mathien, and Heidi Roberts, "Turquoise Trade of the Ancestral Puebloan: Chaco and Beyond," *Journal of Archaeological Science* 45 (2014): 187–95.

64 Stephen Plog and Carrie Heitman, "Hierarchy and Social Inequality in the American Southwest, AD 800–1200," *Proceedings of the National Academy of Sciences* 107 (2010): 19619–26.

65 Kendrick Frazier, *People of Chaco: A Canyon and its Culture* (New York: WW Norton & Company, 1999), 101–2.

66 David Roberts, D. Merriam, and G. Child, "Riddles of the Anasazi," *Smithsonian* 34, no. 4 (2003): 72–81.

67 W. H. Willis, Brandon L. Drake, and Wetherbee B. Dorshow, "Prehistoric Deforestation at Chaco Canyon?" *Proceedings of the National Academy of Sciences* 111 (2014): 11584–91.

68 Tom Dillehay, Alan L. Kolata, and Mario Pino, "Pre-industrial Human and Environment Interactions in Northern Peru during the Late Holocene," *The Holocene* 14 (2004): 272–81; Tom D. Dillehay and Alan L. Kolata, "Long-Term Human Response to Uncertain Environmental Conditions in the Andes," *Proceedings of the National Academy of Sciences* 101 (2004): 4325–30.

69 Gregory Zaro and Adán Umire Alvarez, "Late Chiribaya Agriculture and Risk Management along the Arid Andean Coast of Southern Perú, AD 1200–1400," *Geoarchaeology* 20 (2005): 717–37.

70 Bertil Mächtle and Bernhard Eitel, "Fragile Landscapes, Fragile Civilizations—How Climate Determined Societies in the pre-Columbian South Peruvian Andes," *Catena* 103 (2013): 62–73.

71 B. Mächtle, K. Schittek, M. Forbriger, F. Schäbitz, and B. Eitel, "A See-saw of Pre-Columbian Boom Regions in Southern Peru, Determined by Large-Scale Circulation Changes," *EGU General Assembly Conference Abstracts* 14 (2012): 8867.

72 Peter B. DeMenocal, "Cultural Responses to Climate Change during the Late Holocene," *Science* 292 (2001): 667–73.

73 Paul Coombes and Keith Barber, "Environmental Determinism in Holocene Research: Causality or Coincidence?" *Area* 37 (2005): 303–11.

74 Charles R. Ortloff and Alan L. Kolata, "Climate and Collapse: Agro-Ecological Perspectives on the Decline of the Tiwanaku State," *Journal of Archaeological*

Science 20, no. 2 (1993): 195–221; J. C. Flores, Mauro Bologna, and Deterlino Urzagasti, "A Mathematical Model for the Andean Tiwanaku Civilization Collapse: Climate Variations," *Journal of Theoretical Biology* 291 (2011): 29–32.

75 Clark L. Erickson, "Neo-environmental Determinism and Agrarian 'Collapse' in Andean Prehistory," *Antiquity* 73 (1999): 634–42.

76 Lars Fehren-Schmitz et al., "Climate Change Underlies Global Demographic, Genetic, and Cultural Transitions in pre-Columbian Southern Peru," *Proceedings of the National Academy of Sciences* 111 (2014): 9443–8.

77 Tiffiny A. Tung et al., "Patterns of Violence and Diet among Children during a Time of Imperial Decline and Climate Change in the Ancient Peruvian Andes," in Amber M. VanDerwarker and Gregory D. Wilson, eds., *The Archaeology of Food and Warfare* (Cham: Springer International Publishing, 2016), 193–228; K. Schittek, M. Forbriger, B. Mächtle, F. Schäbitz, V. Wennrich, M. Reindel, and B. Eitel, "Holocene Environmental Changes in the Highlands of the Southern Peruvian Andes (14° S) and Their Impact on Pre-Columbian Cultures," *Climate of the Past* 11, no. 1 (2015): 27–44.

78 Alex J. Chepstow-Lusty et al., "Putting the Rise of the Inca Empire within a Climatic and Land Management Context," *Climate of the Past* 5 (2009): 375–88.

Chapter Five

1 Chantal Camenisch et al., "The 1430s: A Cold Period of Extraordinary Internal Climate Variability during the Early Spörer Minimum with Social and Economic Impacts in North-Western and Central Europe," *Climate of the Past* 12, no. 11 (2016): 2107.

2 C. Camenisch, R. Brázdil, A. Kiss et al., "Extreme Heat and Drought in 1473 and Their Impacts in Europe in the Context of the Early 1470s," *Regional Environmental Change* 20 (2020). https://doi-org.fitchburgstate.idm.oclc.org/10.1007/s10113-020-01601-0.

3 Morgan Kelly and Cormac Ó. Gráda, "The Waning of the Little Ice Age: Climate Change in Early Modern Europe," *Journal of Interdisciplinary History* 44, no. 3 (2014): 301–25. For the Response see Sam White, "The Real Little Ice Age," *Journal of Interdisciplinary History* 44, no. 3 (2014): 351. See also Ulf Büntgen and Lena Hellmann, "The Little Ice Age in Scientific Perspective: Cold Spells and Caveats," *Journal of Interdisciplinary History* 44, no. 3 (2014): 353–68.

4 G. H. Miller et al., "Abrupt Onset of the Little Ice Age Triggered by Volcanism and Sustained by Sea-Ice/Ocean Feedbacks," *Geophysical Research Letters* 39 (2012): L02708, doi:10.1029/2011GL050168.

5 Dagomar Degroot, "Climate Change and Society in the 15th to 18th Centuries," *WIREs Climate Change* 9 (2018): e518. https://doi.org/10.1002/wcc.518.

6 C. MacFarling Meure et al., "Law Dome CO2, CH4, and N2O Ice Core Records Extended to 2000 years BP," *Geophysical Research Letters* 33 (2006): L14810, doi:10.1029/ 2006GL026152.

7 W. S. Broecker, "Was a Change in Thermohaline Circulation Responsible for the Little Ice Age?" *Proceedings of the National Academy of Sciences* 97 (2000): 1339.

8 Jared M. Diamond, *Collapse: How Societies Choose to Fail or Succeed* (New York: Viking, 2005), 222.

9 Ibid., 230.

10 W. J. D'Andrea, Y. Huang, S.C. Fritz, and N.J. Anderson, "Abrupt Holocene Climate Change as an Important Factor for Human Migration in West Greenland," *Proceedings of the National Academy of Sciences* 108 (2011): 9765–9, doi:10.1073/pnas.1101708108.

11 Sofia Ribeiro, Matthias Moros, Marianne Ellegaard, and Antoon Kuijpers, "Climate Variability in West Greenland during the Past 1500 Years: Evidence from a High-Resolution Marine Palynological Record from Disko Bay," *Boreas* 41, no. 1 (2012): 68–83.

12 H.H. Lamb, *Weather, Climate and Human Affairs (Routledge Revivals): A Book of Essays and Other Papers* (New York: Routledge, 2012), 42.

13 Eli Klintisch, "The Lost Norse," *Science* 354 (2016): 696–701; and Steven Hartman, A. E. J. Ogilvie, Jón Haukur Ingimundarson et al., "Medieval Iceland, Greenland, and the New Human Condition: A Case Study in Integrated Environmental Humanities," *Global and Planetary Change* 156 (2017): 123–39.

14 Andrew J. Dugmore et al., "Cultural Adaptation, Compounding Vulnerabilities and Conjunctures in Norse Greenland," *Proceedings of the National Academy of Sciences* 109 (2012): 3658–63.

15 Jette Arneborg, Jan Heinemeier, Niels Lynnerup, Henrik Loft Nielsen, Niels Rud, and Árny E. Sveinbjörnsdóttir, "Change of Diet of the Greenland Vikings Determined from Stable Carbon Isotope and 14C Dating of Their Bones," *Radiocarbon* 41, no. 2 (1999): 157–68; Dugmore et al., "Cultural Adaptation, Compounding Vulnerabilities and Conjunctures in Norse Greenland," 3658–63.

16 Klintisch, "The Lost Norse," 696–701.

17 William Farr, "The Influence of Scarcities and the High Prices of Wheat on the Mortality of the People of England," *Journal of the Statistical Society of London* 9 (1846): 161.

18 William Rosen, *The Third Horseman: Climate Change and the Great Famine of the 14th Century* (New York: Viking, 2014), 122–58 and 180.

19 William C. Jordan, *The Great Famine: Northern Europe in the Early Fourteenth Century* (Princeton, NJ: Princeton University Press, 1996), 112; Brian M. Fagan, *The Little Ice Age: How Climate Made History, 1300–1850* (New York: Basic Books, 2002), 41.

20 Wolfgang Behringer, *A Cultural History of Climate* (Cambridge: Polity, 2010), 99–101.

21 K. L. Kausrud, M. Begon, T.B. Ari et al., "Modeling the Epidemiological History of Plague in Central Asia: Palaeoclimatic Forcing on a Disease System Over the Past Millennium," *BMC Biol* 8 (2010). https://doi.org/10.1186/1741-7007-8-112); and Boris V. Schmid, Ulf Büntgen, W. Ryan Easterday et al.,

"Climate-Driven Introductions of Plague into Europe," *PNAS* 112 (2015): 3020–5; doi:10.1073/pnas.1412887112.

22 R. P. H. Yue and H.F. Lee, "Climate Change and Plague History in Europe," *Science and China Earth Science* 61 (2018): 163–77. https://doi.org/10.1007/s11430-017-9127-x.

23 Norman F. Cantor, *The Medieval Reader* (New York: HarperCollins, 1994), 281.

24 Nils Hybel and Bjørn Poulsen, *The Danish Resources c. 1000–1550: Growth and Recession* (Leiden and Boston: Brill, 2007), 35.

25 Jean M. Grove, *Little Ice Ages: Ancient and Modern* (London: Routledge, 2004), 1: 161.

26 Damian Evans et al., "A Comprehensive Archaeological Map of the World's Largest Preindustrial Settlement Complex at Angkor, Cambodia," *Proceedings of the National Academy of Sciences* 104, no. 36 (2007): 14277–82.

27 Brendan M. Buckley et al., "Climate as a Contributing Factor in the Demise of Angkor, Cambodia," *Proceedings of the National Academy of Sciences* 107, no. 15 (2010): 6748–52.

28 Dan Penny, Tegan Hall, Damian Evans, et al., "Geoarchaeological Evidence from Angkor, Cambodia, Reveals a Gradual Decline Rather Than a Catastrophic 15th-Century Collapse," *PNAS* 116 (2019): 4871–6; https://doi.org/10.1073/pnas.1821460116; Alison K. Carter, Miriam T. Stark, Seth Quintus et al., "Temple Occupation and the Tempo of Collapse at Angkor Wat, Cambodia," *PNAS* 116 (2019): 12226–31. https://doi.org/10.1073/pnas.1821879116.

29 Dan Penny, Cameron Zachreson, Roland Fletcher, et al., "The Demise of Angkor: Systemic Vulnerability of Urban Infrastructure to Climatic Variations," *Science Advances* 4 (2018). doi:10.1126/sciadv.aau4029.

30 Victor Lieberman and Brendan Buckley, "The Impact of Climate on Southeast Asia, Circa 950–1820: New Findings," *Mod Asian Stud* 46, no. 5 (2012): 1069.

31 Ibid., 1072–3.

32 J. M. Russell and T. C. Johnson, "Little Ice Age Drought in Equatorial Africa: Intertropical Convergence Zone Migrations and El Nino-Southern Oscillation Variability," *Geology* 35 (2007): 21–4.

33 James M. Russell, Dirk Verschuren, and Hilde Eggermont, "Spatial Complexity of 'Little Ice Age' Climate in East Africa: Sedimentary Records from Two Crater Lake Basins in Western Uganda," *The Holocene* 17, no. 2 (2007): 183–93.

34 Lonnie G. Thompson et al., "Kilimanjaro Ice Core Records: Evidence of Holocene Climate Change in Tropical Africa," *Science* 298 (2002): 589–93.

35 Peter Robertshaw and David Taylor, "Climate Change and the Rise of Political Complexity in Western Uganda," *Journal of African History* 41 (2000): 25 and 27.

36 Susan Keech McIntosh, "Reconceptualizing Early Ghana," *Canadian Journal of African Studies/La Revue Canadienne des études africaines* 42, no. 2–3

(2008): 350–4; Scott MacEachern, "Rethinking the Mandara: Political Landscape, Enslavement, Climate and an Entry into History in the Second Millennium AD," in J. Cameron Monroe and Akinwumi Ogundiran, eds., *Power and Landscape in Atlantic West Africa: Archaeological Perspectives* (Cambridge: Cambridge University Press, 2012), 325–6.

37 George E. Brooks, *Landlords and Strangers: Ecology, Society, and Trade in Western Africa, 1000–1630* (Boulder: Westview Press, 1993); James C. McCann, "Climate and Causation in African History," *The International Journal of African Historical Studies* 32, no. 2/3 (1999): 268; James L.A. Webb, *Desert Frontier: Ecological and Economic Change along the Western Sahel, 1600–1850* (Madison: University of Wisconsin Press, 1995).

38 Matthew J. Hannaford et al., "Climate Variability and Societal Dynamics in Pre-Colonial Southern African History (AD 900–1840): A Synthesis and Critique," *Environment and History* 20 (2014): 411–45.

39 Thomas N. Huffman and Stephan Woodborne, "Archaeology, Baobabs and Drought: Cultural Proxies and Environmental Data from the Mapungubwe Landscape, Southern Africa," *The Holocene* 26 (2016): 464–70.

40 Thomas Hobbes, *Leviathan* (Auckland: The Floating Press, 2009, first published 1651), 179.

41 Emmanuel Le Roy Ladurie, *The French Peasantry, 1450–1660* (Berkeley: University of California Press, 1987).

42 Rosanne D'Arrigo, Patrick Kliner, Timothy Newfield et al., "Complexity in Crisis: The Volcanic Cold Pulse of the 1690s and the Consequences of Scotland's Failure to Cope," *Journal of Volcanology and Geothermal Research* 389 (2020). https://doi.org/10.1016/j.jvolgeores.2019.106746.

43 Laura Rayner, "The Tribulations of Everyday Government in Williamite Scotland," in Sharon Adams and Julian Goodare, eds., *Scotland in the Age of Two Revolutions* (Woodbridge, Suffolk: Boydell Press, 2014), 206; Karen J. Cullen, *Famine in Scotland: The "ill Years" of the 1690s* (Edinburgh: Edinburgh University Press, 2010), 10.

44 Samuel Pepys, Robert Latham, and William Matthews, *The Diary of Samuel Pepys, vol. 6, 1665* (London: HarperCollins, 2000), 208.

45 J. N. Hays, *Epidemics and Pandemics: Their Impacts on Human History* (Santa Barbara, CA: ABC-CLIO, 2005), 152.

46 Andrew B. Appleby, "Disease or Famine? Mortality in Cumberland and Westmorland 1580–16401," *The Economic History Review* 26, no. 3 (1973): 403–32.

47 Richard H. Steckel, "New Light on the 'Dark Ages': The Remarkably Tall Stature of Northern European Men During the Medieval era," *Social Science History* 28, no. 2 (2004): 211–28.

48 Cullen, *Famine in Scotland-the'Ill Years' of the 1690s*, 66–71.

49 Ibid., 9, 177, and 182.

50 Dagomar DeGroot, *The Frigid Golden Age: Climate Change, the Little Ice Age, and the Dutch Republic, 1560–1720* (Cambridge: Cambridge University Press, 2018), 109–247.

51 Ibid.

52 D. Degroot, K. Anchukaitis, M. Bauch et al., "Towards a Rigorous Understanding of Societal Responses to Climate Change," *Nature* 591 (2021): 544. https://doi.org/10.1038/s41586-021-03190-2.

53 Philipp Blom, *Nature's Mutiny: How the Little Ice Age of the Long Seventeenth Century Transformed the West and Shaped the Present* (New York: Liveright Publishing, 2020).

54 White, "The Real Little Ice Age," 136–42.

55 A. Nicault, "Mediterranean Drought Fluctuations," *Climate Dynamics* 31 (2008): 227–45.

56 Jelena Mrgic, "Wine or Raki-The Interplay of Climate and Society in Early Modern Ottoman Bosnia," *Environment and History* 17 (2011): 621 and 636.

57 Sam White, *The Climate of Rebellion in the Early Modern Ottoman Empire* (Cambridge: Cambridge University Press, 2011), 153.

58 Sam White, "Rethinking Disease in Ottoman History," *International Journal of Middle East Studie*s 42, no. 4 (2010): 559.

59 Ibid; Sam White, "The Little Ice Age Crisis of the Ottoman Empire: A Conjuncture in Middle East Environmental History," in Alan Mikhail, ed., *Water on Sand: Environmental Histories of the Middle East and North Africa* (Oxford: Oxford University Press, 2013), 79–80.

60 Geoffrey Parker, *Global Crisis: War, Climate Change and Catastrophe in the Seventeenth Century* (New Haven: Yale University Press, 2013), 403–5.

61 Jingyun Zheng et al., "How Climate Change Impacted the Collapse of the Ming Dynasty," *Climatic Change* 127, no. 2 (2014): 169–82.

62 Tonio Andrade, *Lost Colony: The Untold Story of China's First Great Victory over the West* (Princeton: Princeton University Press, 2011), 56; Timothy Brook, *The Troubled Empire: China in the Yuan and Ming Dynasties* (Cambridge: Harvard University Press, 2010), 59 and 250.

63 Brook, *The Troubled Empire*, 250.

64 Jingyun Zheng et al., "How Climate Change Impacted the Collapse of the Ming Dynasty," *Climatic Change* 127, (2014): 169–82.

65 Q. Liu, G. Li, D. Kong, et al., "Climate, Disasters, Wars and the Collapse of the Ming Dynasty," *Environmental Earth Science* 77 (2018). https://doi.org/10.1007/s12665-017-7194-4.

66 J. Cui, H. Chang, G. S. Burr, et al., "Climatic Change and the Rise of the Manchu from Northeast China during AD 1600–1650," *Climatic Change* 15 (2019): 405–23 (2019). https://doi.org/10.1007/s10584-019-02471-0.

67 Zheng et al., "How Climate Change Impacted the Collapse of the Ming Dynasty."

68 Parker, *Global Crisis*, 484–506.

69 Sam White, "Shewing the Difference Between Their Conjuration, and Our Invocation on the Name of God for Rayne: Weather, Prayer, and Magic in Early American Encounters," *The William and Mary Quarterly* 72, no. 1 (2015): 33–56.

70 Karen Ordahl Kupperman, "The Puzzle of the American Climate in the Early Colonial Period," *The American Historical Review* 87, no. 5 (1982): 1264, 1266, 1269, 1271, 1276–7, 1288–9.

71 Marcy Rockman, "New World with a New Sky: Climatic Variability, Environmental Expectations, and the Historical Period Colonization of Eastern North America," *Historical Archaeology* 44 (2010): 6.

72 Sam White, *A Cold Welcome: The Little Ice Age and Europe's Encounter with North America* (Cambridge, MA: Harvard University Press, 2017), 240–6.

73 Ibid.

74 Edmund Clarence Stedman, Ellen Mackay Hutchinson, and Arthur Stedman, *A Library of American Literature From the Earliest Settlement to the Present Time* (New York: C. L. Webster, 1889), 1: 126.

75 Thomas Wickman, "'Winters Embittered with Hardships': Severe Cold, Wabanaki Power, and English Adjustments, 1690–1710," *The William and Mary Quarterly* 72, no. 1 (2015): 78.

76 Ibid., 57–98.

77 Ibid., 74.

78 John Demos, *The Unredeemed Captive: A Family Story from Early America* (New York: Alfred Knopf, 1994).

79 Wickman, "Winters Embittered with Hardships," 76.

80 Ibid., 90.

81 Sean P. A. Desjardins, "Neo-Inuit Strategies for Ensuring Food Security during the Little Ice Age Climate Change Episode, Foxe Basin, Arctic Canada," *Quaternary International* 549 (2020): 163–75. https://doi.org/10.1016/j.quaint.2017.12.026.

82 Jason Hall, "Maliseet Cultivation and Climatic Resilience on the Wəlastəkw/St. John River During the Little Ice Age," *Acadiensis* 44 (2015): 3–25.

83 Kelly and Gráda, "The Waning of the Little Ice Age; White, "The Real Little Ice Age."

84 Behringer, *A Cultural History of Climate*, 133.

85 Dean Phillip Bell, "The Little Ice Age and the Jews: Environmental History and the Mercurial Nature of Jewish–Christian Relations in Early Modern Germany," *AJS Review* 32, no. 1 (2008): 11.

86 Ibid., 13–14.

87 Teresa Kwiatkowska, "The Light Was Retreating Before Darkness: Tales of the Witch Hunt and Climate Change," *Medievalia* 42 (2016): 34.

88 Emily Oster, "Witchcraft, Weather and Economic Growth in Renaissance Europe," *The Journal of Economic Perspectives* 18, no. 1 (2004): 215–28; Behringer, *A Cultural History of Climate,* 132; Wolfgang Behringer, "Climatic Change and Witch Hunting: The Impact of the Little Ice Age on Mentalities," *Climatic Change* 43 (1999): 338.

89 Behringer, "Climatic Change and Witch-hunting," 340.

90 Ibid., 344.

91 Ibid., 344–5.

92 Emerson W. Baker, *A Storm of Witchcraft: The Salem Trials and the American Experience* (Oxford: Oxford University Press, 2014), 58–9.

93 Michèle Hayeur Smith, "'Some in Rags and Some in Jags and Some in Silken Gowns': Textiles from Iceland's Early Modern Period," *International Journal of Historical Archaeology* 16, no. 3 (2012): 520–1.

94 Behringer, *A Cultural History of Climate,* 136.

95 Ibid., 136–7.

96 John F. Richards, *The Unending Frontier: An Environmental History of the Early Modern World* (Berkeley: University of California Press, 2005), 463.

97 Ibid., 527.

98 Ibid., 536.

99 James R. Gibson, *Feeding the Russian fur Trade: Provisionment of the Okhotsk Seaboard and the Kamchatka Peninsula, 1639–1856* (Madison: University of Wisconsin Press, 1969), 25.

100 Richards, *Unending Frontier*, 471 and 494.

101 Johan C. Varekamp, "The Historic Fur Trade and Climate Change," *Eos* 87, no. 52 (2006): 593–6.

102 Werner Rösener, *Peasants in the Middle Ages* (Urbana: University of Illinois Press, 1992), 79–81.

103 Robert Jütte, *Poverty and Deviance in Early Modern Europe* (Cambridge: Cambridge University Press, 1994), 70.

104 Paolo Malanima, *Pre-Modern European Economy* (Leiden and Boston: Brill, 2009), 59.

105 Gunhild Eriksdotter, "Did the Little Ice Age Affect Indoor Climate and Comfort?: Re-theorizing Climate History and Architecture from the Early Modern Period," *Journal for Early Modern Cultural Studies* 13, no. 2 (2013): 34 and 36.

106 John Perlin, *A Forest Journey: The Role of Wood in the Development of Civilization* (New York: W.W. Norton, 1989), 245.

107 Jack Temple Kirby, *Poquosin: A Study of Rural Landscape & Society* (Chapel Hill: University of North Carolina Press, 1995), 201.

108 Ibid., 201.

109 Robert Marks, *China: An Environmental History*, 2nd ed. (Lanham, MD: Rowman & Littlefield, 2017), 161.

110 Ji-Hyung Cho, "The Little Ice Age and the Coming of the Anthropocene," *The Asian Review of World Histories* 2, no. 1 (2014): 12.

111 Reid Mitenbuler, "The Stubborn American who Brought Ice to the World," *The Atlantic*, February 5, 2013. https://www.theatlantic.com/national/archive/2013/02/the-stubborn-american-who-brought-ice-to-the-world/272828/.

112 William S. Atwell, "Volcanism and Short-Term Climatic Change in East Asian and World History, c. 1200–1699," *Journal of World History* (2001): 29–98.

113 Jón Steingrímsson, *Fires of the Earth: The Laki Eruption, 1783–1784* (Reykjavík: Nordic Volcanological Institute, 1998), 41.

114 Benjamin Franklin and Epes Sargent, *The Select Works of Benjamin Franklin: Including His Autobiography* (Boston: Phillips, Sampson & Co., 1854), 294; Ricardo M. Trigo, J. M. Vaquero, and R. B. Stothers, "Witnessing the Impact of the 1783–1784 Laki Eruption in the Southern Hemisphere," *Climatic Change* 99, no. 3–4 (2010): 535–46.

115 S. Brönnimann, J. Franke, S. U. Nussbaumer, et al., "Last Phase of the Little Ice Age Forced by Volcanic Eruptions," *Nature and Geoscience* 12 (2019): 650–6. https://doi.org/10.1038/s41561-019-0402-y.

116 Gillen D'Arcy Wood, *Tambora: The Eruption That Changed the World* (Princeton, NJ: Princeton University Press, 2013).

117 William K. Klingaman and Nicholas P. Klingaman, *The Year without Summer: 1816 and the Volcano That Darkened the World and Changed History* (New York: St. Martin's Press, 2013), 192.

118 Ibid., 192.

119 Wolfgang Behringer, *Tambora and the Year without a Summer: How a Volcano Plunged the World into Crisis* (Polity, 2019), 32–3.

120 Rudolf Brázdil et al., "Climatic Effects and Impacts of the 1815 Eruption of Mount Tambora in the Czech Lands," *Climate of the Past* 12, no. 6 (2016): 1361–74.

121 Ibid., 110–12.

122 Ibid., 117–27.

123 Klingaman and Klingaman, *The Year Without Summer*, 193.

124 Jürgen Osterhammel, *The Transformation of the World: A Global History of the Nineteenth Century* (Princeton, NJ: Princeton University Press, 2014), 199.

125 Rüdiger Glaser, Iso Himmelsbach, Iso, and Annette Bösmeier, "Climate of Migration? How Climate Triggered Migration from Southwest Germany to North America during the 19th Century," *Climate of the Past* 13 (2017): 1573–92. https://doi.org/10.5194/cp-13-1573-2017.

126 Wood, *Tambora*, 67.

127 Peter C. Perdue, *China Marches West: The Qing Conquest of Central Asia* (Cambridge, MA: Belknap Press of Harvard University Press, 2005), 283.

128 E. Backhouse and J. O. P. Bland, *Annals and Memoirs of the Court of Peking (from the 16th to the 20th Century)* (Taipei: Ch'eng Wen Pub. Co., 1970), 325.

129 Shuji Cao, Yushang Li, and Bin Yang, "Mt. Tambora, Climatic Changes, and China's Decline in the Nineteenth Century," *Journal of World History* 23, no. 3 (2012): 587– 607.

130 Wood, *Tambora*, 120.

131 Lingbo Xiao, Xiuqi Fang, Jingyun Zheng, and Wanyi Zhao, "Famine, Migration and War: Comparison of Climate Change Impacts and Social Responses in North China between the Late Ming and Late Qing Dynasties," *The Holocene* 25, no. 6 (2015): 900–10.

132 Jørgen Klein et al., "Climate, Conflict and Society: Changing Responses to Weather Extremes in Nineteenth Century Zululand," *Environment and History* (forthcoming).

133 Michael Garstang, Anthony D. Coleman, and Matthew Therrell, "Climate and the Mfecane," *South African Journal of Science* 110, no. 5–6 (2014): 1–6.

Chapter Six

1 W. F. Ruddiman, "The Anthropogenic Greenhouse Era Began Thousands of Years Ago," *Climatic Change* 61 (2003): 261–93.

2 John Walter, Roger Schofield, and Andrew B. Appleby, *Famine, Disease, and the Social Order in Early Modern Society* (Cambridge: Cambridge University Press, 1989), 36; Victor Lieberman and Brendan Buckley, "The Impact of Climate on Southeast Asia, Circa 950–1820: New Findings," *Modern Asian Studies* 46 (2012): 1049–96.

3 Robert Marks, *China: Its Environment and History* (Lanham, MD: Rowman & Littlefield, 2012), 145.

4 Jack A. Goldstone, "Efflorescences and Economic Growth in World History: Rethinking the 'Rise of the West' and the Industrial Revolution," *Journal of World History* 13 (2002): 323–89.

5 R.A. Houston, "Colonies, Enterprises, and Wealth: The Economies of Europe and the Wider World," in Euan Cameron, ed., *Early Modern Europe: An Oxford History* (Oxford: Oxford University Press, 1999), 147.

6 David Levine, *Reproducing Families: The Political Economy of English Population History* (Cambridge: Cambridge University Press, 1987), 99.

7 Rupert Sargent Holland, *Historic Inventions* (Philadelphia: Macrae Smith Company, 1911), 76.

8 Manfred Weissenbacher, *Sources of Power: How Energy Forges Human History*, vol. 1 (Santa Barbara, CA: Praeger, 2009), 202.

9 Stephen Mosley, *The Chimney of the World: A History of Smoke Pollution in Victorian and Edwardian Manchester* (London: Routledge, 2008), 17.

10 Charles Dickens, *Hard Times* (London: Bradbury & Evans), 26.

11 James J. Sheehan, *German History, 1770– 1866* (Oxford: Clarendon Press, 1989), 742.

12 Marius B. Jansen, *The Cambridge History of Japan*, vol. 5 (Cambridge: Cambridge University Press, 1989), 495.

13 Astrid Kander, Paolo Malanima, and Paul Warde, *Power to the People: Energy in Europe Over the Last Five Centuries* (Princeton, NJ: Princeton University Press, 2013), 140.

14 E. Burgess, "General Remarks on the Temperature of the Terrestrial Globe and the Planetary Spaces; by Baron Fourier," *American Journal of Science* 32 (1837): 1–20. Translation From the French, of Fourier, J.B.J., 1824,

"Remarques Générales Sur Les Températures Du Globe Terrestre Et Des Espaces Planétaires," *Annales de Chimie et de Physique*, Vol. 27, 136–67.

15 G. S. Callendar, "The Artificial Production of Carbon Dioxide and its Influence on Temperature," *Quarterly Journal of the Royal Meteorological Society* 64 (1938): 223–40. doi:10.1002/qj.49706427503.

16 R. Revelle and H. E. Suess, "Carbon Dioxide Exchange between Atmosphere and Ocean and the Question of an Increase of Atmospheric CO_2 during the Past Decades," *Tellus* 9 (1957): 18–27.

17 D. M. Etheridge et al., "Natural and Anthropogenic Changes in Atmospheric CO_2 over the Last 1000 Years from air in Antarctic ice and Firn," *Journal of Geophysical Research* 101 (1996): 4115–28.

18 Connie A. Woodhouse, Jeffrey J. Lukas, and Peter M. Brown, "Drought in the Western Great Plains, 1845–56: Impacts and Implications," *Bulletin of the American Meteorological Society* 83 (2002): 1485; Edward R. Cook, Richard Seager, Mark A. Cane, and David W. Stahle, "North American Drought: Reconstructions, Causes, and Consequences," *Earth-Science Reviews* 81 (2007): 93–134.

19 Gerald Michael Greenfield, "The Great Drought and Elite Discourse in Imperial Brazil," *The Hispanic American Historical Review* 72 (1992): 376; Lise Sedrez, "Environmental History of Modern Latin America," in Thomas H. Holloway, ed., *A Companion to Latin American History* (Malden, MA: Blackwell, 2008), 455.

20 Sunil S. Amrith, *Crossing the Bay of Bengal* (Cambridge, MA: Harvard University Press, 2013), 114; Mike Davis, *Late Victorian Holocausts: El Niño Famines and the Making of the Third World* (London: Verso, 2002), 7.

21 Kathryn Edgerton-Tarpley, *Tears from Iron: Cultural Responses to Famine in Nineteenth-Century China* (Berkeley: University of California Press, 2008), 26 and 40–1.

22 "Pictures to Draw Tears from Iron," Visualizing Cultures, MITOPEN COURSEWARE, Massachusetts Institute of Technology, 2010, http://ocw.mit.edu/ans7870/21f/21f.027/tears_from_iron/tfi_essay_06.pdf.

23 Edgerton-Tarpley, *Tears from Iron*, 28; Pierre-Etienne Will, Roy Bin Wong, James Lee, Jean Oi, and Peter Perdue, *Nourish the People: The State Civilian Granary System in China, 1650–1850* (Ann Arbor: University of Michigan Center for Chinese Studies, 1991), 3, 14, and 21.

24 Edgerton-Tarpley, *Tears from Iron*, 31–2.

25 Will et al., *Nourish the People*, 91–2.

26 Ibid., Edgerton-Tarpley, *Tears from Iron*, 91.

27 Edgerton-Tarpley, *Tears from Iron*, 101–2.

28 *Speeches by Babu Surendra Nath Banerjea*, 1876–80, vol. 6 (Calcutta: S.K. Lahiri, 1908), 268.

29 David Silbey, *The Boxer Rebellion and the Great Game in China* (New York: Hill and Wang, 2012), 66.

30 Diana Preston, *The Boxer Rebellion: The Dramatic Story of China's war on Foreigners That Shook the World in the Summer of 1900* (New York: Berkley Books, 2000), 29.

31 Mike Davis, *Late Victorian Holocausts*, 143 and 154.

32 Ibid., 27–8, 31, 44, 51, 142, and 315.

33 Ibid., 199.

34 Ibid, 188–94.

35 Ibid., 201.

36 B. I. Cook, R. L. Miller, and R. Seager, "Dust and sea Surface Temperature Forcing of the 1930s 'Dust Bowl' Drought," *Geophysical Research Letters* 35 (2008): L08710, doi:10.1029/2008GL033486.

37 BBC, "On This Day, 20 July," http://news.bbc.co.uk/onthisday/hi/dates/stories/july/20/ newsid_3728000/3728225.stm.

38 "Air Conditioning: No Sweat," *The Economist* January 5, 2013.

39 Charles Dickens, *Oliver Twist* (Philadelphia: Lea & Blanchard), 86.

40 David Allen Pfeifer, *Eating Fossil Fuels: Oil, Food, and the Coming Crisis in Agriculture* (Gabriola Island, BC: New Society Publishers, 2006).

41 World Resources Institute, "The History of Carbon Dioxide Emissions," http://www.wri.org/blog/2014/05/history-carbon-dioxide-emissions.

Chapter Seven

1 NOAA, "Global Climate Report—September 2016," National Centers for Environmental Information, https://www.ncdc.noaa.gov/sotc/global/201609.

2 National Oceanic and Atmospheric Administration, "2020 was Earth's 2nd-Hottest Year, Just Behind 2016," January 14, 2021, https://www.noaa.gov/news/2020-was-earth-s-2nd-hottest-year-just-behind-2016; National Aeronautics and Space Administration, "2020 Tied for Warmest Year on Record, NASA Analysis Shows," January 14, 2021, https://climate.nasa.gov/news/3061/2020-tied-for-warmest-year-on-record-nasa-analysis-shows/.

3 Cornelia Dean, "As Alaska Glaciers Melt, it's Land That's Rising," *New York Times*, May 17, 2009.

4 J. Richter-Menge, M. L. Druckenmiller, and M. Jeffries, eds., *2019: Arctic Report Card 2019*, https://www.arctic.noaa.gov/Report-Card.

5 Richard Slats et al., "Voices from the Front Lines of a Changing Bering Sea," J. Richter-Menge, M. L. Druckenmiller, and M. Jeffries, eds., *2019: Arctic Report Card 2019*, https://www.arctic.noaa.gov/Report-Card.

6 Lonnie G. Thompson, Ellen Mosley-Thompson, Mary E. Davis, and Henry H. Brecher, "Tropical Glaciers, Recorders and Indicators of Climate Change, Are Disappearing Globally," *Annals of Glaciology* 52, no (2011): 23–34.

7 L. G. Thompson et al., "Annually Resolved Ice Core Records of Tropical Climate Variability over the Past~ 1800 Years," *Science* 340, no. 6135 (2013): 945–50.

8 Mark Carey, *In the Shadow of Melting Glaciers: Climate Change and Andean Society* (Oxford: Oxford University Press, 2010).

9 Simon J Cook et al., "Glacier Change and Glacial Lake Outburst Flood Risk in the Bolivian Andes," *The Cryosphere* 10 (2016): 2399.

10 Barbara Fraser, "Melting in the Andes: Goodbye Glaciers," *Nature* 491 (2012): 180–2.

11 Kenneth R. Weiss, "Drying Lakes: Warming Climates, Drought, and Overuse Are Draining Some of the World's Biggest Lakes, Threatening Habitats and Cultures," *National Geographic* 233, no. 3 (2018): 108.

12 Soumaya Belmecheri et al., "Multi-Century Evaluation of Sierra Nevada Snowpack," *Nature Climate Change* 6, (2016): 2–3.

13 Inside Climate News, "Drought Fears Take Hold in a Four Corners Region Already Beset by the Coronavirus Pandemic," June 8, 2020, https://insideclimatenews.org/news/08062020/drought-southwest-us-agriculture-water-rain-climate-change/.

14 MET Office, "What Caused the Record UK Winter Rainfall of 2013–14?" June 22, 2017, http://www.metoffice.gov.uk/news/releases/2017/record-uk-winter-rainfall-of-2013-14.

15 Nicholas Kristof, "Will Climate Get Some Respect Now?" *New York Times*, November 1, 2012.

16 First Street Foundation, "Highlights From "The First National Flood Risk Assessment," June 29, 2020, https://firststreet.org/flood-lab/published-research/2020-national-flood-risk-assessment-highlights/.

17 Harold L. Platt, *Sinking Chicago: Climate Change and the Remaking of a Flood-Prone Environment.* (Philadelphia: Temple University Press, 2018).

18 Christopher Flavelle et al., "New Data Reveals Hidden Flood Risk Across America," *New York Times*, June 29, 2020.

19 Zeninjor Enwemeka, "Boston's Top 10 Biggest Snowstorms," *WBUR*, January 28, 2015; Andy Rosen, "Last 7 Days Were Boston's Snowiest on Record," *Boston Globe*, February 2, 2015.

20 Stephanie C. Herring, Martin P. Hoerling, Thomas C. Peterson, and Peter A. Stott, "Explaining Extreme Events of 2013 from a Climate Perspective," *Bulletin of the American Meteorological Society* 95, (2014): S1–S104.

21 Stefan Rahmstorf and Dim Coumou, "Increase of Extreme Events in a Warming World," *Proceedings of the National Academy of Sciences* 108, no. 44 (2011): 17905–9.

22 Noah S. Diffenbaugh, Daniel L. Swain, Danielle Touma, "Global Warming Increases California Drought Risk," *PNAS* 112(2015): 3931–6; doi:10.1073/pnas.1422385112.

23 Justin T. Martin, Gregory T. Pederson, Connie A. Woodhouse et al., "Increased Drought Severity Tracks Warming in the United States' Largest River Basin," *PNAS* 117 (2020): 11328–36, doi:10.1073/pnas.1916208117.

24 Global Post, "The Biggest Disaster You've Probably Never Heard of," February 14, 2014.

25 Edward Wong, "Resettling China's 'Ecological Migrants,'" *New York Times*, October 25, 2016.

26 Sam Jones, "Bolivia after the Floods: 'The Climate Is Changing; We Are Living That Change,'" *Guardian*, December 8, 2014.

27 Associated Press, "Malawi Floods Kill 176 People," *Guardian*, January 17, 2015.

28 Chris Funk, Diego Pedreros, Sharon Nicholson et al., "Examining the Potential Contributions of Extreme "Western V" Sea Surface Temperatures to the 2017 March–June East African Drought," *Bulletin of the American Meteorological Society* 100 (2019): S55–S60, https://doi.org/10.1175/BAMS-D-18-0108.1 via Carbon Brief https://www.carbonbrief.org/mapped-how-climate-change-affects-extreme-weather-around-the-world.

29 Oxfam International, "Drought in East Africa," https://www.oxfam.org/en/drought-east-africa-if-rains-do-not-come-none-us-will-survive.

30 M. K. Roxy, S. Ghosh, A. Pathak, A. et al., "A Threefold Rise in Widespread Extreme Rain Events Over Central India," *Nature Communications* 8 (2017), doi:10.1038/s41467-017-00744-9.

31 Michelle C. Mack et al., "Carbon Loss from an Unprecedented Arctic Tundra Wildfire," *Nature* 475, no. 7357 (2011): 489–92.

32 Carbon Brief, "Siberia's 2020 Heatwave Made '600 Times More Likely' by Climate Change," July 15, 2020, https://www.carbonbrief.org/siberia-s-2020-heatwave-made-600-times-more-likely-by-climate-change; and Mark Parrington, https://twitter.com/m_parrington/status/1278243686225190913/photo/3.

33 Julia Hollingsworth, "Over 9,000 Evacuated as Wildfire Rages in Spanish Holiday Hotspot, the Canary Islands, for Second Time This Summer," *CNN*, August 19, 2019, https://www.cnn.com/2019/08/19/europe/canary-islands-gran-canaria-fire-intl-hnk/index.html.

34 Special Climate Statement 73—Extreme Heat and Fire Weather in December 2019 and January 2020. Australian Government Bureau of Meteorology, http://www.bom.gov.au/climate/current/statements/scs73.pdf; and NOAA, State of the Climate, 2019.

35 Maddie Stone, "A Forgotten Forest of Ancient Trees Was Devastated by Bushfires," *The Atlantic*, February 25, 2020.

36 Lauren Bennett, Dr. Sabine Kasel, Dr. Tom Fairman and Ruizhu Jiang, "Why Australia's Severe Bushfires May Be Bad News for Tree Regeneration" January 30, 2020, https://phys.org/news/2020-01-australia-severe-bushfires-bad-news.html.

37 NASA, "Climate Conditions Determine Amazon fi re Risk," June 7, 2013, http://www.nasa.gov/topics/earth/features/ amazon- fire-risk.html#.V4Ub3ZMrLox.

38 Richard Conniff, "Escalator to Extinction," November 13, 2018, *YaleEnvironment360*, https://e360.yale.edu/features/escalator-to-extinction-can-mountain-species-adapt-to-climate-change.

39 J. Rocklöv and R. Dubrow, "Climate Change: An Enduring Challenge for Vector-Borne Disease Prevention and Control," *Nature Immunology* 21(2020): 479–83, https://doi.org/10.1038/s41590-020-0648-y.

40 "As Earth Warms, the Diseases That May Lie within Permafrost Become a Bigger Worry," *Scientific American*, November 21, 2016.

41 K. Hueffer, D. Drown, V. Romanovsky et al., "Factors Contributing to Anthrax Outbreaks in the Circumpolar North," *EcoHealth* 17 (2020): 174–80, https://doi-org.fitchburgstate.idm.oclc.org/10.1007/s10393-020-01474-z.

42 Michael Lemonick, "The Secret of Sea Level Rise: It Will Vary Greatly by Region," *YaleEnvironment360*, March 22, 2010, http://e360.yale.edu/features/the_secret_of_sea_level_rise_it_will_vary_greatly_by_region.

43 *Fiji Times Online*, August 16, 2014, http://www.fijitimes.com/story.aspx?id=277453.

44 Understanding Risk, "Dealing with Coastal Risks in Small Island States," https://understandrisk.org/event-session/dealing-with-coastal-risks-in-small-islandstates-training-session-on-simple-assessments-of-coastal-problems-andsolutions-in-small-island-developing-states/.

45 NOAA, "2019 State of U.S. High Tide Flooding with a 2020 Outlook," NOAA Technical Report NOS CO-OPS 092, 2020. https://tidesandcurrents.noaa.gov/publications/Techrpt_092_2019_State_of_US_High_Tide_Flooding_with_a_2020_Outlook_30June2020.pdf.

46 Christopher Joyce, "Climate Change Worsens Coastal Flooding from High Tides," *NPR*, October 8, 2014, http://www.npr.org/2014/10/08/354166982/climate-change-worsens-coastal-flooding-from-high-tides.

47 Jon Gertner, "Should the United States Save Tangier Island from Oblivion?" *New York Times*, July 6, 2016.

48 Suging Seas, "Louisiana and the Surging Sea," *Climate Central*, http://sealevel.climatecentral.org/research/reports/louisiana-and-the-surging-sea.

49 NOAA, "2019 State of U.S. High Tide Flooding with a 2020 Outlook."

50 Poppy McPherson, "Dhaka: The City Where Climate Refugees Are Already a Reality," *Guardian*, December 1, 2015.

51 S. A. Kulp and B. H. Strauss, "New Elevation Data Triple Estimates of Global Vulnerability to Sea-Level Rise and Coastal Flooding," *Nature Commun*ication 10, 4844 (2019), https://doi.org/10.1038/s41467-019-12808-zZ.

52 Fred Pearce, "On Java's Coast, A Natural Approach to Holding Back the Waters," *YaleEnvironment360*, May 2, 2019, https://e360.yale.edu/features/on-javas-coast-a-natural-approach-to-holding-back-the-waters.

53 Alex Chapman and Van Pham Dang Tri, "Climate Change Is Driving Migration from Vietnam's Mekong Delta," *The Conversation*, January 9, 2018, https://theconversation.com/climate-change-is-triggering-a-migrant-crisis-in-vietnam-88791.

54 M. Rhein et al., eds., *Climate Change 2013: The Physical Science Basis. Contribution of Working Group I to the Fifth Assessment Report of the*

Intergovernmental Panel on Climate Change (Cambridge and New York: Cambridge University Press, 2013).

55 Michael Milstein, "Unusual North Pacific Warmth Jostles Marine Food Chain," *Northwest Fisheries Science Center*, September 2014, https://www.nwfsc.noaa.gov/news/features/food_chain/index.cfm.

56 Alastair Bland, "As Oceans Warm, the World's Kelp Forests Begin to Disappear," *YaleEnvironment360*, November 20, 2017, https://e360.yale.edu/features/as-oceans-warm-the-worlds-giant-kelp-forests-begin-to-disappear.

57 Christopher Free, James T. Thorson, Malin L Pinsky et al., "Impacts of Historical Warming on Marine Fisheries," *Science* 363 (2019): 979–83, doi:10.1126/science.aau1758.

58 Thomas L. Frölicher, Erich M. Fischer, and Nicolas Gruber, "Marine Heatwaves Under Global Warming," *Nature* 560 (2018): 360–4.

59 Knutson et al., "Tropical Cyclones and Climate Change Assessment Part II: Projected Response to Anthropogenic Warming," *Bulletin of the American Meteorological Society* 101 (2020): E303–22.

60 Knutson et al., "Tropical Cyclones and Climate Change Assessment Part II," and Kossin et al., "Global Increase in Major Tropical Cyclone Exceedance Probability over the Past Four Decades," *Proceedings of the National Academy of Sciences* 117 (2020): 11975–80.

61 Andrew H. Altieri and Keryn B. Gedan, "Climate Change and Dead Zones," *Global Change Biology* 21 (2015): 1395–406.

62 J.C. Lehrter et al., "Predicted Effects of Climate Change on Northern Gulf of Mexico Hypoxia," in D. Justic, K. Rose, R. Hetland, and K. Fennel, eds., Modeling Coastal Hypoxia (Cham, Switzerland: Springer, 2017), 173–214.

63 IPCC, "2013: Summary for Policymakers," in T. F. eds., *Climate Change 2013: The Physical Science Basis. Contribution of Working Group I to the Fifth Assessment Report of the Intergovernmental Panel on Climate Change* (Cambridge and New York: Cambridge University Press, 2013).

64 R. A. Feely, S. C. Doney, and S. R. Cooley, "Ocean Acidification: Present Conditions and Future Changes in a High- CO2 World," *Oceanography* 22 (2009): 36–47, doi:10.5670/oceanog.2009.95.

65 Éva Plagányi, "Climate Change Impacts on Fisheries," *Science* 363 (2019): 93–131, doi:10.1126/science.aaw5824.

66 Kendra Pierre-Louis, "Warming Waters, Moving Fish," *New York Times*, November 29, 2019.

67 Antoaneta Roussi, "Row Over Africa's Largest Dam in Danger of Escalating, Warn Scientists," *Nature* 583 (2020): 501–2.

68 Alan Basist and Claude Williams, "Mekong River Wetness Anomalies in the 2019 Monsoon Season," *Eyes on the Earth*, April 20, 2020, https://www.eyesonearth.org/single-post/2020/04/20/Mekong-River-Wetness-Anomalies-in-the-2019-Monsoon-Season.

69 Mekong River Commission Procedures for Notification, Prior Consultation and Agreement, April 2, 2020, https://www.mrcmekong.org/assets/Uploads/Cambodia-Official-Reply-Form.pdf.

70 Paul Hockenos, "Turkey's Dam-Building Spree Continues, At Steep Ecological Cost," *YaleEnvironment360*, October 3, 2019, https://e360.yale.edu/features/turkeys-dam-building-spree-continues-at-steep-ecological-cost.

71 Yale Environment360, "When the Water Ends: Africa's Climate Conflicts," *Yale School of Forestry and Environmental Studies*, October 26, 2010, http://e360.yale.edu/features/when_the_water_ends_africas_climate_conflicts.

72 Wario R Adano., Ton Dietz, Karen Witsenburg, and Fred Zaal, "Climate Change, Violent Conflict and Local Institutions in Kenya's Drylands," *Journal of Peace Research* 49 (2012): 65–80.

73 Cullen S. Hendrix and Idean Salehyan, "Climate Change, Rainfall, and Social Conflict in Africa," *Journal of Peace Research* 49 (2012): 35–50.

74 Human Rights Watch, "'There Is No Time Left' Climate Change, Environmental Threats, and Human Rights in Turkana County, Kenya," https://www.hrw.org/report/2015/10/15/there-no-time-left/climate-change-environmental-threats-and-human-rights-turkana.

75 Ethan L. Hall, *Conflict for Resources: Water in the Lake Chad Basin*. Army Command and General Staff Coll Fort Leavenworth KS School of Advanced Military Studies, 2009.

76 François Gemenne, Jon Barnett, W. Neil Adger, and Geoffrey D. Dabelko, "Climate and Security: Evidence, Emerging Risks, and a New Agenda," *Climatic Change* 123 (2014): 1–9.

77 Solomon M. Hsiang, Kyle C. Meng, and Mark A. Cane, "Civil Conflicts Are Associated with the Global Climate," *Nature* 476 (2011): 438–41.

78 Caitlin E. Werrell, Francesco Femia, and Anne-Marie Slaughter, "The Arab Spring and Climate Change," *Center for American Progress*, February 28, 2013, https://www.americanprogress.org/issues/security/reports/2013/02/28/54579/the-arab-spring-and-climate-change/.

79 World Bank, *East Asia and Pacific Cities: Expanding Opportunities for the Urban Poor* (Washington, DC: World Bank, 2017), https://openknowledge.worldbank.org/bitstream/handle/10986/27614/211093ov.pdf.

80 Tom Di Liberto, "India Heat Wave Kills Thousands," June 9, 2015, https://www.climate.gov/news-features/event-tracker/india-heat-wave-kills-thousands.

81 Gulrez Azhar et al., "Heat Wave Vulnerability Mapping for India," *International Journal of Environmental Research and Public Health* 14, no. 4 (2017): 357, https://doi.org/10.3390/ijerph14040357.

82 NPR All Things Considered, "As Rising Heat Bakes U.S. Cities, The Poor Often Feel It Most," https://www.npr.org/2019/09/03/754044732/as-rising-heat-bakes-u-s-cities-the-poor-often-feel-it-most.

83 An Introduction to Climate Change, Health, and Equity, https://www.apha.org/-/media/files/pdf/topics/climate/apha_climate_equity_introduction.ashx?la=en&hash=B40A6A0109D9C5474B7C7362176BEA9E9DFC16CC.

84 Bruce Bekkar, Susan Pacheco, and Rupa Basu, "Association of Air Pollution and Heat Exposure with Preterm Birth, Low Birth Weight, and Stillbirth in

the USA Systematic Review," *Journal of the American Medical Association* 3, no. 6 (2020): e208243. doi:10.1001/jamanetworkopen.2020.8243.

85 Rachel Morello-Frosch, Manuel Pastor, James Sadd, and Seth B. Shonkoff, *The Climate Gap Inequalities in How Climate Change Hurts Americans & How to Close the Gap*, https://dornsife.usc.edu/assets/sites/242/docs/The_Climate_Gap_Full_Report_FINAL.pdf.

86 Adrianna Quintero & Juanita Constible, "Nuestro Futuro: Climate Change and U.S. Latinos," *NRDC*, October 13, 2016, https://www.nrdc.org/resources/nuestro-futuro-climate-change-and-us-latinos.

87 D. C. Mitchell, J. Castro, T. L. Armitage et al., "Recruitment, Methods, and Descriptive Results of a Physiologic Assessment of Latino Farmworkers: The California Heat Illness Prevention Study," *Journal of Occupational and Environmental Medicine* 59 (2017): 649–58, https://doi.org/10.1097/JOM.0000000000000988; and Ethel V. Taylor, Ambarish Vaidyanathan, W. Dana Flanders, et al., "Differences in Heat-Related Mortality by Citizenship Status: United States, 2005–2014," *American Journal of Public Health* 108 (2018): S131–6, https://doi.org/10.2105/AJPH.2017.304006.

88 Diane J. Nelson, "Protecting California's Farmworkers as Temperatures Climb," *UC Davis Science & Climate*, August 31, 2017, https://climatechange.ucdavis.edu/news/protecting-californias-farmworkers-as-temperatures-climb/.

89 Extreme Weather Events: Perspectives & Safety Impacts in Agriculture, UC Davis Western Center for Agricultural Health and Safety, https://indd.adobe.com/view/0405c2bc-c95e-438d-b5af-3c80c73d00eb.

90 Tessa Toumbourou, "Empty Pocket Season': Dayak Women Farmers Grapple with the Impacts of Oil Palm Plantations," *Mongabay*, August 20, 2018, https://news.mongabay.com/2018/08/empty-pocket-season-dayak-women-farmers-grapple-with-the-impacts-of-oil-palm-plantations/.

91 Abrahm Lustgarten, "Palm Oil Was Supposed to Help Save the Planet. Instead It Unleashed a Catastrophe," *New York Times Magazine*, November 20, 2018.

92 Human Rights Watch, *When We Lost the Forest, We Lost Everything" Oil Palm Plantations and Rights Violations in Indonesia*, September 22, 2019, https://www.hrw.org/report/2019/09/23/when-we-lost-forest-we-lost-everything/oil-palm-plantations-and-rights-violations#.

93 Lorena Allam and Nick Evershed, "Too Hot for Humans? First Nations People Fear Becoming Australia's First Climate Refugees," *Guardian*, https://www.theguardian.com/australia-news/2019/dec/18/too-hot-for-humans-first-nations-people-fear-becoming-australias-first-climate-refugees.

94 G. Wako, M. Tadesse, and A. Angassa, "Camel Management as an Adaptive Strategy to Climate Change by Pastoralists in Southern Ethiopia," *Ecol Process* 6 (2017), https://doi.org/10.1186/s13717-017-0093-5); and E.E. Watson, H.H. Kochore, and B. H. Dabasso, "Camels and Climate Resilience: Adaptation in Northern Kenya," *Human Ecology* 44 (2016): 701–13, https://doi.org/10.1007/s10745-016-9858-1.

95 T. W. Ng'ang'a, J. Y. Coulibaly, T. A. Crane et al., "Propensity to Adapt to Climate Change: Insights from Pastoralist and Agro-Pastoralist Households of Laikipia County, Kenya," *Climatic Change* 161 (2020), https://doi.org/10.1007/s10584-020-02696-4.

96 Jacinta Mukulu Waila, Michael Wandanje Mahero, Shamilah Namusisi et al., "Outcomes of Climate Change in a Marginalized Population: An Ethnography on the Turkana Pastoralists in Kenya," *American Journal of Public Health* 108 (2018): S70–1, https://doi.org/10.2105/AJPH.2017.304063.

97 Ben Walker, "Climate Change is Making This Bolivian Village a Ghost Town," *Inside Climate News*, August 25, 2017, https://insideclimatenews.org/news/25082017/climate-change-shocks-bolivia-rural-poor-migration-agriculture-quinoa.

98 Jenny Gonzales, "Green Alert: How Indigenous People Are Experiencing Climate Change in the Amazon," *Mongabay*, May 27, 2020, https://news.mongabay.com/2020/05/green-alert-how-indigenous-have-been-experiencing-climate-change-in-the-amazon/.

99 Wayne S. Walker, Seth R. Gorelik, Alessandro Baccini et al., "The Role of Forest Conversion, Degradation, and Disturbance in the Carbon Dynamics of Amazon Indigenous Territories and Protected Areas," *Proceedings of the National Academy of Sciences* 117 (2020): 3015–25, doi:10.1073/pnas.1913321117.

100 Jonathan Blitzer, "How Climate Change Is Fuelling the U.S. Border Crisis," *The New Yorker*, April 3, 2019.

101 Charlie Fidleman, "When Climate Change Drove All the Men Away," *Canada's National Observer*, https://www.nationalobserver.com/2020/02/24/features/when-climate-change-drove-all-men-away.

102 Sonia Narang, "Navajo Women Struggle to Preserve Traditions as Climate Change Intensifies," *Global Post*, May 25, 2018, https://www.pri.org/stories/2018-05-25/navajo-women-struggle-preserve-traditions-climate-change-intensifies.

103 Valerie Volcovici, "A Tribe's Uphill Battle Against Climate Change," *Eos*, April 20, 2020, https://eos.org/articles/a-tribes-uphill-battle-against-climate-change; and Bob Berwyn, "Global Warming Is Pushing Pacific Salmon to the Brink, Federal Scientists Warn," *Inside Climate News*, July 29, 2019, https://insideclimatenews.org/news/29072019/pacific-salmon-climate-change-threat-endangered-columbia-river-california-idaho-oregon-study/.

104 Fourth National Climate Assessment, Chapter 22, Northern Great Plains, https://nca2018.globalchange.gov/chapter/22/#:~:text=Indigenous%20peoples%20of%20the%20Northern,melting%20glaciers%20and%20reduction%20in.

105 Kathryn Norton-Smith, Kathy Lynn, Karletta Chief et al., "Climate Change and Indigenous Peoples: A Synthesis of Current Impacts and Experiences," *USDA, 25*, https://www.climatehubs.usda.gov/content/climate-change-and-indigenous-peoples-synthesis-current-impacts-and-experiences.

106 US Climate Resilience Toolkit, Tribal Nations, https://toolkit.climate.gov/topics/tribal-nations; and Norton-Smith et al., "Climate Change and Indigenous Peoples," 25, 29.

107 Kenyon Wallace, "Beyond Frozen," *The Star*, July 24, 2019, https://projects.thestar.com/climate-change-canada/nunavut/.

108 Amali Tower, "Shrinking Options: The Nexus between Climate Change, Displacement, and Security in the Lake Chad Basin," *Climate Refugees*, September 18, 2017, https://www.climate-refugees.org/reports/2017/9/18/shrinking-options-the-nexus-between-climate-change-displacement-and-security-in-the-lake-chad-basin.

109 Bond McGillivray, "Climate Change Driven Crop Failure in Central America," November 5, 2019, https://storymaps.arcgis.com/stories/e0d52fa5d69b42bc8ffd2040e4e9971f.

110 Miranda Cady Hallett, "How Climate Change Is Driving Emigration from Central America," September 26, 2019, https://theconversation.com/how-climate-change-is-driving-emigration-from-central-america-121525.

111 Tim Padgett, "Guatemalan Climate Change Refugees Pouring Over U.S. Border – And Into South Florida," *WJCT*, April 8, 2019, https://news.wjct.org/post/guatemalan-climate-change-refugees-pouring-over-us-border-and-south-florida.

112 John Podesta, "The Climate Crisis, Migration, and Refugees," *Brookings Institution*, July 25, 2019, https://www.brookings.edu/research/the-climate-crisis-migration-and-refugees/.

113 M. E. Hauer, E. Fussell, V. Mueller et al., "Sea-Level Rise and Human Migration," *Nature Review of Earth Environment* 1 (2020): 28–39, https://doi.org/10.1038/s43017-019-0002-9; and Matthew E. Hauer, "Migration Induced by Sea-Level Rise Could Reshape the US Population Landscape," *Nature Climate Change* 7(2017): 321–5.

Chapter Eight

1 T. F. Stocker et al., eds., *Climate Change 2013: The Physical Science Basis. Contribution of Working Group I to the Fifth Assessment Report of the Intergovernmental Panel on Climate Change, Summary for Policymakers* (Cambridge and New York: Cambridge University Press, 2013), 24.

2 http://earthobservatory.nasa.gov/ IOTD/ view.php?id=86027.

3 Robert M. DeConto and David Pollard, "Contribution of Antarctica to Past and Future Sea-Level Rise," *Nature* 531 (2016): 591–7.

4 James Hansen et al., "Ice Melt, Sea Level Rise and Superstorms: Evidence from Paleoclimate Data, Climate Modeling, and Modern Observations That 2 C Global Warming Is Highly Dangerous," *Atmospheric Chemistry and Physics Discuss* 15 (2015): 20059–179.

5 IPCC, 2013: Summary for Policymakers.

6 E. A. G. Schuur et al., "Climate Change and the Permafrost Carbon Feedback," *Nature* 520 (2015): 171–9.

7 V. Masson-Delmotte et al., eds., *IPCC, 2021: Climate Change 2021: The Physical Science Basis. Contribution of Working Group I to the Sixth Assessment Report of the Intergovernmental Panel on Climate Change, Summary for Policymakers* (Cambridge University Press, In Press).

8 World Weather Attribution, https://www.worldweatherattribution.org/analysis/rainfall/.

9 P. Stott, D. Stone, and M. Allen, "Human Contribution to the European Heatwave of 2003," *Nature* 432 (2004): 610–4, https://doi.org/10.1038/nature03089.

10 N. Christidis, G. Jones, P. Stott, "Dramatically Increasing Chance of Extremely Hot Summers since the 2003 European Heatwave," *Nature Climate Change* 5, no. 4 (2015): 6–50, https://doi.org/10.1038/nclimate2468.

11 World Weather Attribution, "Stormy January over Western Europe, 2018," March 16, 2018, https://www.worldweatherattribution.org/the-stormy-month-of-january-2018-over-western-europe/.

12 "Attributing Extreme Weather to Climate Change," *CarbonBrief*, https://www.carbonbrief.org/mapped-how-climate-change-affects-extreme-weather-around-the-world.

13 R. G., "FPC's Gloomy Outlook on Gas," *Science* 187 (1975): 151.

14 C. Le Quéré et al., "Global Carbon Budget 2015," *Earth System Science Data*, 7 (2015): 349–96.

15 Zeke Hausfather, "Analysis: How Much 'Carbon Budget' Is Left to Limit Global Warming to 1.5C?" *CarbonBrief*, September 4, 2018, https://www.carbonbrief.org/analysis-how-much-carbon-budget-is-left-to-limit-global-warming-to-1-5c.

16 K. Riahi, S. Rao, V. Krey et al., "RCP 8.5—A Scenario of Comparatively High Greenhouse Gas Emissions," *Climatic Change* 109 (2011), https://doi.org/10.1007/s10584-011-0149-y.

17 Dr. Christine Shearer, "Guest Post: How Plans for New Coal Are Changing Around the World," *CarbonBrief*, August 13, 2019, https://www.carbonbrief.org/guest-post-how-plans-for-new-coal-are-changing-around-the-world.

18 "Mapped, the World's Coal Plants," *CarbonBrief*, March 26, 2020, https://www.carbonbrief.org/mapped-worlds-coal-power-plants.

19 "Choked by Coal: The Carbon Catastrophe in Bangladesh," *Market Forces*, November 26, 2019, https://www.marketforces.org.au/bangladesh-choked-by-coal/.

20 "Analysis: Will China Build Hundreds of New Coal Plants in the 2020s?" *CarbonBrief*, March 24, 2020, https://www.carbonbrief.org/analysis-will-china-build-hundreds-of-new-coal-plants-in-the-2020s.

21 Stockholm Environment Institute, "How the Fossil Fuel 'Production Gap' Hinders Climate Goals," https://www.sei.org/featured/the-fossil-fuel-production-gap-climate-goals/.

22 *Projection Gap Report, 2019*, http://productiongap.org/wp-content/uploads/2019/11/Production-Gap-Report-2019.pdf, 22, 30–6.

23 "Fracking's Secret Problem—Oil Wells Aren't Producing as Much as Forecast," *Dow Jones Institutional News*, March 4 2019.

24 "Our World in Data, Renewables: What Share of Electricity Comes from Renewables?" https://ourworldindata.org/electricity-mix#renewables-what-share-of-electricity-comes-from-renewables.

25 Justin Rowlatt, "Britain Goes Coal Free as Renewables Edge Out Fossil Fuels," *BBC*, June 10, 2020, https://www.bbc.com/news/science-environment-52973089.

26 "Our World in Data, Renewables: What Share of Electricity Comes from Renewables?," https://ourworldindata.org/electricity-mix#renewables-what-share-of-electricity-comes-from-renewables.

27 David Roberts, "More Natural Gas Isn't A 'Middle Ground'—It's a Climate Disaster," *Vox*, May 30, 2019, https://www.vox.com/energy-and-environment/2019/5/30/18643819/climate-change-natural-gas-middle-ground/; and Our World in Data, "Levelized Cost of Energy by Technology, World," https://ourworldindata.org/grapher/levelized-cost-of-energy.

28 Michael E. Mann, Raymond S. Bradley, and Malcolm K. Hughes, "Global-Scale Temperature Patterns and Climate Forcing Over the Past Six Centuries," *Nature* 392 (1988): 779–87.

29 K. E. Trenberth and J. T. Fasullo, "An Apparent Hiatus in Global Warming?" *Earth's Future* 1 (2013): 19–32.

30 Thomas R. Karl et al., "Possible Artifacts of Data Biases in the Recent Global Surface Warming Hiatus," *Science* 348 (2015): 1469–72.

31 NASA Global Climate Change, "Global Temperature," https://climate.nasa.gov/vital-signs/global-temperature/#:~:text=Nineteen%20of%20the%2020%20warmest,National%20Oceanic%20and%20Atmospheric%20Administration.

32 Michael Roppolo, "Americans More Skeptical Of Climate Change Than Others in Global Survey," *CBS News*, July 23, 2014, http://www.cbsnews.com/ news/americans-more-skeptical-of-climate-change-than-others-in-global-survey/.

33 Bruce Stokes, Richard Wike, and Jill Carle, "Global Concern about Climate Change," *Pew Research Center*, November 5, 2015, http://www.pewglobal.org/ 2015/11/05/global-concern-about-climate-change-broad-support-for-limitingemissions/.

34 Cary Funk, Alec Tyson, Brian Kennedy, and Courtney Johnson, "3. Concern over Climate and the Environment Predominates among These Publics," *Pew Research Center*, September 29, 2020, https://www.pewresearch.org/science/2020/09/29/concern-over-climate-and-the-environment-predominates-among-these-publics/.

35 World Weather Attribution, "Attribution of the Australian Bushfire Risk to Anthropogenic Climate Change," January 10 2020, https://www.worldweatherattribution.org/bushfires-in-australia-2019-2020/.

36 United Nations, *United Nations Framework Convention on Climate Change*, 1992, https://unfccc.int/ resource/ docs/ convkp/ conveng.pdf.

37 PBL Netherlands Environmental Assessment Agency, *Trends in Global CO2 Emissions, 2015 Report*, Joint Research Centre, European Commission, http://edgar.jrc.ec.europa.eu/ news_ docs/ jrc- 2015- trends- in- global- co2-emissions-2015- report- 98184.pdf, 28–9.

38 Carbon Brief, "Two Degrees: The History of Climate Change's Speed Limit," December 8 2014, https://www.carbonbrief.org/two-degrees-the-history-of-climate-changes-speed-limit.

39 J. Hansen et al., "Assessing 'Dangerous Climate Change': Required Reduction of Carbon Emissions to Protect Young People, Future Generations and Nature," *PLoS ONE* 8, no. 12 (2013): e81648, doi:10.1371/journal.pone .0081648.

40 Mike Young, "Two Degrees Warmer May Be Past the Tipping Point," *University of Copenhagen*, December 11, 2009, https://uniavisen.dk/en/ two-degreeswarmer-may-be-past-the-tipping-point/.

41 United Nations, *Paris Agreement*, 2015, United Nations Framework Convention on Climate Change, 3, https://unfccc.int/files/essential_background/convention/application/pdf/english_paris_agreement.pdf.

42 CAIT, Climate Data Explorer, "Historical Emissions," *World Resources Institute*, http://cait.wri.org/ historical.

43 Mengpin Ge, Johannes Friedrich, and Thomas Damassa, "6 Graphs Explain the World's Top 10 Emitters," *World Resources Institute*, November 25, 2014, https://wri.org/blog/2014/11/6-graphs-explain-world%E2%80%99s-top-10-emitters.

44 http://www.statista.com/statistics/263265/top-companies-in-the-world-byrevenue/.

45 Sander van der Linden, Edward Maibach, and Anthony Leiserowitz, "Improving Public Engagement with Climate Change Five 'Best Practice' Insights from Psychological Science," *Perspectives on Psychological Science* 10 (2015): 758–63.

Chapter Nine

1 Peter Wang Hjemdahl, "The Climate Movement's Diversity Problem," *Philadelphia Inquirer*, December 10, 2019.

2 Beth Gardiner, "Unequal Impact: The Deep Links between Racism and Climate Change," *Yale Environment360*, June 9, 2020, https://e360.yale.edu/features/unequal-impact-the-deep-links-between-inequality-and-climate-change.

3 Luke Tryl, "Extinction Rebellion's Tactics Are Working Like a Charm—Even if You Don't Happen to Like Them," *The Independent*, September 7, 2020.

4 Anmar Frangoul, "Swedish Pension Fund with Billions of Assets under Management to Stop Fossil Fuel Investments," *CNBC*, March 17, 2020.

5 Joana Setzer and Rebecca Byrnes, *Global Trends in Climate Change Litigation: 2020 Snapshot*, July 2020, https://www.lse.ac.uk/granthaminstitute/wp-content/uploads/2020/07/Global-trends-in-climate-change-litigation_2020-snapshot.pdf.

6 Climate Watch, "Historical Greenhouse Gas Emissions," https://www.climatewatchdata.org/ghg-emissions?end_year=2017&start_year=1990.

7 Transform to Net Zero, https://transformtonetzero.org/.

8 David Chandler, "New Approach Suggests Path to Emissions-Free Cement," *MIT News*, September 16, 2019.

9 "Plastic & Climate: The Hidden Costs of a Plastic Planet, Executive Summary," *Center for International Environmental Law*, https://www.ciel.org/wp-content/uploads/2019/05/Plastic-and-Climate-Executive-Summary-2019.pdf.

10 David Roberts, "Pulling CO2 out of the Air and Using it Could Be a Trillion-Dollar Business," *Vox*, November 22, 2019, https://www.vox.com/energy-and-environment/2019/9/4/20829431/climate-change-carbon-capture-utilization-sequestration-ccu-ccs.

11 Sarah Zielinski, "Iceland Carbon Capture Project Quickly Converts Carbon Dioxide into Stone," *Smithsonian.com*, June 9, 2016, http://www.smithsonianmag.com/science-nature/iceland-carbon-capture-project-quicklyconverts-carbon-dioxide-stone-180959365/.

12 Robert F. Service, "Industrial Waste Can Turn Planet-Warming Carbon Dioxide into Stone," *Science*, September 3, 2020, https://www.sciencemag.org/news/2020/09/industrial-waste-can-turn-planet-warming-carbon-dioxide-stone#:~:text=Industrial%20waste%20can%20turn%20planet%2Dwarming%20carbon%20dioxide%20into%20stone,-By%20Robert%20F&text=Over%202%20days%2C%20the%20slurry,into%20to%20carbon%2Dbased%20minerals.

13 Tona Kunz, "Questions Rise about Seeding for Ocean CO2 Sequestration," *Argonne National Laboratory*, June 12, 2013, http://www.aps.anl.gov/News/APS_News/Content/APS_NEWS_20130612.php.

14 Paul Voosen, "Geoengineers Inch Closer to Sun-Dimming Balloon Test," *Science*, December 15, 2020.

15 "Durban Adaptation Charter for Local Governments," https://unfccc.int/files/meetings/durban_nov_2011/statements/application/pdf/111209_cop17_hls_iclei_charter.pdf.

16 NWS Phoenix, "Phoenix Heat by the Numbers," https://twitter.com/NWSPhoenix/status/1318381423011663873/photo/1 and Tonya Mosley and Allision Hagan, "Phoenix Residents Will Need to Adapt to an Even Hotter Climate," *Here and Now*, September 18, 2019.

17 Gilbert M. Gaul, "Fortress Charleston: Will Walling Off the City Hold Back the Waters?" *Yale Environment360*, May 5, 2020, https://e360.yale.edu/features/fortress-charleston-will-walling-off-the-city-hold-back-the-waters.

18 Nathaniel Rich, "Destroying a Way of Life to Save Louisiana," *New York Times Magazine*, July 21, 2020.

19 *Louisiana's Comprehensive Master Plan for a Sustainable Coast*, http://coastal.la.gov/wp-content/uploads/2017/04/2017-Coastal-Master-Plan_Web-Book_CFinal-with-Effective-Date-06092017.pdf.

20 Kevin Sack and John Schwarz, "Left to Louisiana's Tides, A Village Fights for Time," *New York Times*, February 24, 2018.

21 Philip Sherwell, "$40bn to Save Jakarta: The Story of the Great Garuda," *Guardian*, https://www.theguardian.com/cities/2016/nov/22/jakarta-great-garuda-seawall-sinking.

22 Christopher Flavelle, "Rising Seas Threaten an American Institution: The 30-Year Mortgage," *New York Times*, June 19, 2020.

23 Al Shaw and Elizabeth Weil, "New Maps Show How Climate Change Is Making California's 'Fire Weather' Worse," *ProPublica*, October 14, 2020, https://projects.propublica.org/california-fire-weather/.

24 "Climate Change Could Make Insurance Too Expensive for Most People—Report," *Guardian*, https://www.theguardian.com/environment/2019/mar/21/climate-change-could-make-insurance-too-expensive-for-ordinary-people-report.

25 "Norway Bets on Global Warming to Thaw Arctic Ice for Oil and Gas Drive," *Reuters*, May 13, 2014, http://www.reuters.com/article/us-energy-arctic-idUSKBN0DT13220140513.

26 Thomas Nilsen, "Norway Proposes to Open 125 New Oil Exploration Blocks in the Barents Sea," *Barents Observer*, June 24, 2020, https://thebarentsobserver.com/en/about-us.

27 Michael E. Mann, *The New Climate War: The Fight to Take Back the Planet* (New York: Public Affairs, 2021).

28 Naomi Klein, *This Changes Everything: Capitalism vs. the Climate* (New York: Simon & Schuster, 2014).

29 P. J. Crutzen and E. F. Stoermer, "The 'Anthropocene'," *Global Change Newsletter* 41 (2000): 17–18.

30 William F. Ruddiman, "The Anthropocene," *Annual Review of Earth and Planetary Sciences* 41, no. 1 (2013): 45–68.

31 P.J. Crutzen and C. Schwagerl, "Living in the Anthropocene: Toward a New Global Ethos," *YaleEnvironment360*, January 24, 2001. https://e360.yale.edu/features/living_in_the_anthropocene_toward_a_new_global_ethos.

32 J. Syvitski et al., "Extraordinary Human Energy Consumption and Resultant Geological Impacts Beginning Around 1950 CE Initiated the Proposed Anthropocene Epoch," *Communications Earth and Environment* 1, 32 (2020). https://doi.org/10.1038/s43247-020-00029-y.

33 "The Term 'Anthropocene' Is Popular—and Problematic," *Scientific American* 319 (2018), doi:10.1038/scientificamerican1218-10.

FURTHER READING

Bartlett, Robert, *The Making of Europe: Conquest, Colonization and Cultural Change 950–1350* (Princeton: Princeton University Press, 1994).
Bar-Yosef, Ofer, “The Natufian culture in the Levant, threshold to the origins of agriculture.” *Evolutionary Anthropology: Issues, News, and Reviews* 6, no. 5 (1998): 159–77.
Behringer, Wolfgang, *A Cultural History of Climate* (Cambridge: Polity, 2010).
Behringer, Wolfgang, *Tambora and the Year without a Summer: How a Volcano Plunged the World into Crisis* (Cambridge: Polity, 2019).
Bernard Knapp, A. and Stuart W. Manning, “Crisis in Context: The End of the Late Bronze Age in the Eastern Mediterranean.” *American Journal of Archaeology* 120 (2016): 99–149. doi:10.3764/aja.120.1.0099.
Brook, Timothy, *The Troubled Empire: China in the Yuan and Ming Dynasties* (Cambridge, MA: Harvard University Press, 2010).
Brooke, John L., *Climate Change and the Course of Global History: A Rough Journey* (New York: Cambridge University Press, 2014).
Buckley, Brendan M. et al. “Climate as a contributing factor in the demise of Angkor, Cambodia.” *Proceedings of the National Academy of Sciences* 107, no. 15 (2010): 6748–52.
Bulliet, Richard W., *Cotton, Climate, and Camels in early Islamic Iran: A Moment in World History* (New York: Columbia University Press, 2009).
Büntgen, Ulf et al. “Cooling and societal change during the Late Antique Little Ice Age from 536 to around 660 AD.” *Nature Geoscience* 9 (2016): 231–6.
Callendar, G. S., “The artificial production of carbon dioxide and its influence on temperature.” *Quarterly Journal of the Royal Meteorological Society* 64 (1938): 223–40. doi:10.1002/qj.49706427503.
Crutzen, P. J. and E. F. Stoermer, “The ‘Anthropocene’.” *Global Change Newsletter* 41 (2000): 17–18.
DeGroot, Dagomar, *The Frigid Golden Age: Climate Change, the Little Ice Age, and the Dutch Republic, 1560–1720* (Cambridge: Cambridge University Press, 2018).
deMenocal, Peter B. and Jessica E. Tierney, “Green Sahara: African Humid Periods Paced by Earth’s Orbital Changes.” *Nature Education Knowledge* 3, no. 10 (2012): 12.
Diamond, Jared, *Collapse: How Societies Choose to Fail or Succeed* (Penguin, 2005).
Douglas, Peter M. J. et al., “Drought, agricultural adaptation, and sociopolitical collapse in the Maya Lowlands.” *Proceedings of the National Academy of Sciences* 112, no. 18 (2015): 5607–12.
Fagan, Brian M., *The Little Ice Age: How Climate Made History, 1300–1850* (New York: Basic Books, 2002).

Fagan, Brian M., *The Great Warming: Climate Change and the Rise and Fall of Civilizations* (New York: Bloomsbury, 2008).

Gammage, Bill, *The Biggest Estate on Earth: How Aborigines Made Australia* (Sydney: Allen & Unwin, 2011).

Goldstone, Jack A., "Efflorescences and economic growth in world history: rethinking the 'Rise of the West' and the Industrial Revolution." *Journal of World History* 13 (2002): 323–89.

Harper, Kyle, *The Fate of Rome: Climate Disease and the End of an Empire* (Princeton: Princeton University Press, 2018).

Hansen, James, *Storms of My Grandchildren: The Truth about the Coming Climate Catastrophe and Our Last Chance to Save Humanity* (Bloomsbury, 2009).

Hodell, David A., Jason H. Curtis, and Mark Brenner, "Possible role of climate in the collapse of Classic Maya civilization." *Nature* 375 (1995): 391–4.

Hsiang, Solomon M., Kyle C. Meng, and Mark A. Cane, "Civil conflicts are associated with the global climate." *Nature* 476 (2011): 438–41.

IPCC (2014), *Climate Change 2014: Synthesis Report*. Contribution of Working Groups I, II and III to the Fifth Assessment Report of the Intergovernmental Panel on Climate Change [Core Writing Team, R.K. Pachauri and L.A. Meyer (eds.)]. IPCC, Geneva, Switzerland, 151 pp.

Klein, Naomi, *This Changes Everything: Capitalism vs. the Climate* (New York: Simon & Schuster, 2014).

Lamb, H. H., *Climate, History, and the Modern World* (London: Routledge, 1995).

Lieberman, Victor and Brendan Buckley, "The Impact of Climate on Southeast Asia, circa 950–1820: New findings." *Modern Asian Studies* 46, no. 5 (2012): 1056–59.

Mann, Michael E., *The New Climate War: The Fight to Take Back the Planet* (New York: Public Affairs, 2021).

Mann, Michael E. et al., "Global Signatures and Dynamical Origins of the Little Ice Age and Medieval Climate Anomaly." *Science* 326 (2009): 1256–60.

Mann, Michael E., Raymond S. Bradley, and Malcolm K. Hughes, "Global-scale temperature patterns and climate forcing over the past six centuries." *Nature* 392 (1988): 779–87.

McAnany, Patricia A. and Norman Yoffee, eds, *Questioning Collapse: Human Resilience, Ecological Vulnerability, and the Aftermath of Empire* (Cambridge: Cambridge University Press, 2009).

McCormick, Michael et al., "Climate change during and after the Roman Empire: Reconstructing the past from scientific and historical evidence." *Journal of Interdisciplinary History* 43 (2012): 169–220.

McKibben, Bill, ed., *The Global Warming Reader: A Century of Writing about Climate Change* (New York: Penguin Books, 2011).

Rahmstorf, Stefan and Dim Coumou, "Increase of extreme events in a warming world." *Proceedings of the National Academy of Sciences* 108, no. 44 (2011): 17905–9.

Raymo, M. E. and W. F. Ruddiman, "Tectonic forcing of late Cenozoic climate," *Nature* 359 (1992): 117–122.

Reich, David, *Who We Are and How We Got Here: Ancient DNA and the New Science of the Human Past* (New York: Pantheon, 2018).

Revelle, R. and H. E. Suess, "Carbon Dioxide Exchange Between Atmosphere and Ocean and the Question of an Increase of Atmospheric CO_2 during the Past Decades." *Tellus* 9 (1957): 18–27.
Rich, Nathaniel, *Losing Earth: The Decade We Could have Stopped Climate Change* (New York: MCD Farrar, Strauss and Giroux, 2019).
Ruddiman, W.F. "The anthropogenic greenhouse era began thousands of years ago." *Climate Change* 61 (2003), 261–93.
Ruddiman, William, *Earth's Climate: Past and Future* (New York: W.H. Freeman, 2014).
Sessa, Kristina, "The New Environmental History of the Fall of Rome: A Methodological Consideration." *Journal of Late Antiquity* 12 (2019): 211–55.
Stinger, Chris, *Lone Survivors: How We Came to be the Only Humans on Earth* (New York: Times Books, Henry Holt, 2012).
Thompson, L. G. et al., "Annually resolved ice core records of tropical climate variability over the past~ 1800 years." *Science* 340, no. 6135 (2013): 945–50.
USGCRP (2018), Impacts, Risks, and Adaptation in the United States: Fourth National Climate Assessment, Volume II: Report-in-Brief [Reidmiller, D.R., C.W. Avery, D.R. Easterling, K.E. Kunkel, K.L.M. Lewis, T.K. Maycock, and B.C. Stewart (eds.)]. U.S. Global Change Research Program, Washington, DC, USA, 186 pp.
Wallace-Wells, David, *The Uninhabitable Earth: Life after Warming* (New York: Tim Duggan Books, 2019).
White, Sam, *The Climate of Rebellion in the Early Modern Ottoman Empire* (Cambridge: Cambridge University Press, 2011).
White, Sam, *A Cold Welcome: The Little Ice Age and Europe's Encounter with North America* (Cambridge, MA: Harvard University Press, 2017).
Wood, Gillen D'Arcy, *Tambora: The Eruption that Changed the World* (Princeton, NJ: Princeton University Press, 2013).
World Weather Attribution, https://www.worldweatherattribution.org/analysis/rainfall/.

INDEX